Population Dynamics of Forest Insects

Population Dynamics of Forest Insects

the
UNIVERSITY
of
GREENWICH

Intercept

Andover, Hampshire

British Library CIP Data available.
ISBN 0 946707 28 6

Published in July 1990 by Intercept Limited,
PO Box 716, Andover, Hants SP10 1YG, England

Laserset in 'Times' by Reachword Limited, Sudbrooke, Lincoln.

Printed in Great Britain by Athenaeum Press, Newcastle upon Tyne.

Contributors

I. ABBOTT, *Department of Conservation and Land Management, PO Box 104, Como WA 6152, Australia*

D. A. BARBOUR, *Forest Insect Surveys, 62 North Gyle Loan, Edinburgh, EH12 8LD, Scotland , UK*

A. M. BARKER, *Department of Biology, Building 44, University of Southampton, Southampton, SO9 5NH, UK*

A. A. BERRYMAN, *Department of Entomology, Washington State University, Pullman, WA 99164-6432, USA*

C. I. CARTER, *Forestry Commission Research Station, Alice Holt Lodge, Wrecclesham, Farnham, Surrey, GU10 4LH, UK*

R. G. CATES, *Chemical Ecology Laboratory, Department of Botany and Range Science, 495 WIDB, Brigham Young University, Provo, Utah 84602, USA*

M. F. CLARIDGE, *School of Pure and Applied Biology, University of Wales, PO Box 915, Cardiff, CF1 3TL, UK*

W. A. COPPER, *Division of Biological Control, University of California, Berkeley 94720, USA*

S. CRUTE, *Department of Biological and Biomedical Sciences, University of Ulster, Coleraine, BT52 1SA, Northern Ireland, UK*

D. L. DAHLSTEN, *Division of Biological Control, University of California, Berkeley 94720, USA*

K. R. DAY, *Department of Biological and Biomedical Sciences, University of Ulster, Coleraine, BT52 1SA, Northern Ireland, UK*

A. F. G. DIXON, *School of Biological Sciences, University of East Anglia, Norwich NR4 7TJ, UK*

P. J. EDWARDS, *Department of Biology, Building 44, University of Southampton, Southampton, SO9 5NH, UK*

J. S. ELKINTON, *Department of Entomology, University of Massachusetts, Amherst, MA 01003, USA*

H. F. EVANS, *Forestry Commission Research Station, Alice Holt Lodge, Wrecclesham, Farnham, Surrey, GU10 4LH, UK*

C. S. FERGUSON, *Department of Entomology, University of Massachusetts, Amherst, MA 01003, USA*

R. A. FLEMING, *Forest Pest Management Institute, Forestry Canada, Box 490, Sault Ste. Marie, Ontario P6A 5M7, Canada*

M. A. FOSTER, *Department of Entomology, Penn State University, University Park, PA 16802, USA*

J. R. GOULD, *Department of Entomology, University of Massachusetts, Amherst, MA 01003, USA*

I. HANSKI, *Department of Zoology, University of Helsinki, P. Rautatiekatu 13, SF-00100, Helsinki, Finland*

S. E. HARTLEY, *Department of Biology, University of York, Heslington, York YO1 5DD, UK*

E. HAUKIOJA, *Laboratory of Ecological Zoology, Department of Biology; and Kevo Subarctic Research Station; University of Turku, SF-20500 Turku, Finland*

K. HELIÖVAARA, *Finnish Forest Research Institute, Vantaa, SF-001301, Finland*

G. P. HOSKING, *Forest Research Institute, Rotorua, New Zealand*

M. D. HUNTER, *Department of Zoology, University of Oxford, South Parks Road, Oxford OX1 3PS, UK*

J. A. HUTCHESON, *Forest Research Institute, Rotorua, New Zealand*

N. A. C. KIDD, *School of Pure and Applied Biology, University of Wales, Cardiff, CF1 3TL, UK*

D. KLIMETZEK, *Forstzoologisches Institut der Universität, Bertoldstrasse 17, D-7800 Freiburg/Br., Federal Republic of Germany*

J. H. LAWTON, *Centre for Population Biology, Department of Pure and Applied Biology, Imperial College, Silwood Park, Ascot, SL5 7PY, UK*

S. R. LEATHER, *Forestry Commission Northern Research Station, Roslin, Midlothian, EH25 9SY, UK*

A. M. LIEBHOLD, *USDA Forest Service, Northeastern Forest Experimental Station, PO Box 4360, Morgantown, WV 26505, USA*

G. B. LEWIS, *School of Pure and Applied Biology, University of Wales, Cardiff, CF1 3TL, UK*

G. E. LONG, *Department of Entomology, Washington State University, Pullman WA 99164-6432, USA*

R. M. J. MacKENZIE, *Forest Research Institute, Rotorua, New Zealand*

S. McNEIL, *Department of Pure and Applied Biology, Imperial College, Silwood Park, Ascot, Berkshire, SLB 7PY, UK*

E. J. MAJOR, *Forestry Commission Research Station, Alice Holt Lodge, Wrecclesham, Farnham, Surrey, GU10 4LH, UK*

R. R. MASON, *Forestry and Range Sciences Laboratory, La Grande, OR 97859, USA*

J. A. MILLSTEIN, *Applied Biomathematics Inc., 100 North Country Road, Setauket, NY 11733, USA*

N. J. MILLS, *CAB International Institute of Biological Control, Silwood Park, Ascot, Berkshire SL5 7TA, UK*

M. E. MONTGOMERY, *USDA Forest Service, Northeastern Forest Experimental Station, 51 Mill Pond Road, Hamden, Connecticut 06514, USA*

P. W. PRICE, *Department of Biological Sciences, Northern Arizona University, Flagstaff, Arizona 86011-5640, USA*

J. ROLAND, *Boreal Institute for Northern Studies and Department of Botany, University of Alberta, Edmonton, Alberta, T6G 2E9, Canada*

D. L. ROWNEY, *Division of Biological Control, University of California, Berkeley 94720, USA*

J. C. SCHULTZ, *Department of Entomology, Penn State University, University Park, PA 16802, USA*

S. D. J. SMITH, *School of Pure and Applied Biology, University of Wales, Cardiff, CF1 3TL, UK*

S. M. TAIT, *Division of Biological Control, University of California, Berkeley 94720, USA*

R. VÄISÄNEN, *Water and Environment Research Institute, Helsinki, Finland*

W. E. WALLNER, *USDA Forest Service, Northeastern Forest Experimental Station, 51 Mill Pond Road, Hamden, Connecticut 06514, USA*

P. J. WALSH, *Forestry Commission Northern Research Station, Roslin, Midlothian, Scotland,UK*

A. D. WATT, *Institute of Terrestrial Ecology, Bush Estate, Penicuik, Midlothian, EH26 0QB, Scotland UK*

J. M. WENZ, *Stanislaus National Forest, USDA Forest Service, Sonora, California 95370, USA*

R. M. WESELOH, *Connecticut Agricultural Experiment Station, New Haven, Connecticut,USA*

C. WEST, *Department of Zoology, University of Oxford, South Parks Road, Oxford OX1 3PS, UK*

J. B. WHITTAKER, *Division of Biological Sciences, Institute of Environmental and Biological Sciences, University of Lancaster, Bailrigg, Lancaster LA4 YQ, UK*

S. D. WRATTEN, *Department of Biology, Building 44, University of Southampton, Southampton, SO9 5NH, UK*

R. ZONDAG, *Forest Research Institute, Roturua, New Zealand*

J. ZOU, *Chemical Ecology Laboratory, Department of Botany and Range Science, 495 WIDB, Brigham Young University, Provo, Utah 84602 USA*

Introduction

Foliage-feeding insects on trees are often very abundant: species such as the gypsy moth, the pine beauty moth and the spruce budworms pose a major threat to forestry and can have a major impact on the ecology of forest habitats. Management of these insect pests is impossible without an understanding of their population dynamics. This book considers many aspects of the population ecology of a range of foliage feeding insects. Most of the chapters deal with pest species, but several look at species without pest status: advances in insect population ecology have come from both pest and 'non-pest' species.

The book is divided into four sections. The first deals mainly with long-term population studies; the second with insect–plant relationships; the third with natural enemies; and the fourth with the use of models in understanding population dynamics and in pest management.

Long-term studies form the backbone of population ecology. Careful monitoring of insect numbers or damage over many years (e.g. *Dixon, Klimetzek*) yields more than just data for analysis, it also provides an essential perspective from which general theories of population ecology emerge. For example, long-term trends in the abundance of pine feeding defoliators (Schwerdtfeger, 1941; Klomp, 1966; *Klimetzek*) are often cited in support of new theories (e.g. Varley, 1949; Andrewartha and Birch, 1954; White, 1974; Rhoades, 1983). In recent years the extensive monitoring approach has largely given way to shorter, intensive, localised studies (e.g. *Dahlsten et al., Day and Crute, Dixon, Leather*). This intensive approach requires careful consideration of sampling techniques (*Dahlsten et al.*) but these studies often incorporate methods to evaluate the roles of the host plant (e.g. *Dixon*), natural enemies (e.g. *Leather*) and the abiotic environment (*Day and Crute*). While these general population studies tend not to consider individual factors in great depth, their strength is that they are often able to relate the roles of one or more factor, albeit often circumstantially, to insect abundance because they are based on the study of real populations. General observations on the occurrence of insects on different tree species can also be used within the framework of island biogeography to predict the occurrence of future pest problems (*Claridge and Evans*).

The second section looks more closely at a range of studies which have attempted to bring a greater understanding of the many ways in which the host plant can affect insect abundance. The argument that because the world is green, natural enemies must be limiting the abundance of phytophagous insects (Hairston *et al.*, 1960) now seems naive. The host plant is no longer considered to be a static resource, but one whose quality varies in time and space. Of particular interest to forest pest management is the association between site conditions (species composition, forest age structure, topography and soil type) and pest outbreaks (*Abbott, Hosking et al., Montgomery*). Unfortunately this association is usually poorly documented and there is no substitute for critical analysis of insect and site data (*Montgomery*). The ecological basis for the association between insect outbreaks and site conditions has been the subject of controversy for some time (e.g. White, 1984; Larsson, 1989). White (1974) proposed that particular combinations of weather and site conditions lead to an increase in the suitability of tree foliage for insect herbivores and this, through an increase in insect survival (or other changes in insect performance), gives rise to insect outbreaks. Support for this plant stress-insect performance hypothesis comes from research on the green spruce aphid (*Major*) but not from related work on the pine beauty moth (*Watt*).

Plant stress may sometimes act to trigger insect outbreaks. Similarly outbreaks may be terminated or their severity reduced by insect damage-induced changes in plant suitability (Haukioja, 1980). Research has shown that these changes are complex: they can affect both the defoliating insect and other species (*Hunter and West*); they may become less significant as the tree ages (*Watt*); the effects range from strongly negative to strongly positive (*Haukioja, Hunter and West*); and they may involve subtle changes in plant chemistry (*Cates and Zou, Kidd et al.*). This is another area of research with controversial aspects: *Wratten and Edwards* consider the evidence supporting the hypothesis that the behavioural avoidance of damaged areas is the result of an evolved strategy in plants to achieve higher competitiveness instead of an evolved strategy on the part of the insects to avoid predation. In contrast, *Hartley and Lawton* argue that there is no convincing evidence that these short-term induced responses affect the performance and abundance of birch feeding insects and that responses observed in plants after insect attack may be primarily an anti-fungal or anti-microbial rather than an anti-insect response.

A rather different way in which insects may be affected by host plant quality is through the impact of air pollution: indeed, if herbivorous insects are positively affected by pollution they may be contributing towards the 'forest decline' in central and northern Europe. *McNeill and Whittaker* and *Heliövaara and Väisänen* consider interactions between air pollution and herbivory for a range of foliage feeding insects.

The major problem in assessing the role of the host plant in insect population dynamics is replacing circumstantial evidence with direct evidence. The chapters in the insect–plant section show how much progress has been made in this area, particularly in measuring the impact of several host plant influences on insect performance (*Haukioja, Hunter and West, Watt, McNeill and Whittaker*) and in assessing the role of particular plant chemicals (*Cates and Zhou, Kidd et al.*). These chapters also point the way towards future research by identifying areas where our knowledge is weak. Perhaps the most serious gap is in assessing the significance of variability in insect performance on insect abundance. This not only requires more quantitative research on insect performance but also an understanding of the role of natural enemies.

The third section considers the natural enemies of forest insects. Predators, parasites and pathogens act in complex ways, but techniques such as host exposure and predator exclusion are helping to unravel this complexity (*Elkinton et al., Mills,*

Roland, Walsh, Weseloh). The knowledge that different predator (*Weseloh*) and parasite (*Mills*) species have been found to specialize at low and high prey densities opens the way towards more successful biological control programmes (*Mills*). The notion, if it ever existed, that biological control by planned introductions is a simple matter must be erased by the case of the introduction of *Cyzenis albicans* to Canada to control the winter moth. The success of the introduction may not have been so much due to a rise in parasitism than to an increase in predation of unparasitised pupae (*Roland*). An understanding of the action of natural enemies has also been of value in explaining why outbreaks of forest insects occur where they do: sites which were thought to have experienced outbreaks because of host plant susceptibility may in fact have done so because they have a limited natural enemy fauna (*Hanski, Walsh*). Research on the influence of predators and parasites in different types of site and at different spatial scales has led to the view that natural enemies may regulate prey populations at a 'metapopulation' scale (*Hanski, Elkinton et al.*). Much can be learned about pest species by comparing them with species which never achieve pest status. In doing so, and in particular by considering the role of natural enemies in latent and eruptive species, *Price* concludes that natural enemies tend to respond to host population change rather than cause it; by constructing life tables to include the behaviour of foraging females, the plant–herbivore interaction is found to be more important than the role of natural enemies. Undoubtedly, the insect–host plant interaction dominates the population dynamics of many forest insects but the action of natural enemies clearly dominates the dynamics of many other species, sometimes in very subtle ways (*Roland*). However, it is interesting that many authors in the natural enemy section also include a consideration of host plant factors and vice versa: the key to understanding the dynamics of most forest insects may be the interaction of host plant and natural enemies. This is particularly well illustrated by the way in which the effect of pathogens on the gypsy moth is mediated by host plant chemistry (*Schultz et al.*). As gypsy moth populations rise the concentration of leaf phenolics increases. This causes a reduction in moth fecundity but it also greatly increases larval resistance to a baculovirus.

The best hope for understanding the complexities of insect population dynamics is modelling. Models can examine the role of individual factors such as predation (*Crute and Day*), parasitism (*Long*) and dispersal (*Barbour*), they can be used to compare the roles of different factors acting alone or together (*Kidd, Liebhold and Elkinton*) and they can be used in pest management (*Fleming*). As research on insect population models develops, one major problem is their degree of complexity. However, it may be that simple models are as good or better than more complex simulation models at understanding and predicting the fluctuations in the numbers of forest insects (*Berryman et al., Kidd*).

In summary, this book presents a range of approaches to the study of forest insect population dynamics: long-term population studies, reductionist studies on the insect–plant and insect–natural enemy relationships which underly insect population behaviour, and theoretical population models. It is our belief that none of these approaches can by themselves adequately describe population dynamics; major advances in the future will be made where these different approaches are combined.

This book is based on the Population Dynamics of Forest Insects conference which was held in Edinburgh, September 1989. The conference steering committee comprised the editors of this book, plus Alan Berryman, Melvin Cannell, Ilkka Hanski and John Lawton, and the organization and running of the conference would have been impossible without Carol Morris, Elma Lawrie, Marjorie Ferguson, Anne McFarlane, James Aegerter and Robbie Wilson. The editors would also like to thank June Nelson and the many referees who made the task of producing this book much easier.

References

ANDREWARTHA, H. G. and BIRCH, L. C. (1954). *The Distribution and Abundance of Animals*. Chigago: University of Chicago Press.

HAIRSTON, N. G., SMITH, F.E. and SLOBODKIN, L. B. (1960). Community structure, population control and competition. *American Naturalist* **94**, 421–425.

HAUKIOJA, E. (1980). On the role of plant defences in the fluctuation of herbivore populations. *Oikos* **35**, 202–213.

KLOMP, H. (1966). The dynamics of a field population of the pine looper, *Bupalas piniarius* (Lep., Geom.). *Advances in Ecological Research* **3**, 207–305.

LARSSON, S. (1989). Stressful times for the plant stress-insect performance hypothesis. *Oikos* **56**, 277–283.

RHOADES, D. F. (1983). Herbivore population dynamics and plant chemistry. In: *Variable Plants and Herbivores in Natural and Managed Systems*, pp. 155–220 (eds R. F Denno and M. S. McClure). New York: Academic Press.

SCHWERDTFEGER, F. (1941). Über die Ursachen des Massenwelchsels der Insekten. *Zeitschrift für Angewandte Entomologie* **28**, 254–303.

VARLEY, G. C. (1949). Population changes in German forest pests. *Journal of Animal Ecology* **18**, 117–122.

WHITE, T. C. R. (1974). A hypothesis to explain outbreaks of looper caterpillars, with special reference to populations of *Selidosema suavis* in a plantation of *Pinus radiata* in New Zealand. *Oecologia* **16**, 279–301.

WHITE, T. C. R. (1984). The abundance of invertebrate herbivores in relation to the availability of nitrogen in stressed food plants. *Oecologia* **63**, 90–105.

Allan D. Watt, Simon R. Leather, Mark D. Hunter and Neil A. C. Kidd
February 1990

Contents

SECTION TWO: INSECT–PLANT INTERACTIONS

SECTION ONE
General Population Studies

1
Population Dynamics of Pine-feeding Insects: a Historical Study

D. KLIMETZEK

Forstzoologisches Institut der Universität, Bertoldstrasse 17, D-7800 Freiburg/Br., Federal Republic of Germany

Introduction
Methods
Results
Discussion
Acknowledgements
References

Introduction

The pine looper *Bupalus piniaria* (L.), pine beauty moth *Panolis flammea* (D. & S.), nun moth *Lymantria monacha* (L.) and common pine sawfly *Diprion pini* (L.) have been forest pests in Germany since at least 1800. This paper describes their occurrence through an analysis of historical data for state-owned forests in Northern Bavaria (administrative areas Oberfranken, Mittelfranken and Oberpfalz) and Southwest Germany (Pfalz) for the years 1811–1988. The starting point of these investigations was the hypothesis that since the beginning of regulated forestry at about 1800, the frequency and intensity of outbreaks in these areas was similar for about 100 years; but thereafter, different patterns evolved.

Methods

In order to evaluate the frequency of outbreaks, an infestation index i was calculated for each year and forestry district, using data available for the intensity of attack and economical damage for each species: $i = 0$ not infested; $i = 1$ insects conspicuously present, no noticeable damage; $i = 2$ little damage only; $i = 3$ extensive damage. The sum of values in a given forestry district for one year was considered the population index for that year (Klimetzek, 1979). This index was then compared for different years to give a chronological representation of occurrence, importance and distribution of each pest species.

Population Dynamics of Forest Insects
© Intercept Ltd, PO Box 716, Andover, Hampshire, SP10 1YG, UK

Results

The two areas which harboured the heaviest infestations of insect pests are both situated in Northern Bavaria (*Figure 1*) and have the following similarities: each is a large, continuous forested area with a high percentage of pine trees (*Table 1*) and their underlying geology has resulted in soils deficient in nutrient chemicals and water-holding capacity. The edges of the areas susceptible to pest attack coincided closely with the borders of geological formations. These areas occurred primarily in low lands characterized by low rainfall and high temperatures; at higher altitudes, they normally occurred on the side of mountain ranges where precipitation was low (Klimetzek, 1979).

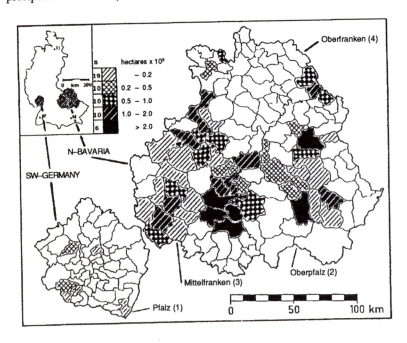

Figure 1.   Location of test areas Southwest Germany (1) and Northern Bavaria (2–4). Hatching denotes location and number (n) of forestry districts and area (ha. x 10^3) treated with chemical insecticides (1925–88) against the pine looper, pine beauty moth, nun moth and common pine sawfly. Administrative areas Pfalz (1), Oberpfalz (2), Mittelfranken (3), Oberfranken (4); boundary lines of forestry districts as of 1930; F: Freiburg, H: Hamburg, M: München.

The chronology of outbreaks since 1811 (since 1801 for the Bavarian administrative areas) is presented in *Figures 2–5*. The most common species in all regions was the pine looper (*Table 1*). Severe outbreaks of this insect occurred in the years 1892–96 and 1924–28 (*Figure 2*). The pine beauty moth (*Figure 3*) and the nun moth (*Figure 4*) were exceptionally abundant in 1837–40, 1889–90,

Table 1. Extent of state-owned forests, percentage of pine trees (pine%) and number of state-owned forestry districts (f.d.) as of 1930 and index values 1811–1988 for pine looper (Bp), pine beauty moth (Pf), nun moth (Lm) and common pine sawfly (Dp).

Area (no.)	Forests (ha.)	Pine (%)	f.d.	Sum of index values				
				Bp	Pf	Lm	Dp	All
Pfalz (1)	110 400	47	46	356	108	79	124	667
Oberpfalz (2)	119 200	56	40	456	280	350	44	1130
Mittelfranken (3)	82 400	54	39	555	577	567	129	1828
Oberfranken (4)	107 650	36	48	400	194	294	136	1024
Total	419 650	48	173	1767	1159	1290	433	4649

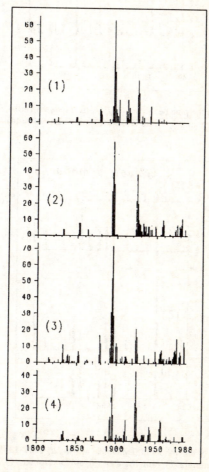

Figure 2. Outbreaks of the pine looper (*Bupalus piniaria*) 1801–1988 in state-owned forests of Pfalz (1), Oberpfalz (2), Mittelfranken (3) and Oberfranken (4).

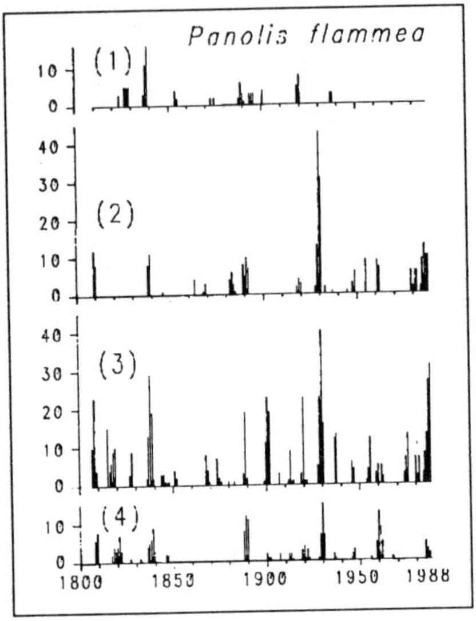

Figure 3. Outbreaks of the pine beauty moth (*Panolis flammea*) 1801–1988 in state-owned forests of Pfalz (1), Oberpfalz (2), Mittelfranken (3) and Oberfranken (4).

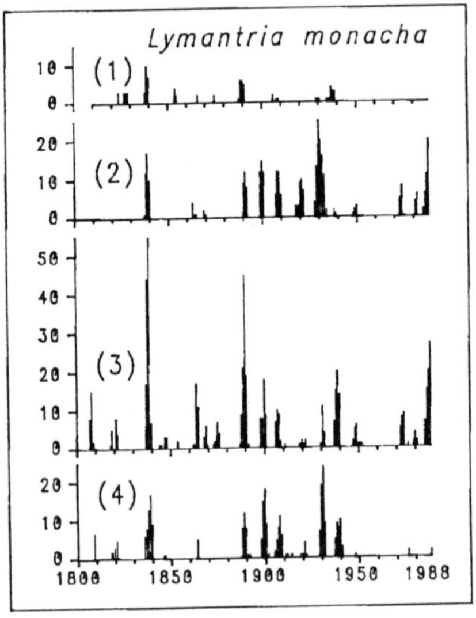

Figure 4. Outbreaks of the nun moth (*Lymantria monacha*) 1801–1988 in state-owned forests of Pfalz (1), Oberpfalz (2), Mittelfranken (3) and Oberfranken (4).

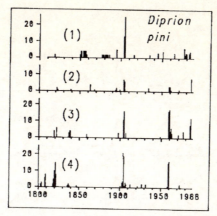

Figure 5. Outbreaks of the common pine sawfly *(Diprion pini)* 1801–1988 in state-owned forests of Pfalz (1), Oberpfalz (2), Mittelfranken (3) and Oberfranken (4).

1928–32, 1937–40 and 1985–88, when both species caused extensive damage in numerous forestry districts. The common pine sawfly *(Figure 5)* had major outbreaks in the years 1904–05 and 1959–60, but generally this insect was only locally important *(Table 1)*.

The occurrence of outbreaks of the four insect species was generally poorly correlated. Exceptions were the nun moth and the pine beauty moth, whose outbreaks usually peaked with a one-year time lag. For individual species, however, there was a significant synchronization in the timing of heavy insect attack in all four administrative areas. This synchronization was most apparent between Mittelfranken and Oberfranken and always stronger among the three administrative areas of Northern Bavaria than between these and Pfalz.

Discussion

The study of historical data in forestry provides one way of analysing insect outbreaks and their possible connections with environmental variables. Statistical tests of the indices of insect occurrence showed with a high probability that the observed pattern of infestation was not random (Klimetzek and Klimetzek, 1991). The original hypothesis of a difference in infestation patterns before and after 1930 was supported. From 1810 to 1930, the occurrence of pine insects in all four administrative areas was similar; since then, the intensity of their outbreaks in Northern Bavaria has been four to eight times higher than in Pfalz. Accordingly, the logarithmic numbers of overwintering pupae, which are comparable to the index values (Klimetzek, 1975), showed damped oscillations and their values were always lower in Southwest Germany than Bavaria *(Figure 6)*. This difference was most pronounced in Mittelfranken and was mostly due to outbreaks of the pine beauty moth and the nun moth.

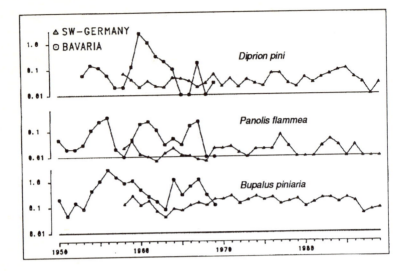

Figure 6.   Population dynamics ($\log_{10}$ overwintering pupae/m^2) of pine looper, pine beauty moth and common pine sawfly in Southwest Germany and Bavaria 1949/50–1988/89 (data from Gauss, 1974; König, 1989; Reindl, 1989).

The distinct spatial synchronization among the administrative districts indicates that, at certain times, one or more environmental factors created conditions favourable for an outbreak (Barbour, this volume, Chapter 30). It is probable that at the turn of the century, conditions started to change, causing the decline of pest insects in Southwest Germany. Analysis of these factors could provide valuable information for pest management in pine forests.

During 1925–88, a total of 64 853 hectares were sprayed with insecticides against the pine insects studied in the test area, half of which were state-owned forests. Sixty per cent of the treated areas are situated in only six Northern Bavarian forestry districts (*Figure 1*); these comprise forests particularly susceptible to insect outbreaks (Klimetzek, 1979). Exceptionally large control operations were conducted in the years 1931, 1956 and 1961–2 against the pine beauty moth, in 1960 against the common pine sawfly and in 1988 against the nun moth (*Figure 7*). The insecticide control area was largest in Mittelfranken (40 555 hectares). In Pfalz, only a total of 1138 hectares were sprayed to control these insects and, since 1952, no further treatment has been considered necessary. Chemical control therefore cannot explain the different development of infestations in Northern Bavaria and Pfalz.

Removal of forest litter is considered to be a major factor causing susceptibility to pest insects in pine forests (Schwenke *et al*., 1970; Krapfenbauer, 1983). The main reason is that this process leads to a withdrawal of nitrogen, on average 10–20 kg/year/ha. Removal of forest litter was traditionally very extensive in Bavaria; in Pfalz, on the other hand, it always remained significantly below

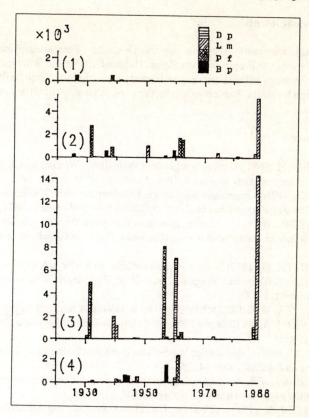

Figure 7. Area (ha. x 10^3) sprayed with chemical insecticides to control the pine looper (Bp), pine beauty moth (Pf), nun moth (Lm) and common pine sawfly (Dp) 1925–1988 in Pfalz (1), Oberpfalz (2), Mittelfranken (3) and Oberfranken (4).

average (Dümlein 1955). At present, for example, in Oberpfalz, pine forests with a long history of forest litter removal contain approximately 1000–2000 kg/ha nitrogen less than stands at comparable sites in which the litter had not been removed (Kreutzer,1972).

A second factor which may have been responsible for the decline in forest pests in Pfalz since the turn of this century was the increased planting of hardwoods instead of pine trees (Lüdge, 1971; Klimetzek, 1979). This hypothesis is supported by the recent decrease in insect pest problems in Oberfranken, where this silvicultural practice has been applied since the 1930s. In conclusion, a comparision of forest management systems in different areas of Germany has demonstrated that the inclusion of hardwoods in pine forests — as a mixture or understorey — seems to be the most promising way to prevent or significantly reduce insect pest outbreaks.

Acknowledgements

This work was partly supported by the Deutsche Forschungsgemeinschaft. Thanks are due to Forest directors Bauer, Haubold, Maier and Stitzinger for permission to study official records at Forestry commissions Neustadt/Pfalz, Regensburg/Oberpfalz, Bamberg/Oberfranken and Ansbach/Mittelfranken.

References

DÜMLEIN, H. (1955). Untersuchungen über Verbreitung, Wirkung und Ablösung der Waldstreunutzung in Bayern. Diss., University of München.

GAUSS, R. (1974). Ergebnisse langjähriger Parasitenstudien an Kieferninsekten des südwestdeutschen Raumes. *Zeitschrift für angewandte Entomologie* 77, 429–438.

KLIMETZEK, D. (1975). Umfang und Auswirkung von Begiftungsaktionen gegen Kiefernraupen in Nordbayern. *Allgemeine Forst- und Jagdzeitung* 146, 186–191.

KLIMETZEK, D. (1979) Insekten-Großschädlinge an Kiefer in Nordbayern und der Pfalz: Analyse und Vergleich 1810–1970. *Freiburger Waldschutz Abhandlungen* 2, 1–173.

KLIMETZEK, D. and KLIMETZEK, F. R. Jr. (1991). A new look at old data: pine insects in South Germany 1801–1988. *Zeitschrift für Angewandte Entomologie* (in press).

KÖNIG, E. (1989) Gegenwärtige Waldschutzsituation in Südwestdeutschland. *Allgemeine Forstzeitschrift* 44, 337–343.

KRAPFENBAUER, A. (1983). Von der Streunutzung zur Ganzbaumnutzung. *Centralblatt für das Gesamte Forstwesen* 100, 143–174.

KREUTZER, K. (1972) . Über den Einfluß der Streunutzung auf den Stickstoffhaushalt von Kiefernbeständen. *Forstwissenschaftliches Centralblatt* 91, 263–270.

LÜDGE, W. (1971). Der Einfluß von Laubholzunterbau auf die Schädlingsdichte in den Kiefernbeständen der Schwetzinger Hardt. *Allgemeine Forst- und Jagdzeitung* 142, 173–178.

REINDL, J. (1989). Situation und Prognose des Forstschädlingsbefalls in Bayern. *Allgemeine Forstzeitschrift* 44, 344–348.

SCHWENKE, W., BÄUMLER, W., KOSCHEL, H., MATSCHEK, M. and ROOMI, M.W.(1970). Über die Verteilung, Biologie und Ökologie von Enchytraeiden, Lumbriciden, Oribatiden und Collembolen im Boden schädlingsdisponierter und nicht disponierter Nadelwälder. *Anzeiger für Schädlingskunde* 43, 33–41.

2
Population Dynamics and Abundance of Deciduous Tree-dwelling Aphids

A. F. G. DIXON

School of Biological Sciences, University of East Anglia, Norwich, NR4 7TJ, UK

Introduction

Although aphid life cycles are undoubtedly complicated, the concern over overlapping generations has, I think, obscured the analysis of aphid population dynamics. One measure of the success (fitness) of an aphid clone is the number of eggs it gives rise to at the end of a season. The problem of detecting the causes of regulation in aphid populations then becomes: what factors affect this relationship? In this paper I shall use this approach, and by reference to my own results and those of the Rothamsted Insect Survey (RIS), show that the abundance of deciduous tree-dwelling aphids is regulated by negative feedback mechanisms. I offer suggestions for the marked differences in abundance of an aphid often observed between trees, even adjacent ones, and why some species of aphids are more abundant than others.

Long-term population trends

The changes in abundance of the beech (*Phyllaphis fagi* (L.)), sycamore (*Drepanosiphum platanoidis* (Schr.)) and Turkey oak (*Myzocallis boerneri* Stroyan) aphids from year to year (*Figures 1, 2*) reveal that, although each species seems to be regulated at a particular density, they fluctuate to varying degrees around these levels. The relationships between log (N_{t+1}/N_t) (i.e. -*k* value of Varley and

Gradwell, 1960) and log (N_t) where N_t and N_{t+1} represent numbers in years t and $t+1$, respectively, were calculated for the beech, lime (*Eucallipterus tiliae* (L.)), oak (*Thelaxes dryophila* (Schr.)) and sycamore aphids from up to 20 years of suction trap catches, and for the average number of Turkey oak aphids per leaf each year over a period of 14 years (*Table 1*). This and other studies (Dixon, 1970b; Barlow and Dixon, 1980; Wellings *et al.*, 1985) reveal that in all species studied there is a strong to overcompensated density-dependent factor or factors operating between years. However, this analysis has two statistical problems in that (i) there are errors in both N_{t+1} and N_t and (ii) the points are not independent. These difficulties can be overcome by following the method employed by Wellings *et al.* (1985) who found that similar significant relationships are obtained for the sycamore aphid even when more rigorous statistical tests are applied.

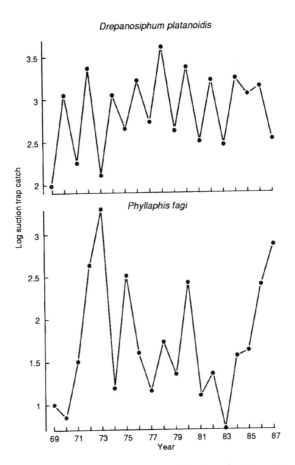

Figure 1.   The total numbers of sycamore aphids, *Drepanosiphum platanoidis*, and beech aphids, *Phillaphis fagi*, caught each year in the suction trap at Dundee, Scotland 1969–1987.

In some species like the sycamore aphid, the changes in abundance from year to year tend to follow a similar pattern throughout the UK (Dixon, 1979), whereas this is not the case in other species like the beech aphid. Also the year-to-year

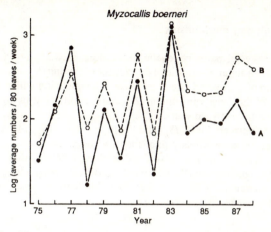

Figure 2. The year-to-year changes in the average numbers of Turkey oak aphids, *Myzocallis boerneri*, per leaf observed on two adjacent Turkey oak trees in Earlham Park, Norwich, England from 1975–88.

Table 1. The relationships between log (N_{t+1}/N_t) and log (N_t) for five species of deciduous tree-dwelling aphids. The relationships result from an analysis of the suction trap catches of beech, lime, oak and sycamore aphids taken at Rothamsted, England, and beech and sycamore aphids at Dundee, Scotland. There was no significant difference in the relationships for the beech aphids caught at Dundee and Rothamsted so the catches were pooled. The relationships for the Turkey oak aphid were obtained from an anlysis of the 14-year population census of the numbers of this aphid present on the leaves of two mature Turkey oak trees in Earlham Park, Norwich. (*$P<0.005$; **$P<0.01$; ***$P<0.001$)

Aphid	Tree	Relationship Log (N_{t+1}/N_t) and log N_t	n	r
Suction trap data				
E. tiliae 1967–87	Lime	$y = 2.44 - 1.48\overset{*}{x}$	20	0.86***
T. dryophila 1967–87	Oak	$y = 0.95 - 0.72x$	20	0.59**
D. platanoidis 1969–87	Sycamore			
Dundee		$y = 4.60 - 1.59\overset{**}{x}$	19	0.92***
Rothamsted		$y = 1.82 - 0.71x$	19	0.59**
P. fagi 1969–87	Beech	$y = 1.38 - 0.86$	38	0.63***
Population census data				
M. boerneri 1975–88	Turkey oak	$y_1 = 3.53 - 1.5\overset{*}{x}$	14	0.90***
		$y_2 = 3.07 - 1.5\overset{*}{x}$	14	0.90***

changes in suction trap catches of beech, lime, oak and sycamore aphids do not appear to be synchronized. This tends to indicate that the disturbing factor is different for each species of aphid.

The timing of regulatory mechanisms

Although there are several parthenogenetic generations between egg hatch in spring and egg laying in autumn, the ratio of the densities in spring (S_n) and autumn (A_n) gives a measure of the overall rate of increase of a population within a year. Similarly, the ratio of the density in autumn (A_n) and the following spring (S_{n+1}) gives the rate of increase between years.

The relationships between log (A_n/S_n) and log (S_n), and log (S_{n+1}/A_n) and log (A_n) in *Table 2*, and those previously published for the alder (*Pterocallis alni* (Deg.)), lime and sycamore aphids (Dixon and Barlow, 1980; Wellings *et al.*, 1985; Gange, 1985) all indicate that the density-dependent factor operating within years tends to be overcompensating and that operating between years close to perfectly density dependent.

Table 2.   The relationships between log (A_n/S_n) and log (S_n), and log (S_{n+1}/A_n) and log (A_n) derived from suction trap and population census data for two species of deciduous tree-dwelling aphids. These relationships were derived from the same data as used for *Table 1* plus a long term population census of the sycamore aphid carried out at Glasgow, Scotland 1966–73. (c.f. Wellings *et al.* 1985)

		Relationships				
		Within years			Between years	
	n	Log (A_n/S_n) and log (S_n)	r	n	Log (S_{n+1}/A_n) and log (A_n)	r
Suction trap data						
D. platanoidis						
Dundee	19	$y = 2.71 - 1.23x$	*** 0.78	18	$y = 2.04 - 0.84x$	** 0.62
Rothamsted	19	$y = 1.86 - 1.11x$	*** 0.74	18	$y = 1.11 - 0.47x$	* 0.48
Population census data						
D. platanoidis						
Tree G_1	8	$y = 3.68 - 1.32x$	** 0.85			
Tree N_2	8	$y = 2.69 - 1.32x$				
M. boerneri						
Tree 1	14	$y = 1.83 - 1.59x$	*** 0.90	13	$y = 2.55 - 0.72x$	** 0.73
Tree 2	14	$y = 2.82 - 1.59x$				

Regulatory mechanisms

The characteristic feature of the within-year population dynamics of these aphids is that high populations in spring are followed by low populations in autumn and *vice versa* (*Figure 3*; Dixon, 1970b) Predators and parasites undoubtedly play a role in determining the magnitude of the dramatic crash in numbers that occurs in early summer in those years when aphids are very abundant in spring, especially in species like the lime and Turkey oak aphids. However, they do not account for the lack of recovery of the population in autumn. For example, from week 13 onwards the autumnal population of the Turkey oak aphid in 1982 and 1983 follow markedly very different trends but both experienced similar slight predation pressure from week 19 onwards. That a similar overcompensated response is observed in the within-year population dynamics of caged populations (see *Figure 6*) also argues against natural enemies having an important role in determining the response.

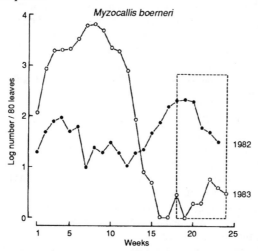

Figure 3. The Turkey oak aphid population trends observed in Earlham Park in 1982 and 1983. The dashed box marks the period of sexual production, with eight times more sexual females produced in 1982 than in 1983.

In the Turkey oak aphid poor performance in autumns of those years when the aphid is abundant in spring is associated with the small size of the aphid. For example, in the autumn of 1983 this aphid was less than half the size, and therefore less fecund, than those present in 1982. Similarly, studies of the sycamore aphid have revealed marked delayed density-dependent effects on fecundity and adult size (quality) (Dixon, 1975; Chambers *et al.* 1985).

Initially, these changes in aphid quality were thought to be driven by aphid induced changes in the quality of their hosts because these aphids undoubtedly have a marked affect on the growth and senescence of the leaves of their host trees (Dixon, 1970b, 1971b, c). However, sycamore saplings subject to very heavy aphid infestations in spring or the previous autumn and spring were no less suitable as hosts for aphids in autumn than previously uninfested saplings (Dixon, 1975 and unpublished results). The concentration of amino acids in the autumnal leaves of previously infested and uninfested saplings is also similar (Wellings and Dixon, 1987). There is therefore no evidence to support the suggestion (Dixon, 1970b) that a delayed aphid-induced change in host-plant quality is important in determining the overcompensated response, at least in the sycamore aphid.

These experiments are open to criticism because they were done on saplings rather than mature trees. The observation that poplar saplings are less suitable than mature trees for the aphid *Pemphigus betae* Doane (Kearsley and Whitham, 1989) tends to reinforce this criticism. However, in the case of sycamore, the indications are that in terms of the effects of the aphid on the growth of trees, and the quality of aphids, saplings are very similar to mature trees (Dixon, 1971b, c).

In contrast, lime trees appear to show an induced response as aphids reared in autumn on lime saplings that were heavily infested with aphids in the spring do worse than aphids reared on control saplings (Barlow and Dixon, 1980). This may be due to the damage inflicted on the phloem system and chloroplasts by feeding aphids. Such damage has been reported for the three aphids infesting pecan (*Carya illinoensis* Koch), which can reach densities that result in 100 000 stylet punctures per leaflet (Tedders and Thompson, 1981).

In the case of the sycamore aphid the strong density-dependent response between years is mainly due to the effect of crowding on dispersal and adult quality, both of which have a marked effect on recruitment (Dixon, 1969; Chambers *et al.*, 1985). Thus, as with the within-year dynamics, intraspecific competition appears to be a major regulatory factor.

To summarize, the studies reported above and those of Gange (1985) on the alder aphid and those of Sluss (1967) and Dixon (1977) on the walnut aphid (*Chromaphis juglandicola* (Kalt.)) all indicate that changes in aphid quality caused by intraspecific competition is a major factor in the population dynamics of deciduous tree-dwelling aphids. Natural enemies tend to be most active at the time when aphid recruitment is declining because of intense intraspecific competition. Thus, the activity of natural enemies is likely to reduce numbers to a lower level than would result if intraspecific competition were operating alone. The marked delay between the cause, high abundance in spring, and its effect on egg laying in autumn possibly accounts for the overcompensated density-dependent response in the within year population dynamics. High abundance in spring has an immediate effect on recruitment and hence the strong density-dependent response between years.

Difference in aphid abundance between trees

Some trees show some regularity in being either heavily infested with aphids and yet others lightly infested, from year to year (*Figure 4*). There has been a tendency to attribute this to either plant resistance or exposure to the elements, which affect a tree's suitability for aphids.

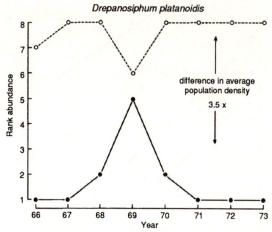

Figure 4. The rank abundance of sycamore aphids on two trees in Glasgow, Scotland 1966–73 (from Wellings *et al.* 1985).

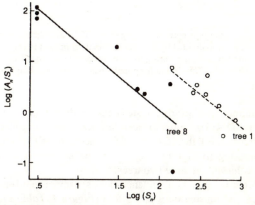

Figure 5. The relationships between the logarithm of the average numbers of sycamore aphids present in autumn (A_n) divided by the average numbers present in spring (S_n) and the logarithm of the average number present in spring for two trees (1+8) in Glasgow, Scotland 1966–73.

For both the sycamore and Turkey oak aphids, the within-year population dynamics on the trees most heavily and lightly infested, show the same overcompensated response (*Figures 5, 6*). However, the elevation of the relationships is significantly higher for the heavily infested trees. How can we account for this?

In the case of the Turkey oak aphid (*Figure 6*), the two trees are similar in terms of the success of the aphids in switching from asexual to sexual reproduction and survival of the overwintering eggs. However, the leaves of the more heavily infested tree start to senesce earlier and are shed some two weeks earlier than those of the other tree. The consequence of this is that although the reproductive rate in the first half of the year is the same on both trees, the reproductive rate is very much

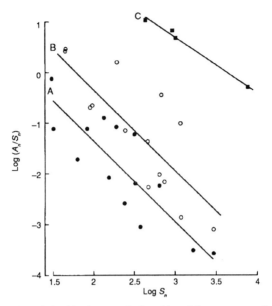

Figure 6. The relationships between the logarithm of the average numbers of Turkey oak aphids present in autumn divided by the average numbers present in spring and the logarithm of the average number present in spring for two adjacent Turkey oak trees (A and B) in Earlham Park, Norwich 1975–88 and for populations on four caged saplings (C).

higher in autumn on the heavily infested tree. The difference in average abundance on the two trees (1.4x) is similar to the difference in the average fecundity achieved on the two trees (1.5x), therefore, the difference in the elevations of the relationships between $\log (A_n/S_n)$ and $\log S_n$ can be accounted for in terms of the recruitment achieved on these two trees.

The difference in average abundance of the sycamore aphid on the most heavily and least heavily infested trees is 3.5-fold (*Figure 4, Table 3*). This cannot be attributed to the reproductive rate achieved on the two trees, or the success in switching from asexual to the sexual egg-laying forms. The most striking difference between the two trees is in how successful the aphid is in establishing itself each spring from overwintering eggs. Bud burst on the lightly infested tree occurs very late and few of the aphids that hatch survive to colonize the unfurling leaves. Although this is undoubtedly a major factor keeping the average level of abundance low on this tree, it does not account for the significant difference in the elevations of the relationships between $\log (A_n/S_n)$ and $\log S_n$ for the two trees (*Figure 5*). This indicates that there is another important factor.

The population trends of Turkey oak aphids on caged saplings show the same within year dynamics but the elevation of the relationship is very much higher (*Figure 6*). The explanation offered for the difference in abundance of the aphids on the two trees sampled in the field can be used to account for the generally higher level of abundance on the caged saplings. The aphid populations on the caged

Table 3. The average abundance (A), recruitment (B), success in switching from asexual to sexual reproduction in autumn (C) and in recruiting aphids [F] to the population in spring relative to the numbers [A] present the previous autumn (D) for the aphids on the two sycamore trees referred to in Figures 4 and 5.

	A	B	C	D
Tree	Av. logarithm of the total numbers recorded each year 1966–73	Av. recruitment/adult/day 1968–73	Av. success in switching from asexual to sexual reproduction 1966–73	Av. log (F_{n+1}/A_n) 1966–73
1	4.09] ***	0.45	0.66	−0.35] *
8	3.52	0.44	0.67	−0.98

saplings will have experienced very low losses due to dispersal and the realized recruitment on the saplings will be very much higher than on the trees sampled in the field. Losses due to dispersal could therefore be a major factor determining the average level of abundance on a tree. Sycamore tree 8 (*Table 3*) is different from all the other trees sampled in being some distance from another sycamore tree and completely surrounded by oaks. Thus losses of adults by dispersal are likely to greatly outnumber gains by recruitment of adults from other trees.

To summarize, tree phenology, in particular both the time of leaf senescence relative to the switch from asexual to sexual reproduction in autumn and the time of bud burst relative to egg hatch, is implicated in determining the difference in abundance of aphids between trees. In addition, it is suggested that isolated trees support fewer aphids because they sustain much higher losses from dispersal than do trees growing in clumps.

Abundance of aphids

The equilibrium population density of a species is seen as the outcome of the interaction between its rate of increase and the strength of the density-dependent factor acting upon it. Natural enemies do not appear to have a great role in regulating the abundance of deciduous tree-dwelling aphids. Moreover, there is no evidence that the efficiency of the natural enemies of the various aphid species differ; therefore, differences between aphid species in r_m, or more particularly realized rate of increase, R, appear to be the most likely cause of differences in abundance between species. The theoretical grounds for this is presented in Dixon and Kindlmann (1990).

Analysis of the empirical data in the previous sections has given the following difference equation for the between-year dynamics

$$\text{Log } X_{t+1} = \log X_t + \log R - M \log X_t \tag{1}$$

where X_t and X_{t+1} are the peak numbers in spring of year t and $t+1$, R the realized rate of increase and M the density-dependent factor. After antilogging, equation

(1) gives:

$$X_{t+1} = RX_t^{1-M}$$ (2)

The equilibrium density ($X*$) of which is

$$X* = R^{1/M}$$ (3)

What factors are likely to affect the degree to which r_m is realized? Dixon *et al.* (1987) argued that one such factor is the probability of finding a host plant [$P(C)$] assuming aphids disperse regularly. Thus, the realized rate of increase, R, which includes the losses incurred in dispersal is

$$R = r_m P(C)$$ (4)

Given that the probability of finding a host plant [P(C)] after D trials is (Dixon *et al.*, 1987):

$$P(C) = 1 - (1 - C)^D$$ (5)

then the equilibrium density is given by:

$$X^* = \left\{ r_m [1 - (1 - C^D)] \right\}^{1/M}$$ (6)

This indicates that, all other things being equal, the proportional cover of the host plant through its effect on realized r_m can markedly affect the abundance of an aphid.

The indigenous deciduous tree-dwelling aphids of Britain all belong to the same family, the Callaphididae. They are either highly host-specific or live on at most two species of a particular genus of tree. It is possible to obtain a qualitative estimate of the abundance of these aphids from Stroyan's (1977) key to this family of aphids. Similar information for their host trees is not available. However, Dr O Rackham, Botany Department, Cambridge University, who has a great knowledge of British trees and woodlands based on extensive field studies, kindly agreed to rank the trees in terms of their abundance in Britain. The relation between the ranked abundance of the aphids and that of their host plants is given in *Figure* 7. This lends support to the idea that plant abundance is a major factor determining aphid abundance.

Discussion

Ignoring the complexity of aphid life cycles and looking for relationships between the numbers of aphids present at the beginning and end of each year and between years has simplified the analysis of the population dynamics of host-specific deciduous tree-dwelling aphids. The seven species so far studied show similar population dynamics: overcompensated density dependence within years, and strong density dependence between years. The strong density-dependent factor operating between years tends to dampen the disturbing effect of the overcompensated density dependent factor operating within years. However, unlike other insects feeding on trees (Edmunds and Alstad, 1981) predators and parasites do not

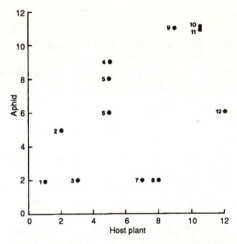

Figure 7. The relationship between the rank order of abundance of 12 species of deciduous tree-dwelling aphids and the rank order of abundance of their host plants ($r_s = 0.52$, $P <$ 0.05). (1) oak aphid, *Tuberculoides annulatus* (Hart.); (2) beech aphid, *P. fagi;* (3) birch aphid, *Euceraphis punctipennis* (Zett.); (4) alder aphid, *P. alni*; (5) sweet chestnut aphid, *Myzocallis castanicola* Baker; (6) hornbeam aphid, *Myzocallis carpini* (Koch); (7) sycamore aphid, *D. platanoidis;* (8) lime aphid, *E. tiliae*; (9) field maple aphid, *D. aceris*; (10) Turkey oak aphid, *M. boerneri*; (11) holm oak aphid, *Myzocallis schreiberi* H.R.L. and Stroyan; (12) walnut aphid, *C. juglandicola*.

play an overwhelming role in the population dynamics of the four species of tree-dwelling aphids that have been studied in detail (e.g. Barlow and Dixon, 1980; Chambers *et al.*, 1985; Wellings *et al.*, 1985). Changes in aphid quality caused by intraspecific competition appears to be the major factor regulating their numbers.

It is generally assumed that the differences in abundance of herbivores between trees, like the difference in abundance of aphids between leaves, can be attributed to either differences in microclimate (Dixon and McKay, 1970) or level of defence (nutritive quality) (Edmunds and Alstad, 1981). With the notable exception of the study by Crawley and Akhteruzzaman (1988) the role of host-plant phenology in the population dynamics of herbivorous insects has been ignored. Although much remains to be done to find other determinants of between-tree differences in aphid abundance, there are good grounds for claiming that plant phenology, in particular the time of bud burst and leaf senescence, plays a major role, which is understandable in terms of aphid reproduction and survival

There is no evidence to support the idea that natural enemies account for the marked differences in abundance observed between similar-sized closely related species of tree-dwelling aphids, even those belonging to the same genus, e.g. the field maple aphid (*D. aceris* Koch) is uncommon and the sycamore aphid (*D. platanoidis*) common, yet they share the same parasitoids and aphid specific predators. However, there are good theoretical grounds and some empirical evidence for the idea that host plant abundance plays a major role in determinig the differences in abundance of tree-dwelling aphids (Dixon and Kindlmann, 1990).

Acknowledgements

I am greatly indebted to Mark Tatchell for details of the Rothamsted Insect Survey suction trap catches, to Sue Mitchell for drawing the figures, Geoff Cleveland for his help with sampling Turkey oak aphids and to Nigel Barlow and Aulay Mackenzie for reading and commenting on an early draft of the manuscript.

References

BARLOW, N.D. and DIXON, A.F.G (1980). *Simulation of Lime Aphid Population Dynamics*. Wageningen: Pudoc.

CHAMBERS, R.J., WELLINGS, P.W. and DIXON, A.F.G. (1985). Sycamore aphid numbers and population density II. Some processes. *Journal of Animal Ecology* **54**, 345–442.

CRAWLEY, M.J. and AKHTERUZZAMAN, M. (1988). Individual variation in the phenology of oak trees and its consequences for herbivorous insects. *Functional Ecology* **2**, 409–415.

DIXON, A.F.G. (1969). Population dynamics of the sycamore aphid *Drepanosiphum platanoides* (Schr.) (Hemiptera:Aphididae) : migratory and trivial flight activity. *Journal of Animal Ecology* **38**, 585–606.

DIXON, A.F.G. (1970a). Quality and availability of food for a sycamore aphid population In: *Animal Populations in Relation to their Food Resources*, pp. 271–287 (ed. A. Watson). Oxford: Blackwell,

DIXON, A.F.G. (1970b). Stabilization of aphid populations by an aphid induced plant factor. *Nature* **227**, 1368–1369.

DIXON, A.F.G. (1971a). The role of intra-specific mechanisms and predation in regulating the numbers of the lime aphid *Eucallipterus tiliae* L. *Oecologia* (Berl.) **8**, 179-193.

DIXON, A.F.G. (1971b). The role of aphids in wood formation 1. The effect of the sycamore aphid *Drepanosiphum platanoides* (Schr.) (Aphididae) on the growth of sycamore *Acer pseudoplatanus* (L.). *Journal of Applied Ecology* **8**, 165–79.

DIXON, A.F.G. (1971c). The role of aphids in wood formation II. The effect of the lime aphid *Eucallipterus tiliae* L. (Aphididae) on the growth of lime *Tilia* x *vulgaris* Hayne. *Journal of Applied Ecology* **8**, 393–399.

DIXON, A.F.G. (1975). Effect of population density and food quality on autumnal reproductive activity in the sycamore aphid, *Drapanosiphum platanoides* (Schr.). *Journal of Animal Ecology* **44**, 297–304.

DIXON, A.F.G. (1977). Aphid ecology: life cycles, polymorphism and population regulation. *Annual Review of Ecology and Systematics* **8**, 329–353.

DIXON, A.F.G. (1979). Sycamore aphid numbers : the role of weather, host and aphid. In: *Population Dynamics*, pp.105–121 (eds R. M. Anderson, B.D. Turner and L.R. Taylor). Oxford: Blackwell.

DIXON, A.F.G. and BARLOW, N.D. (1980). Population regulation in the lime aphid. *Zoological Journal of the Linnean Society* **67**, 225–237.

DIXON, A.F.G. and McKAY, S. (1970). Aggregation in the sycamore aphid

Drepanosiphum platanoides (Schr.) (Hemiptera : Aphididae) and its relevance to the regulation of population growth. *Journal of Animal Ecology* 39, 439–454.

DIXON, A.F.G. and KINDLMANN, P. (1990). Role of plant abundance in determining the abundance of herbivorous insects. *Oecologia* (Berl.)(in press).

DIXON, A.F.G., KINDLMANN, P., LEPS, J. and HOLMAN, J. (1987). Why there are so few species of aphids, especially in the tropics. *American Naturalist* 129, 580–582.

EDMUNDS, G.F. and ALSTAD, D.N. (1981). Responses of black pineleaf scales to host variability. In: *Insect Life History Patterns: Habitat and Geographic Patterns*. pp. 29–38 (eds R.F. Denno and H. Dingle). New York: Springer.

GANGE, A.C. (1985). The ecology of the alder aphid (*Pterocallis alni* (Degeer)) and its role in integrated orchard pest management. PhD Thesis, University of London. 556 pp.

KEARSLEY, M.J.C. and T.G. WHITHAM (1989). Developmental changes in resistance to herbivory : implications for individuals and populations. *Ecology* 70, 422–434.

SLUSS, R.R. (1967). Population dynamics of the walnut aphid *Chromaphis juglandicola* (Kalt) in northern California. *Ecology* 48, 41–58.

STROYAN, H.L.G. (1977). Homoptera Aphidoidea (Part). Chaitophoridae and Callaphididae. *Royal Entomology Society of Lond. Handbooks for the Indentification of British Insects* 11 (Part 4a) 130 pp.

TEDDERS, W.L. and THOMPSON, J.M. (1981). Histological investigation of stylet penetration and feeding damage to pecan foliage by three aphids (Hemiptera (Homoptera): Aphididae). *Southeastern Fruit and Tree Nut Research Laboratory, Science and Education Administration, USDA Miscellaneous Publication* 12, 69-83.

VARLEY, G.C., and GRADWELL, G.R. (1960). Key factors in population studies. *Journal of Animal Ecology* 29, 399-401.

WELLINGS, P.W. and DIXON, A.F.G. (1987). Sycamore aphid numbers and population density III. The role of aphid-induced changes in plant quality. *Journal of Animal Ecology* 56, 161-170.

WELLINGS, P.W., CHAMBERS, R.J., DIXON, A.F.G. and AIKMAN, D.P. (1985). Sycamore aphid numbers and population density 1. Some patterns. *Journal of Animal Ecology* 54, 411-424.

3

The Abundance of Spruce Aphids under the Influence of an Oceanic Climate

KEITH DAY AND STEPHEN CRUTE

Department of Biological and Biomedical Sciences, University of Ulster, Coleraine, BT52 1SA, Northern Ireland, UK

Introduction

High population densities of the green spruce aphid (*Elatobium abietinum*) may cause substantial needle loss in Sitka spruce *(Picea sitchensis)*. In continental regions of Europe holocyclic populations of the aphid reduce the risk of winter mortality by entering the egg stage. However, it is generally agreed that in more maritime climates anholocyclic populations are strongly governed by low winter temperatures (Powell and Parry, 1976). Furthermore, Ohnesorge (1961), Bejer-Petersen (1962) and Carter (1972) have found relationships between mild winters and extensive defoliation the following summer.

Forests studied in Northern Ireland are characterized by the mild winters of a maritime climate. Clare Forest, near the Co. Antrim coast is typical in this respect (Day, 1984, 1986). Only limited, mainly western and coastal regions of the United Kingdom share a similar climate in the coldest month of the year (Anon., 1975). It should follow that in most years *E. abietinum* will survive the winter so well that it is a constant threat to these coastal forests, and there are certainly years when defoliation is locally acute, but these are exceptional and more often needle retention is moderately good.

Winter temperatures most likely to kill aphids by freezing are those below $-5°C$ which occur after a period of much higher mean temperature to which aphids are acclimatized (Carter, 1972). A number of factors moderate the ability of aphids to

Population Dynamics of Forest Insects
© Intercept Ltd, PO Box 716, Andover, Hampshire, SP10 1YG, UK

avoid freezing (Bevan and Carter, 1980) and the definition of a lethal temperature varies accordingly. There is, however, broad consensus among authors that severe frosts kill substantial numbers of overwintering aphids and greatest emphasis has been placed upon this in interpretations of population patterns.

These, and other interpretations of the processes governing aphid abundance from year to year in a forest under maritime influence, can be examined now that population estimates are available for several consecutive years.

Methods

The field studies were undertaken at Clare forest, Co. Antrim. Estimates of aphid population density were made for the same 15 trees for eight years from 1982. This formed part of a wider study commencing in 1981 and reported elsewhere (Day, 1984, 1986).

Trees were sampled once annually at the end of May when the mean aphid population density was near its maximum level (Day, 1984 and *Figure 1*). To provide further assurance that population maxima were being represented by the sampling date, three censuses were made on 25 May, 31 and 7 June in 1988.

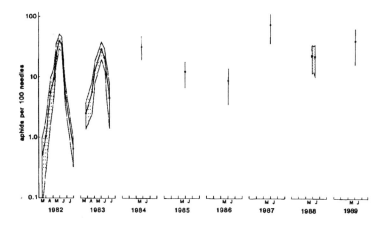

Figure 1.   Estimates of mean population density of aphids from 15 trees at Clare forest. Shaded areas or vertical bars represent 95% confidence limits of the mean values.

Foliage samples were cut in the forest and examined for aphids in the laboratory within a two-day period. Samples awaiting examination were kept in a cold room at approximately 4°C. All needles were counted and population estimates expressed as aphids per 100 needles; mean needle number per sample shoot was normally in excess of 200 needles.

Records of daily maximum and minimum air temperature were obtained from the University of Ulster meteorological station at Coleraine. Aphid physiological time (Hughes, 1972) was calculated as accumulated day °C above the threshold of

4°C for each daily mean temperature from January 1st. Thresholds for development appear to be inversely age-dependent (Crute, unpublished) and the adoption of a single figure is a compromise about a range of values for which 4°C is the approximate mean. From the beginning of May each year the forest was visited regularly to check on foliar phenology. The date of bud burst was estimated as the day when half of the lateral buds on half of the trees reached the stage when green needles first break free from the bud scales (Lines and Mitchell, 1966; Cannell and Smith, 1983).

Results

POPULATION LEVELS

Mean aphid population density from fifteen sample trees is recorded in *Figure 1.* The three 1988 estimates, covering the normal census period, are very similar and provide support for the choice of the end of May as a period representing the population maximum. During eight years, peak aphid numbers only once (in 1987) reached a level where needle loss eventually became generally apparent in the forest, although in other years, notably 1989, some of the sample trees were severely defoliated. Peak aphid numbers were lowest in 1986. Net population change between consecutive years is presented alongside an annual temperature profile in *Figure 2.*

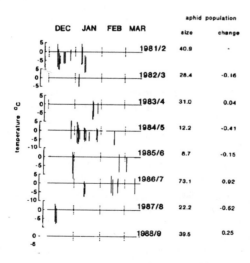

Figure 2. Profiles of low winter temperatures and aphid population indices. Each vertical bar denotes a day when minimum temperature fell to or below –5°C. The lower level of a bar indicates the minimum temperature reached, and the upper level indicates the mean daily temperature over the previous five days. Aphid population size for each summer maximum is recorded as the mean aphid number per 100 needles, and population change between consecutive years is the difference between $\log_{10}$ population estimates.

REGULATION

A plot of population change, log (N_{t+1}/N_t), against population size (N_t) is significant and, although departing from the strictest assumptions of regression, is nevertheless strongly indicative of the existEnce of density dependent processes operating between years (*Figure 3*). The predictive value of log (N_t) is high, 61% of the variation in the rate of population change being explained by population levels in the previous year.

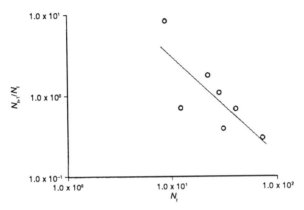

Figure 3.   The density dependent relationship between the rate of change in population size between successive years and the population maximum in the current year. $\text{Log}(N_{t+1}/N_t) = 1.69 - 1.21 \log(Nt)$, $P_{b=0} = 0.04$, $r^2 = 0.61$.

TEMPERATURE EFFECTS

The days when temperature dropped to or lower than $-5°C$ are indicated in *Figure 2*. Temperatures above this value are unlikely to freeze aphids and since, it is a sudden temperature fall which is most likely to effect mortality when they are physiologically unprepared (Carter, 1972), the drop in temperature from the mean during a preceding period of five days is also given. The overall patterns of cold temperature are not easily interpreted in relation to population size. In this form *Figure 2* contains little evidence for a simple relationship between either the severity of temperature reduction or the frequency with which such cold days occurred during a winter, and either subsequent summer maximum aphid density or the net population change taking place between years. The winters of 1982/3 and 1987/8 were relatively mild and yet were followed by only moderate aphid levels. Relatively severe winters in 1981/2 and 1985/6 were followed by high and low aphid populations respectively.

However, since there is a strong suggestion of regulation between years, it is the residuals from the density-dependent relationship (*Figure 3*) which can be examined for any sign that departures from equilibrium are determined by winter weather. Three main patterns of temperature are considered here. Although they are not quantitatively independent they affect aphids in different ways.

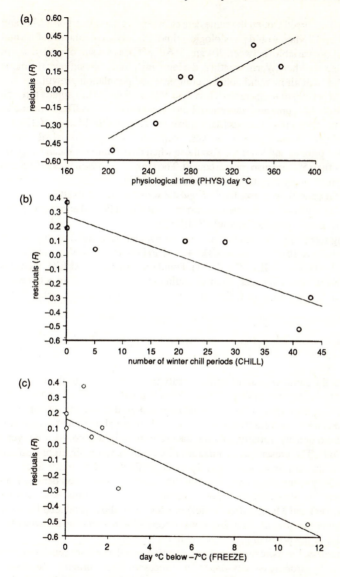

Figure 4.   The relationships between the residuals from the density dependent relationship (*Figure 3*) and climatic variables. (*a*) The physiological time accumulated from 1 January to bud burst. Physiological time (PHYS) is in units of day °C above a threshold of 4°C, accumulated from 1 January. $R = 0.0048 \, PHYS - 1.383$, $P_{b=0} = 0.001$, $r^2 = 0.77$. (*b*) The number of winter periods where mean daily temperature did not exceed 5°C for 14 or more consecutive days (*CHILL*). $R = 0.2742 - 0.0139 \, CHILL$, $P_{b=0} = 0.016$, $r^2 = 0.72$. (*c*) Low winter temperatures liable to kill aphids by freezing (FREEZE). Temperature data are accumulated day °C below a daily minimum of −7°C during the preceding winter. $R = 0.1586 - 0.063 \, FREEZE$, $P_{b=0} = 0.019$, $r^2 = 0.70$.

Bud burst occurred around the same date each year, at the earliest on 8 May and at the latest on 18 May. Aphid physiological time (PHYS) accumulated before bud burst, was more variable (between 204 and 366 day °C), as a result of cool or warm winter/spring weather. Warm weather, during which more physiological time is accumulated, will allow rapid aphid recruitment and population growth.

Extended periods at temperatures less than 5°C are liable to kill aphids through a syndrome of chill coma and starvation (Powell and Parry, 1976). The frequencies of moderately cool winter periods (daily mean at 5°C or less for 14 days) (CHILL), were calculated for each year. However, years when less physiological time is accumulated prior to bud burst are also those when days without a mean temperature greater than 5°C are also most frequent. The temperature variables PHYS and CHILL were strongly negatively correlated ($r^2 = 0.86$).

Very cold conditions presented in *Figure 2* were further indexed by accumulated days °C below –7°C (frequent occurrences of which Powell and Parry, 1976, suggest will keep a population low) (FREEZE).

A strong relationship exists between actual population size (N) each year and the two variables, PHYS ($\ln N = 0.315 + 0.01$ PHYS, $P = 0.0065$) and CHILL ($\ln N = 3.92 - 0.0314$ CHILL, $P = 0.011$). Population size and FREEZE were not significantly related. However, all three could independently explain between 70 and 80% of the variation in residuals from the density dependent relationship (*Figure 4*).

Discussion

Elsewhere, the summer peak of aphid density has been related to the density of aphids surviving the winter (Parry, 1969; Powell and Parry, 1976; Dixon, 1988). However, the data from which this relationship is derived are drawn from observations on some trees that had experienced chemical treatment designed to depress aphid density initially, and the data refer to only two consecutive years (Parry, 1969). The present study includes observations spanning eight years on aphids from the same trees where aphid population fluctuated naturally. Such data are more likely to highlight any processes which govern population density from year to year, but are deficient in not including population estimates in October (before winter) and March (after winter) in addition to those produced for May/June (summer maximum). Such additional information would have allowed clear comparison of the population trends within each year.

Nevertheless, the indications that spruce aphid populations are regulated from year to year by strongly density-dependent processes is encouraging. Regulation at this level and with no additional disturbance would bring peak aphid populations to a stable equilibrium around 20 or 30 aphids per 100 needles. Density dependent processes which might account for this have not yet been highlighted in the ecology of spruce aphids, except in a spatial sense (Day, 1986). Population growth on individual trees was inversely density dependent and could be explained by centrifugal dispersal and loss of older aphids (Day, 1986), but it is not known whether such effects are translated into temporal density dependence of sufficient strength. Other biotic processes are currently being investigated through simulation modelling (Crute and Day, this volume, Chapter 29).

Further departures from equilibrium may be attributable to winter weather. There can be no doubt that aphids are killed by freezing. Powell and Parry (1978) have observed this under field conditions, while it has been demonstrated experimentally by Carter (1972), Powell, (1974) and Parry (1979). It is less certain whether temperatures reached in the field are low enough to be effective in generating large changes in aphid population from year to year, particularly in forests experiencing mild maritime winters. The falls in temperature illustrated in *Figure 3,* if compared with Carter's (1972) data, were probably not great enough to effect high percentage aphid mortalities by freezing. The greatest fall in temperature, a range of 14°C to −12.5°C, occurred during the winter of 1985/6 and might well have generated significant population change, but for the eight years studied such temperature change was exceptional.

The winters studied varied also in the frequency of periods when mean daily temperature remained below °5C. Such temperatures are at or below the developmental threshold for spruce aphids (Crute, unpublished) and have been described as a threshold for population increase or decrease (Dixon, 1985). Powell and Parry (1976) observed reductions in aphid numbers through chill coma and starvation when temperatures did not rise above 5 or 6°C for two to three weeks. Coma starvation or simply failure to develop may be important factors, but field studies are necessary to quantify them further. Pre-freeze processes at low temperature are also known to be a dominant lethal effect in the overwintering biology of anholocyclic *Myzus persicae* (Bale, Harrington and Clough, 1988). As with the spruce aphid, the advantages to predominantly anholocyclic clones of the peach-potato aphid in areas with mild winters are partially offset by the increased risks of experiencing occasionally lethal winter weather.

When daily mean temperatures rise above 4°C during the winter and spring, aphid development and reproduction is expected. Developmental rates are temperature related (Fisher, 1982), so that higher temperatures will generally result in faster population growth (Dixon, 1985). This aspect has not been ignored in previous work (Powell and Parry, 1976) but might now be emphasized more as a determinant of population maxima each year at mild-winter sites.

Population growth is particularly pronounced when needle sap nutrient levels are high, before bud burst. It is known that amino acid levels in previous year's needles decline from early May onward (Parry, 1974). Bud burst is a convenient reference point for the rapidly changing nutritional conditions available to aphids (Carter and Cole, 1977). Patterns of fecundity in field populations of aphids also suggest that numbers not influenced by other factors will grow substantially faster before bud burst than after it (Parry, 1974; Parry and Powell, 1977; Day, 1984).

The total physiological time (accumulated day °C > 4°C) available for population growth will be greater in years with mild winters and warm springs, but such conditions do not advance bud burst accordingly. Studies of foliar phenology (Cannell and Smith, 1983) show that thermal time to bud burst is prolonged when fewer winter chill days are experienced. Normally, so few chill days occur in coastal Northern Ireland that the relationship developed by Cannell and Smith (1983) to predict bud burst is inappropriate (Day, 1984). Bud burst occurs at about the same time each year at Clare forest but is preceded by very different periods of physiological time. It is these that generate the greatest deviations from aphid equilibrium density from year to year. Milder winters followed by warm springs will enhance aphid population growth by extending the aphid physiological time elapsed before bud burst and the onset of deleterious nutritional conditions.

In conclusion, the absence of huge fluctuations in density from year to year are testimony to the existence of regulatory phenomena, while the major density disturbing factors in the population dynamics of spruce aphids are liable to occur during the winter and spring. Although all temperature variables could account for much of the additional variation in aphid abundance, physiological time and chill periods were thought to be most influential, and mortality through freezing less so, since deep frosts are less frequent than in other parts of the British Isles and Europe.

Acknowledgements

We are grateful to the Northern Ireland Forest Service for access to the forest site and to the Department of Agriculture for Northern Ireland for a studentship to S. J. C. Professor A. F. G. Dixon was kind enough to provide some useful insights into the data.

References

ANON, (1975). Maps of mean and extreme temperature over the United Kingdom 1941–1970. *Climatological Memorandum* 73, 53 pp. Bracknell: Meteorological Office.

BALE, J. S., HARRINGTON, R. and CLOUGH, M.S. (1988). Low temperature mortality of the peach-potato aphid *Myzus persicae*. *Ecological Entomology* 13, 121–129.

BEJER-PETERSEN, B. (1962). Peak years and the regulation of numbers in the aphid *Neomyzaphis abietina* Walker. *Oikos* 16, 155–168.

BEVAN, D. and CARTER, C.I. (1980). Frost proofed aphids. *Antenna* 4, 6–8.

CANNELL, M.G.R. and SMITH, R.I. (1983). Thermal time, chill days and prediction of budburst in *Picea sitchensis*. *Journal of Applied Ecology* 20, 951–964.

CARTER, C.I. (1972). Winter temperatures and survival of the green spruce aphid, *Elatobium abietinum* (Walker). *Forest Record* (Forestry Commission, UK) 84, 1–10.

CARTER, C.I. and COLE, J. (1977). Flight regulation in the green spruce aphid *(Elatobium abietinun)*. *Annals of Applied Biology* 89, 9–14.

DAY, K.R. (1984). The growth and decline of a population of the green spruce aphid *Elatobium abietinum* during a three year study, and the changing pattern of fecundity, recruitment and alary polymorphism in a Northern Ireland forest. *Oecologia* 64, 118–124.

DAY, K.R. (1986). Population growth and spatial patterns of spruce aphids (*Elatobium abietinum*) on individual trees. *Journal of Applied Entomology* 102, 505–515.

DIXON, A.F.G. (1985). *Aphid Ecology*. 157 pp. Glasgow and London: Blackie.

FISHER, M. (1982). Morph determination in *Elatobium abietinum* (Walk.) the green spruce aphid. Unpublished PhD thesis, University of East Anglia.

HUGHES, R.D. (1972). Population dynamics. In: *Aphid Technology*, pp. 275–293 (ed. H.F. van Emden). London and New York: Academic Press.

LINES, R. and MITCHELL, A.F. (1966). Differences in phenology of Sitka spruce provenances. *Report on Forest Research for 1965* pp. 173–184. London: HMSO.

OHNESORGE, B. (1961). Wann sind Schaden durch die Sitkalaus zu erwarten? *Allgemeine Forstzeitschrift* **16**, 408–410.

PARRY, W.H. (1969). A study of the relationship between defoliation of Sitka spruce and population levels of *Elatobium abietinum* (Walker). *Forestry* **42**, 69–82.

PARRY, W.H. (1974). The effects of nitrogen levels in Sitka spruce needles on *Elatobium abietinum* populations in northeastern Scotland. *Oecologia* **15**, 305–320.

PARRY, W.H. (1979). Acclimatisation in the green spruce aphid, *Elatobium abietinum*. *Annals of Applied Biology* **92**, 299–306.

PARRY, W.H. and POWELL, W. (1977). A comparison of *Elatobium abietinum* populations on Sitka spruce trees differing in needle retention during aphid outbreaks. *Oecologia* **27**, 239–252.

POWELL, W. (1974). Supercooling and the low-temperature survival of the green spruce aphid *Elatobium abietinum*. *Annals of Applied Biology* **78**, 27–37.

POWELL, W. and PARRY, W.H. (1976). Effects of temperature on overwintering populations of the green spruce aphid *Elatobium abietinum*. *Annals of Applied Biology* **82**, 209–219.

4

The Role of Host Quality, Natural Enemies, Competition and Weather in the Regulation of Autumn and Winter Populations of the Bird Cherry Aphid

SIMON R. LEATHER

Forestry Commission, Northern Research Station, Roslin, Midlothian, EH25 9SY, UK

Introduction
    Life cycle
    Sampling
Autumn populations
    Host selection
    Natural enemies
Host quality
    Population development in roadside and woodland environments
    Induced defences and interspecific competition
Overwintering studies
    Weather
    Predation
    Competition
Conclusions
Acknowledgements
References

Introduction

The bird cherry aphid, *Rhopalosiphum padi,* is a common host-alternating aphid found throughout northern Europe, although in some parts of its range it is anholocyclic (Leather, Walters and Dixon, 1989). It spends the summer on graminaceous hosts, including cereals, and the rest of the year on its primary host, bird cherry, *Prunus padus* (Leather, 1990). Although *R. padi* has been the subject of a number of studies (e.g. Rogerson, 1947; Dixon, 1971, 1976; Dedryver, 1978; Leather and Dixon, 1981; Wiktelius and Pettersson, 1985), these have mainly addressed that part of the life cycle occurring on cereals and grasses (but see Dixon, 1976 and Leather, 1990). This part of the life cycle, although important, has been reviewed recently (Leather, Walters and Dixon, 1989) and will only be touched on slightly in this paper.

    This paper focuses on the biology and ecology of *R. padi* on *P. padus* in Scotland, during the autumn and winter months. New information is presented and

Population Dynamics of Forest Insects
© Intercept Ltd, PO Box 716, Andover, Hampshire, SP10 1YG, UK

the role of host quality, natural enemies, competition and weather at this time of year in the population regulation of *R. padi* assessed.

LIFE CYCLE

Rhopalosiphum padi hatches from the egg to produce an apterous (non-winged), highly fecund fundatrix, which gives rise to an even more fecund apterous second generation, the fundatrigeniae (Leather and Dixon, 1981), which in turn gives rise to an alate emigrant third generation which migrates from *P. padus* to the secondary graminaceous host (Dixon, 1971). In autumn, in response to shortening day lengths and decreasing temperatures (Dixon and Dewar, 1974), the gynopara, a non-feeding parthenogenetic female, returns from its graminaceous host to *P. padus* (Leather, 1981a, 1982), where it gives rise to the egg-laying sexual females (oviparae). These mate with the males, which also originate on Gramineae but at a later point in the progeny sequence (Dixon and Dewar, 1974), and lay their eggs in the axils of the buds of current year shoots during September and October (Leather, 1980, 1981b).

SAMPLING

Six *P. padus* trees have been sampled for *R. padi* at Roslin Glen, at least once a week since 1982. In 1987, four further trees, also in Roslin Glen but growing within 3 m of a road have also been sampled for *R. padi*. All of the trees are surrounded by similar vegetation, are at an almost identical altitude and have the same degree of exposure to the sky (Leather, unpublished). Sampling varies according to the stage of the life cycle and is described in Leather and Lehti (1981, 1982).

Autumn populations

In autumn, *P. padus* is recolonized by the gynoparae of *R. padi* flying from the secondary graminaceous hosts (Leather, 1990). The number of colonizing gynoparae is dependent on population development on grasses and cereals and is linked with weather; mild wet summers resulting in large numbers of gynoparae (A'Brook, 1981). The summer populations on grasses and cereals are dependent on the number of aphids leaving *P. padus* in early summer and arriving on the secondary host (Leather, 1990). All things being equal, a large spring population should result in a large autumn population. However, this is not the case. In fact, only in 1985 did this happen, and it is interesting to note that 1985 was characterized as being one of the wettest summers on record.

The date of first arrival of gynoparae of *R. padi* on *P. padus* at Roslin Glen, has shown great variability ranging from 7 August in 1988 to 18 September in 1983. This has been shown to be an adaptation to the relationship existing between the timing of leaf fall and August weather (Ward, Leather and Dixon, 1984). Autumn populations have also shown great variations in size ranging from 19.9 aphids/10 leaves in 1983 to 1318.2 aphids/10 leaves in 1986.

The population developing on *P. padus* in autumn is not so much a result of the number of gynoparae arriving, but a consequence of the fecundity of those gynoparae once they arrive. As the gynoparae of *R. padi* do not feed (Leather, 1982), the differences in fecundity (ranging from 3.23 oviparae/gynopara in 1983 to 9.2 oviparae/gynopara in 1986) must be attributable to either their pre-flight nutrition or the weather (in particular temperature) at the time of settling on *P. padus*. The mean fecundity of gynoparae at Roslin Glen or elsewhere, has never been shown to reach the potential maximum (24 oviparae/gynopara) (Leather, 1981a).

HOST SELECTION

Although all the trees sampled at Roslin Glen are very close together and growing under virtually identical conditions, the aphid populations that they support are far from identical. This is a result either of differential colonization or differences in survival and development brought about by differing host quality within the same species. As differences between trees tend to be consistent from year to year (Leather, 1986), it would appear that one of the major differences must be in terms of host quality. Gynoparae do not settle on *P. padus* trees at random and the number of gynoparae landing on each tree is positively and significantly correlated with the fecundity of the preceding spring generations on the same trees (Leather, 1986). Thus it is apparent that the gynoparae of *R. padi* are not only able to detect differences in host quality between individual *P. padus* trees but to modify their reproductive behaviour accordingly (Leather, 1986).

NATURAL ENEMIES

The numbers of natural enemies in autumn on *P. padus* vary little from year to year and have very little bearing on aphid population densities. The most common predators are coccinellids, anthocorids, syrphid larvae and spiders. The numbers of coccinellids, anthocorids and syrphids are much lower than those found in the spring but spiders are more common in the autumn. This is a result of both their role as non-specific predators and their migratory behaviour. Although spring and autumn populations of *R. padi* reach similar levels, the numbers of predators, even when spiders are included, are lower in autumn than in spring. This could be a result of the fact that natural enemies in general are frequently in the early stages of hibernation in autumn. Alternatively, as autumn morphs of *R. padi* are smaller than those occurring in the spring (Dixon, 1976; Leather and Dixon, 1981; Leather, 1988) they provide less food per aphid and thus support a lower natural enemy population, despite occurring in similar numbers to their spring counterparts. There is little or no evidence to suggest that natural enemies on *P. padus* have a regulatory effect on *R. padi* numbers in the autumn.

Host quality

Although there are intrinsic differences in host quality within closely related *P. padus* growing in the same vicinity (Leather, 1986), other factors can affect the host quality of individual trees, even those that have suckered from a common parent and are thus of the same clonal material.

POPULATION DEVELOPMENT IN ROADSIDE AND WOODLAND ENVIRONMENTS

Roadside *Prunus padus* trees appear to be preferred hosts for *R. padi*, with greater numbers of autumn migrants settling on them than on comparable trees in woodland areas. Preliminary analysis and work in progress appears to indicate that the nutrient status of roadside trees is different from that of woodland trees, total nitrogen content of leaves being significantly higher in roadside trees in autumn (2.34% compared to 1.86%, $P<0.001$, $n = 32$). Spring populations also develop at a greater rate on roadside trees than they do on woodland trees, the fecundity of the fundatrices and fundatrigeniae being higher on roadside trees than on woodland trees. The duration of the populations in both spring and autumn are longer on roadside trees than on woodland trees and egg mortality rates are quite different (*Table 1*). Motorway pollution has been shown to increase aphid populations on

Table 1. The influence of environment on life history parameters of *R. padi* on *P. padus*. (SU = sample units.)

	1987		1988	
	Roadside	Woodland	Roadside	Woodland
Peak autumn count (aphids/10 SU)	450.0	84.7	1148.2	457.1
Eggs/100 buds	17.0	0.8	108.2	12.1
% overwintering mortality	48.1	87.5	51.1	81.0
Peak spring count (aphids/10 SU)	2321.7	484.3	125.9	6.3
Duration of autumn population (days)	81	67	57	46
Duration of spring population (days)	77	69	105	91

plants (Bolsinger and Fluckiger, 1987) as has SO_2 gas (Warrington, 1987). It appears that the suitability of *P. padus* for aphid growth and reproduction are favourably altered by low levels of pollution. The reduced overwintering mortality on roadside trees is marked and the magnitude of it (30–40% less than that suffered by eggs on woodland trees) would suggest that it is predation that has been reduced by a considerable amount. Predators, as would be expected, are more susceptible to the effects of pollution than their aphid prey. Full details of this work will be presented at a later date.

INDUCED DEFENCES AND INTERSPECIFIC COMPETITION

Early season defoliation of *Prunus padus* trees, such as that caused by the bird cherry ermine moth *Yponomeuta evonymellus*, has a marked effect on the subsequent herbivore damage experienced by those trees later in the season and also reduces the number of eggs laid on them by *R. padi* (Leather, 1990). Work in progress suggests that this could be the result of chemical changes in the leaves: nitrogen levels are significantly increased by severe defoliation (Leather, unpublished data) and marked morphological changes occur to the host trees (Leather, 1988) which make the tree more suitable to colonization by the moth. A significantly negative correlation has been shown to exist between overwintering *R. padi* egg numbers and overwintering *Y. evonymellus* numbers on *P. padus* (Leather, 1988). As both these insects share a similar phenology and are both host specific to *P. padus*, there is every reason to suppose that interspecific competition does take place, but until further analysis is performed this will have to remain an area of speculation.

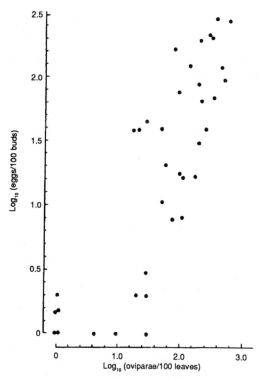

Figure 1. Relationships between numbers of oviparae of *Rhopalosiphum padi* and eggs laid on different *Prunus padus* trees, 1983–88.

Overwintering studies

Rhopalosiphum padi overwinters as an egg in the axils of the buds of *P. padus* (Leather and Lehti, 1981) and is thus exposed to the vagaries of winter weather. Egg populations vary greatly from year to year. Over the study period initial egg populations have ranged from 3.5 eggs/100 buds (1987–88) to 215.1 eggs/100 buds (1985–86) and this is closely correlated with the number of oviparae present (*Figure 1*).

Eggs of *R. padi* suffer a considerable loss over the winter. At Roslin since 1982, this has on average ranged from 47 to 76%, but in Finland it can be as high as 84%. Overwintering mortality is closely correlated with the length of the winter (Leather, 1990) and tends to occur at an almost constant rate in both Britain (Leather, 1980, 1981b) and Finland (Leather and Lehti, 1981). Mortality does increase in the spring (Leather, 1980) and this could be due to the increase in activity of the major predators, arthropods and birds (Leather, 1981b).

WEATHER

Although the eggs are exposed to the full harshness of winter conditions, there is little evidence to suggest that low temperatures *per se* have any effect on egg mortality. What evidence there is seems to suggest that cold weather would have little effect on the eggs at all, as they are extremely cold hardy and are able to survive temperatures as low as $-37°C$ (Somme, 1969). Certainly egg mortality in Finland, where temperatures are much colder than in Britain, is no greater than would be expected from the length of the winter (Leather, 1990). Although the rate of loss in mild winters does not appear to be greater or less than in severe winters (Leather, 1990), it is possible that the spring mortality at egg hatch is greater in mild conditions due to increased predator activity.

PREDATION

Little work has been done on the effects of predators on overwintering aphid egg populations, although Way and Banks (1964) considered that birds and anthocorids were important in *Aphis fabae* egg mortality. Gange and Llewellyn (1988) found little evidence for bird predation of eggs of *Pterocallis alni* and ascribed the majority of predation to insects. In *R. padi*, egg mortality due to predation by birds can be as high as 15% and that due to arthropods 31% (Leather, 1981b). Eggs that are totally protected from predators suffer only 35% mortality, presumably from weather effects or natural intrinsic mortality. Thus predation over the winter is an important feature of the life cycle.

COMPETITION

There are three areas in which competition can occur at the overwintering stage: (a) competition between oviparae for mates, (b) interference competition between oviparae at oviposition, and (c) competition for egg-laying sites.

Competition for mates would result in an increase in the proportion of infertile eggs present. This has never exceeded 5% at Roslin and does not appear to be correlated in any way with the number of oviparae present. Interference between oviparae should result in reduced fecundities. Achieved fecundity within cohorts is very low, ranging from as low as 0.005 eggs/oviparae to 0.14 eggs/oviparae. However, the number of eggs laid is positively correlated to the number of

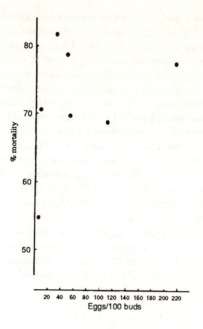

Figure 2. Relationship between eggs laid and overwintering mortality of *Rhopalosiphum padi*.

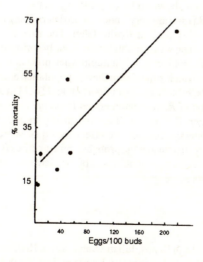

Figure 3. Relationship between initial egg density of *Rhopalosiphum padi* and mortality in the first two weeks of winter ($y = 0.254x + 20.8$, $r = 0.87$, $P<0.01$).

oviparae present (*Figure 1*) and, although fecundity varies significantly between years ($F = 5.39, 5/32\ df, P<0.01$) there is no significant correlation between density of oviparae and fecundity ($r = 0.10, n = 38$). There is thus no evidence for intraspecific interference competition in the oviparae of *R. padi*.

On the other hand competition for egg-laying sites may be more of a problem. Ideal egg-laying sites, i.e. tight buds on current year shoots, are in limited supply. Although the average bud has enough room for about 10 eggs, any more eggs laid in the vicinity of that bud would have to be laid either on the bud, on the bark or on top of other eggs. This will inevitably make the eggs more apparent to predators and more susceptible to weather effects such as rain. Total egg mortality is not density dependent (*Figure 2*) and this has been found to be so for other aphids, e.g. *Cinara pinea* (Kidd and Tozer, 1985). Initial egg mortality, i.e. that occurring in the first two weeks after leaf fall is however significantly correlated with initial egg density (*Figure 3*). This competition for egg-laying sites will cause an increase in the number of poorly attached eggs as well as an increase in their apparency to natural enemies. It is possible that the incidence of fungal disease may increase at very high densities; the oviparae certainly become very vulnerable under these conditions and their corpses are found adhering to the branches on and near the buds for a number of weeks after leaf fall (Leather, personal observation).

Conclusions

One of the most important factors regulating numbers of autumn populations of *R. padi* is host quality. However, the role of natural enemies should not be underestimated as they have been shown to be of great importance in other aphid/tree interactions (Dixon, 1985) and are very important in determining the initial size of the spring populations of *R. padi* at Roslin Glen. The role of natural enemies appears to be enhanced when weather conditions in early spring are more adverse than normal. Intraspecific competition in autumn when populations are extremely high can be important in reducing the size of the egg population entering the winter. Interspecific competition effects are also important and could in some years lead to a significant reduction of *R. padi* numbers on *P. padus* trees in certain areas. Populations of *Y. evonymellus* tend to fluctuate greatly in both time and space (Leather, 1986b) and thus the competitive effect would be sporadic rather than sustained. It is interesting to note that the probable intraspecific effect acts through insect-induced changes in the nutritional status of the plant rather than from direct insect–insect competition.

Acknowledgements

It is a pleasure to thank Hugh Evans, Julian Evans, Derek Redfern, John Stoakley and Allan Watt for their helpful criticism of this paper. I am indebted to the Scottish Wildlife Trust for permission to sample in their reserve.

References

A'BROOK, J. (1981). Some observations in west Wales on the relationships between numbers of alate aphids and weather. *Annals of Applied Biology* 97, 11–15.

BOLSINGER, M. and FLUCKIGER, W. (1987). Enhanced aphid infestation at motorways: the role of ambient air pollution. *Entomologia Experimentalis et Applicata* 45, 237–243.

DEDRYVER, C. A. (1978). Biologie des pucerons des céréales dans l'Ouest de la France. 1. Repartition et évolution des populations de *Sitobion avenae* F., *Metopolophium dirhodum* (Wlk.), et *Rhopalosiphum padi* L. de 1974 à 1977 sur blé d'hiver dans le bassin de Rennes. *Annales de Zoologique. Ecologie des Animales* 10, 483–505.

DIXON, A. F. G. (1971). The life cycle and host preferences of the bird cherry-oat aphid, *Rhopalosiphum padi* (L.) and their bearing on the theories of host alternation in aphids. *Annals of Applied Biology* 68, 135–147.

DIXON, A. F. G. (1976). Reproductive strategies of the alate morphs of the bird cherry-oat aphid, *Rhopalosiphum padi* (L.). *Journal of Animal Ecology* 45, 817–830.

DIXON, A. F. G. (1979). Sycamore aphid numbers: the role of weather, host and aphid. In: *Population Dynamics,* pp. 105–121 (eds R. M. Anderson, B. D. Turner and L. R. Taylor). Oxford: Blackwell.

DIXON, A. F. G. (1985). *Aphid Ecology.* Glasgow and London: Blackie.

DIXON, A. F. G. and DEWAR, A. M. (1974). The time of dermination of gynoparae and males in the bird cherry-oat aphid, *Rhopalosiphum padi* (L.). *Annals of Applied Biology* 78, 1–6.

GANGE, A. C. and LLEWELLYN, M. (1988). Egg distribution and mortality in the alder aphid, *Pterocallis alni*. *Entomologia Experimentalis et Applicata* 48, 9–14.

KIDD, N. A. C. and TOZER, D. J. (1985). Distribution, survival and hatching of overwintering eggs in the large pine aphid, *Cinara pinea* (Mordv.) (Hom., Lachnidae). *Zeitschrift für Angewandte Entomologie* 100, 17–23.

LEATHER, S. R. (1980). Egg survival in the bird cherry-oat aphid, *Rhopalosiphum padi*. *Entomologia Experimentalis et Applicata* 27, 96–97.

LEATHER, S. R. (1981a). Reproduction and survival: a field study of the gynoparae of the bird cherry-oat aphid, *Rhopalosiphum padi* (Homoptera: Aphididae) on its primary host *Prunus padus*. *Annales Entomologici Fennici* 47, 131–135.

LEATHER, S. R. (1981b). Factors affecting egg survival in the bird cherry-oat aphid, *Rhopalosiphum padi*. *Entomologia Experimentalis et Applicata* 30, 131–135.

LEATHER, S. R. (1982). Do gynoparae and males need to feed? An attempt to allocate resources in the bird cherry-oat aphid, *Rhopalosiphum padi*. *Entomologia Experimentalis et Applicata* 31, 386–390.

LEATHER, S. R. (1986). Host monitoring by aphid migrants: do gynoparae maximise offspring fitness? *Oecologia* 68, 367–369.

LEATHER, S. R. (1988). Consumers and plant fitness: coevolution or competition? *Oikos* 53, 285-288.

LEATHER, S. R. (1990). Two case studies: the pine beauty moth and the bird cherry aphid. In: *Pests, Pathogens and Plant Communities,* pp. 145–167 (eds J. J. Burdon and S. R. Leather). Oxford: Blackwell.

LEATHER, S. R. and DIXON, A. F. G. (1981). Growth, survival and reproduction of the bird cherry aphid, *Rhopalosiphum padi,* on its primary host. *Annals of Applied Biology* **99**, 115–118.

LEATHER, S. R. and LEHTI, J. P. (1981). Abundance and survival of eggs of the bird cherry-oat aphid, *Rhopalosiphum padi* in southern Finland. *Annales Entomologici Fennici* **47**, 125-130.

LEATHER, S. R. and LEHTI, J. P. (1982). Field studies on the factors affecting the population dynamics of the bird cherry-oat aphid, *Rhopalosiphum padi (L.)* in Finland. *Annales Agriculture Fenniae* **21**, 20–31.

LEATHER, S. R., WALTERS, K. F. A. and DIXON, A. F. G. (1989). Factors determining the pest status of the bird cherry-oat aphid, *Rhopalosiphum padi* (L.) (Hemiptera: Aphididae) in Europe: a study and review. *Bulletin of Entomological Research* **79**, 345–60.

ROGERSON, J. P. (1947). The oat bird-cherry aphis, *Rhopalosiphum padi* L., and comparison with *R. crataegellum,* Theo. (Hemiptera, Aphididae). *Bulletin of Entomological Research* **38**, 157–176.

SOMME, L. (1969). Mannitol and glycerol in overwintering aphid eggs. *Norske Entomologie Tidsskrift* **16**, 107–111.

WARD, S. A., LEATHER, S. R. and DIXON, A. F. G. (1984). Temperature prediction and the timing of sex in aphids. *Oecologia* **62**, 230–233.

WARRINGTON, S. (1987). Relationship between SO2 dose and growth of the pea aphid, *Acyrthosiphon pisum,* on peas. *Environmental Pollution Series A* **43**, 155–162.

WAY, M. J. and BANKS, C. J. (1964). Natural mortality of eggs of the black bean aphid, *Aphis fabae* Scop., on the spindle tree, *Euonymus europaeus* L. *Annals of Applied Biology* **54**, 255–267.

WIKTELIUS, S. and PETTERSSON, J. (1985). Simulations of bird cherry-oat aphid population dynamics; a tool for developing strategies for breeding aphid resistant plants. *Agriculture, Ecosystems and Environment* **14**, 159–170.

5

Long-term Population Studies of the Douglas-fir Tussock Moth in California

D. L. DAHLSTEN,* D. L. ROWNEY,* W. A. COPPER,*
S. M. TAIT* AND J. M. WENZ**

*Division of Biological Control, University of California,
Berkeley 94720, USA
**Stanislaus National Forest, USDA, Forest Service, Sonora,
California 95370, USA

Introduction

Douglas-fir tussock moth (DFTM), *Orgyia pseudotsugata* (McDunnough) (Lepidoptera: Lymantriidae) populations fluctuate from extremely low numbers for many years to massive outbreaks that cause serious defoliation and tree mortality of several conifers in western North America. There is good evidence that outbreak cycles are synchronized and occur approximately every nine years (Shepherd *et al.*, 1984). The primary host varies by region. In California, the primary host is white fir, *Abies concolor* (Gord. and Glend.) Lindl.

    The tussock moth is univoltine and overwinters as eggs. Larvae hatch in the spring shortly after fir buds begin to burst in June. The first instar larvae are the major dispersing stage since adult females are wingless and lay eggs on their cocoons. Larvae develop through five to six instars and begin pupating in late July.

Population Dynamics of Forest Insects
© Intercept Ltd, PO Box 716, Andover, Hampshire, SP10 1YG, UK

Adults emerge between August and November. Details of the biology and host preference can be found in Brookes, Stark and Campbell (1978).

In this paper the development of sampling procedures, useful at low populations, for defoliators and associated arthropods on white fir is described. A monitoring technique was developed which may be useful for long-term population studies, predicting outbreaks, and evaluating control programmes (Dahlsten *et al.*, 1977; Dahlsten *et al.*, 1985). Other workers have developed methods for sampling tussock moth egg masses and larvae on Douglas-fir in British Columbia (Shepherd, Otvos and Chorney, 1984; Shepherd, 1985) and larvae on white fir in the western United States (Mason, 1970, 1977, 1987).

Methods

INTENSIVE TREE SAMPLING 1976-77

Two areas in California were selected for sampling, based on DFTM activity in previous years. Three plots were established in each area: at Yellowjacket Springs, Tom's Creek, and Roney Flat in Modoc County; and at Iron Mountain, Plummer Ridge, and Baltic Ridge in El Dorado County.

A road ran lengthwise through each plot, and plot lengths were measured by odometer; they were approximately 2–5 km long. Each plot was divided into quarters, and two spots were randomly selected in each quarter. At each spot, the nearest white fir between 9 and 12 m in height became the first sample tree. Sample spots were permanent and were revisited each subsequent sampling period; since the sampling was destructive, on each subsequent visit the 9 to 12m white fir nearest to the originally selected sample tree was chosen. Eight trees, one from each of the eight spots in a plot, were sampled in each of the six plots during each sample period, giving a sample size of 48 trees per period. Five periods during the DFTM generation were sampled in 1976. The five trees in each spot, therefore, spanned the development of the DFTM generation and gave phenological information for the white fir defoliator guild, and for predators and parasites.

Computer-generated random number lists were used to select a random sample of one-third of all branches on each sample tree. Sample branches were caught in large canvas bags and beaten over a large canvas on the ground. Foliage area and all insects and spiders found were recorded. A crew of three or four, processing from two to four trees per day, was needed for the intensive sampling procedure. In spring and fall the above sampling was supplemented by a search of the entire tree for DFTM cocoons and egg masses, since these occurred in relatively low numbers.

Because cocoons and egg masses were rare in 1976, in 1977 only the two plots with the most 1976 cocoons and egg masses were sampled during the first and fifth periods. No sampling was done during the third period.

Field data sheets were designed for direct keyboard entry, and computer programs were written to produce summaries of each insect species' density by whole trees, plots, areas and by each of 12 equal crown levels. Another program was written to simulate sampling in different ways. This program gave variance, bias and cost figures necessary to sample a plot at any level of precision for each sampling method.

EL DORADO COUNTY PERMANENT STUDY PLOTS

The El Dorado Plots — Iron Mountain (IM), Plummer Ridge (PR) and Baltic Ridge (BR) were established in 1971 in co-operation with the Pacific Northwest Forest and Range Experiment Station, USDA Forest Service, Corvallis, Oregon (Dahlsten *et al.*, 1977). These plots have been maintained to the present with some replacement of plots due to tree mortality followed by logging. Each of the areas is a mountain ridge running west to east and with elevations ranging from 1700 to 1830 m. Iron Mountain is the lowest ridge and Baltic Ridge is the highest. The ridges consist of mixed west side Sierra conifer stands dominated by white fir. Each ridge was divided into eight plots of approximately 4 ha (10 acres). Initially 10 tussock moth larval sample white fir trees were selected per plot, giving 80 sample trees per ridge. Variations from this scheme are described below.

CRYPTIC SHELTERS 1978–86

After DFTM were observed pupating in nest boxes used to study cavity-nesting birds (Dahlsten and Copper, 1979), five other types of cryptic shelters were studied in 1977 (Dahlsten *et al.*, 1985). In 1978, two types of shelters were used based on cost and ease of use, since there appeared to be no major preference differences among the five types. Twenty paper carton cryptic shelters, 10 on host white firs and 10 on adjacent non-host trees, were installed in each of eight plots at each of the three areas (total of 24 plots with 480 shelters). In addition, 10 wood block shelters (10 x 10 x 5 cm with two 2.5 cm holes) were installed on the same host trees on the IM plot (total of 80). Since numbers of cocoons were similar for paper cartons and wood blocks on the same trees at IM (*Figure 1*), in 1979 all paper cartons were replaced with the more durable wood blocks. These were left in place for subsequent seasons as a monitoring tool. In 1986 each established cryptic plot was supplemented with 10 additional host trees with two 4-hole wood shelters. As the 2-hole shelters deteriorate, they are being replaced by 4-hole shelters on all trees. Egg masses found in shelters are reared and dissected each year to determine parasitism.

LARVAL SAMPLING 1971–86

Mid-crown sampling of the trees in the Eldorado plots began in 1971. All the white firs were approximately 12 m high (40 ft.) and relatively open grown. Samples were taken according to a scheme developed by Mason (1970), and consisted of three twigs, two from the inside crown and one from the outside, each approximately 38 cm (15 in.) long. Samples were taken on several dates in some of the years from 1971 to 1977 but only one date for the early instar data is reported here. Densities of larvae were recorded per 0.64 m² (1000 sq. in.) of foliage. The foliage area was calculated by assuming a foliated twig approximated a triangle whose base was the widest point of the twig and whose height was the length of the foliated twig.

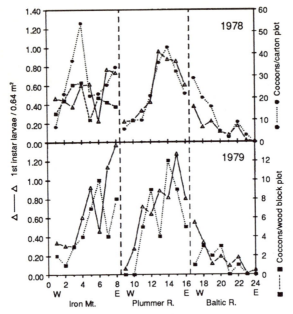

Figure 1.   Douglas-fir tussock moth cocoon densities from paper carton and wooden block cryptic shelters compared to three crown level early instar larval estimates on 24 plots on three ridges in 1978–79 in El Dorado County, California.

Lower crown beating (LCB) (Mason, 1977) small larval estimates were made on six of the 24 plots in 1978–79 and on the nine pheromone plots from 1984–86 (*Figure* 2). The closest 100 white fir trees of any size, including three crown level (3CL) sample trees in 1978 and 1979, to each of the plots were sampled. Sampling consists of beating three branches in the lower crown over a drop cloth hand-held beneath at least 56 cm (22 in.) of branch, and recording only the presence or absence of larvae.

In 1978 and 1979 3CL first instar larval sampling was carried out in the same areas and plots as cryptic shelters. Ten white firs in the 9–12 m class were chosen per plot, and two branch tips about 0.7 m long (approx. 0.64 m² or 1000 sq. in.) were cut with pole pruners from the lower, middle and upper crown. The branches were caught in a basket attached to a telescoping pole, and each branch was beaten over a white sheet. DFTM larvae and associated arthropods were counted. Mean larvae per 0.64 m² of foliage was calculated for each tree from the mean of all branch samples, and means per plot were calculated from the tree means.

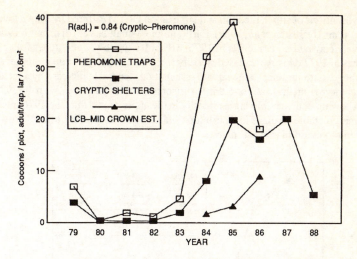

Figure 2. Comparison of pheromone traps (1979–86), cryptic shelters (1979–88), and lower crown beating (1984–86) population estimates for the Douglas-fir tussock moth on all plots in El Dorado County, California.

PHEROMONE TRAPS 1978–86

Pheromone traps were used in each plot in 1978 and 1979. Traps, made from a half gallon milk carton, were lined with an adhesive to trap moths that entered. There were 10 traps per plot, baited with synthetic pheromone (Z-6-heneicosen-11-one) formulated in plastic pellets. One pellet was placed in each trap by impaling it on a pin, sticking the pin through an interior wall of the trap, and bending the tip of the pin on the outside of the trap to hold it in place. Each trap was hung near the end of a relatively open-grown host tree branch. Traps were placed a minimum of 23 m apart at the beginning of flight season and monitored weekly throughout the flight period. From 1980 to 1986 the pheromone traps were spread out along each ridge and did not necessarily coincide with the cryptic shelter plots. However trap population estimates from each ridge can be compared with the cryptic shelter estimates by ridge (*Figure 2*).

Results

INTENSIVE TREE SAMPLING 1976–77

Foliage distributions of DFTM and associated insects were calculated by twelfths of the live crown from the whole-tree sample of 48 trees per period with both areas combined. Numbers of egg masses and cocoons were too low to estimate meaningful distributions.

Distributions of small, medium and large larval DFTM differed by crown level and by years (Dahlsten *et al.*, 1989). Early summer (small larvae) distribution was relatively constant across levels in 1976, except for the lower and upper foliage, whereas in 1977 density increased steadily from the lower to the upper quarter of the foliage. Late summer (large larvae) distributions tended to increase by a factor of 10 or more from the lower one-third to upper one-third of the trees, with the 1977 trees showing considerably higher density in the upper crown than the 1976 trees. The unpredictable changes in distribution indicate the need for multilevel crown sampling to avoid biased estimates.

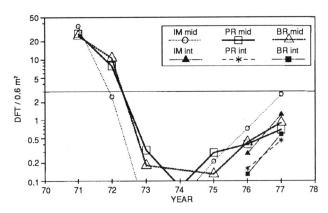

Figure 3. Douglas-fir tussock moth early instar larval estimates from midcrown sampling of white fir, 1971-77, and intensive tree sampling, 1976-77, on three ridges in El Dorado County, California.

Early instar DFTM larval estimates for 1976 and 1977 on the three El Dorado County plots (IM, PR, BR) are shown in *Figure 3*. Note that the intensive sample estimates are lower than the mid-crown estimates on each plot and in each year.

If the objective of sampling is to estimate total larvae in foliage, a multilevel sample will be required for relatively precise, unbiased estimates. To illustrate this, a computer- generated subsampling of the original data from all trees was used, under a variety of sampling rules, to compare the DFTM density estimates they produced to density obtained using the complete intensive sample. Estimated cost figures were then used to determine the most efficient methods for given total error levels.

The computer sampling program simulated these sampling methods: two branches taken at random from the lowest 2 m; two branches from the middle third of the crown (mid-crown sampling method); two, three, or four branches from the whole crown at random; and two, three and four equal crown levels, with sets of two, three or four branches from each level. For each of these methods (a total of 14) the program calculated tree mean densities using means per level weighted by the average proportion of foliage per level.

Within-tree sampling error (*WSE*) was the square root of the variance of the density estimates for all possible samples. Between-tree errors (*BSE*) were calculated from the mean squared differences between area means and individual tree means. Bias was found by subtracting the density mean (*SM*) of the samples chosen by the program from the 'actual' (intensive sample) tree mean density (*AM*). Total standard error (*TSE*) for a sampling method with n sample trees was then calculated as:

$$TSE = \sqrt{((BSE^2 + WSE^2)/n + BIAS^2)}$$

where $BIAS = AM - SM$

When choosing a sampling method, it is important to use a method with low and stable bias, because bias cannot be reduced by increasing sample size. For the sampling methods tried above, the methods using two, three or four branches from two or three levels generally yielded the lowest percentage bias figures (Dahlsten *et al.*, 1989). The percentage bias for the lower 2m method and the mid-crown method were both high and unstable.

Comparisons between sampling methods may be made by selecting an acceptable level for *TSE* and calculating the number of trees and total branches required for a given mean density and its associated *BSE*, *WSE* and *BIAS*. Labour costs for a method may then be calculated from the estimated time to locate a tree and sample a branch. A conservative estimate is 15 min per tree, plus 3 min per branch for a crew of three people.

For example, in 1977 the mean density of small larvae was 0.434/0.6 m2, the *WSE* varied from 0.181 to 0.804, and the *BSE* varied from 0.409 to 0.441, depending on the sampling method. Total trees and person-hours needed to determine the mean with a *TSE* of 20%, 40% or 60% of the mean were calculated for each method, and trees were plotted vs. person-hours for different methods at

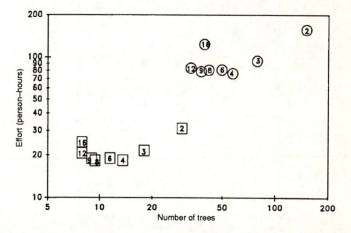

Figure 4. Time in hours to sample DFTM on different numbers of trees with varying numbers of branches per tree (numbers are the number of branches from 2 to 16) with 20% standard error (circles) and 40% standard error (squares). Based on 1977 Period 2 small larvae sampling, El Dorado and Modoc counties, California (from Dahlsten *et al.* 1989).

two error levels (*Figure 4*: example for Period 2, 1977). Only the low bias methods and more efficient of any two methods that used the same number of branches per tree are shown.

For any error rate, the minimum point for curves in terms of effort (person-hours) indicates the most efficient sampling for the time assumptions used. The three-level, two-branch-per-level method is a good choice as it is easy for field crews to divide a crown by eye into three levels, and it ensures a relatively representative sample, even if the branches chosen in each level are not random. Methods taking greater numbers of branches per tree are more likely to cause significant damage to the tree.

Tree and effort figures were calculated for all the sample periods in both years. Relationships between methods for other periods were similar to those for Period 2, 1977 (*Figure 4*). However, the numbers of trees necessary for a given proportional sample error increased significantly for sample periods with lower mean densities. Using the three-level, two-branch-per- level method, the number of trees necessary for standard errors of 20%, 40% and 60% of the mean were calculated

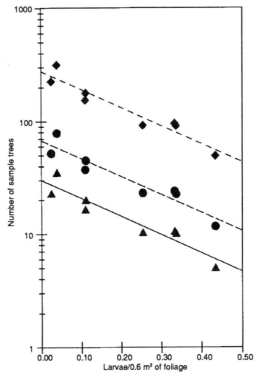

Figure 5.   Number of sample trees needed for sampling DFTM larvae at different densities for standard errors = 20% of mean (diamonds), 40% of mean (circles), and 60% of mean (triangles), using the three-crown-level, two-branch per-level method (from Dahlsten *et al.*, 1989).

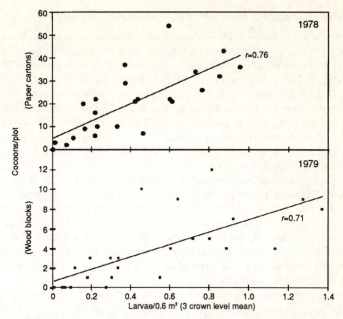

Figure 6.   Regression of three crown level larval estimates per plot in 1978-79 on cocoon densities from cryptic shelters per plot (paper cartons in 1978 and wooden blocks in 1979) on three ridges in El Dorado County, California.

and plotted vs. density, along with least squares regression lines for each error level (*Figure 5*). Using this figure, it is possible to plan a low-level population sampling programme, given the degree of precision required and an estimate of the populations in an area, perhaps from the previous year's population or a pilot study.

MID-CROWN SAMPLING OF EARLY INSTAR DFTM 1971-1977

Populations declined on all three ridges from 1971 to 1974 and then began to increase gradually from 1975 through 1977 (*Figure 3*). The whole-tree intensive sampling densities in 1976 and 1977 were consistently lower than the mid-crown estimates at the same locations. Multilevel sampling would give more accurate estimates at DFTM early larval densities below $3/0.64$ m^2. Above this level, as in 1971 and 1972, the mid-crown sampling is adequate and the upper and lower levels can be deleted, as the mid-crown branches will generally give densities representative of the overall tree population density. Defoliation is detectable at 20 larvae/0.64 m^2 (Mason and Luck, 1978), and economic damage occurs at early instar densities in excess of 50/0.64 m^2 (Wickman, 1978).

CRYPTIC SHELTERS AND LARVAL SAMPLING, 1978–79

Means of cryptic cocoon totals by ridge reflect differences seen between ridges in the 3CL larval sample (i.e. IM and PR are about the same, while BR is significantly lower). Numbers of cocoons in the paper cartons and wood blocks in 1978 were much higher than in the wood blocks in 1979 for similar larval populations in all areas (*Figure 1*). Assuming that paper cartons and wood blocks are similarly attractive as cocoon sites, the difference between 1978 and 1979 is probably due to greater larval mortality in later instars in 1979. The limited lower crown beating samples (two plots per ridge) gave estimates an order of magnitude higher than the more intensive 3CL sampling done at the same time, so they were considered too unreliable to be used in analysis.

A graph of cocoon sums and larval density means by plots along the ridges (*Figure 1*) shows a large variation between plots. In general, the geographic trend in cocoon counts along the ridges is similar to that of the larval density, although variation between two or three adjacent plots may be very high. Differences in cocoon trends compared to larval density could be due to both random variation and differential larval mortality in later larval instars. The 1979 cocoon counts, although averaging about one third of the 1978 counts, still show a geographic trend similar to the larval density.

Correlations were made between cryptic shelter plot totals and corresponding larval density and pheromone trap counts and pheromone trap numbers for the 24

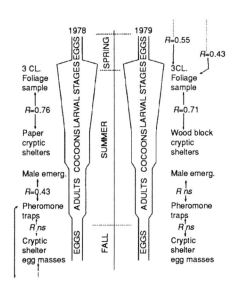

Figure 7   Summary of sampling procedures used for the Douglas-fir tussock moth life stages in 1978–79 in El Dorado County, California on three different ridges showing correlations between the different methods.

plots separately in 1978 and 1979. Correlations were regarded as significant if they exceeded $r = 0.404$, the 5% probability level for 22 degrees of freedom. The highest correlations were for total cocoon counts vs. larval density for both 1978 ($r = 0.76$) and 1979 ($r = 0.71$) (*Figure 6*). Some other variables are significantly correlated but with more variation. *Figure 7* shows a schematic of life stages with sampling methods and meaningful correlations tested. The correlation (0.55) between 1978 egg masses and 1979 larval density is not strong enough for making a prediction model, but does indicate a significant relationship and should be studied further.

Host tree shelters accounted for about three-quarters of the cocoons and egg masses found. Egg masses were about 20% of total cocoons in 1978 for both host and non-host trees; numbers of cocoons, and especially egg masses, were too low for reliable estimates in 1979.

CRYPTIC SHELTERS 1979–86

Cryptic data for each plot were pooled because DFTM populations were very low for most of these years. Total tussock moth count (egg mass count, etc.) for each plot was one sample, with eight samples (plots) per ridge. Analyses were not done on data for 1980–82 because the counts were near zero for all the ridges during these years. Data were summarized as totals per plot and per ridge, and the mean of the plot totals for each ridge. Chi-squared analyses were done comparing plots within each ridge for each year and between ridge totals for each year. Plot means and confidence intervals were plotted by year for all ridges combined and for each ridge.

No significant differences were found for between-ridge comparisons, except for the first year, 1979, when total tussock moth cocoons, and total tussock moth cocoons on host trees, were significantly lower on Baltic Ridge compared to the other two ridges. Within-ridge (between-plot) analyses showed significant differences in the following cases:

> 1985, IM ($P = 0.001$): high counts in upper plots
> 1985, BR ($P = 0.001$): high counts in lower plots
> 1984, IM ($P = 0.01$): high counts in middle plots
> 1984, BR ($P = 0.001$): high counts in lower-middle plots

Numbers from 1979 to 1983 were too low in many plots for valid tests.

Overall totals decreased from a low 1979 level to near zero in 1980–83. From 1984 to 1986 the means increased steadily to about five times the 1979 level. The same trend held for egg mass numbers. Variation increased more within ridges (between plots) than between ridges. Ridges were primarily different only in the location of their high and low plots along the east–west length of the ridges. When comparing by years (*Figure 2*), we can disregard ridge differences and use a sample size of 24 plots for each year.

For populations with ridge means higher than about 0.5 to 0.8 cocoon/shelter the 4-hole wood shelter type attracted more DFTM than did the 2-hole shelters. At lower populations, counts do not seem to be affected by the shelter type.

COMPARISON OF CRYPTIC SHELTERS WITH PHEROMONE TRAPS 1979–86

Means of pheromone trap catches were calculated by ridge for each year. Correlations were calculated using the pheromone means and the corresponding cryptic plot means. For all ridge-year combinations, r was 0.864; for IM only, where most of the pheromone data was collected, r was 0.916. For the relatively low populations of DFTM during these years, the cryptic counts were a significant predictor of both pheromone and larval samples, but the pheromone and larval samples were weakly or not correlated with each other.

Discussion

These methods have been tested from 1978 to 1986 over nearly an entire DFTM population cycle, from declining post-outbreak numbers through near-zero years to rising populations. Computer comparison of various larval sampling schemes indicates that sampling should be adjusted with population density for maximum efficiency and accuracy.

The wooden cryptic shelters are a promising, simple method to detect increasing DFTM populations. Cocoon counts appear to be a good indicator for the current year's larval density, while egg mass numbers should be studied as a potential index of increasing or decreasing future larval densities. Detection in the fall provides ample time for resource managers to plan control strategies. Placing a pair of 4-hole wooden blocks on 10 host trees in an area that is in the forest type susceptible to DFTM should be adequate. A number of plots should be evaluated in each area as there is considerable variation even among closely spaced plots. The technique may also be useful for long-term population studies.

The shelters should be examined in the fall after DFTM adult flight has terminated. Cocoons should be collected for rearing of parasitoids and determination of male emergence. Egg masses can be left until the following spring if information on total egg parasitization is desirable.

The cryptic shelters provide information valuable in predicting population trends for applied or basic studies: sex ratio, number of egg masses, number of eggs per egg mass, number of cocoons, late larval-pupal (cocoon) parasitism and egg parasitism. Shelter data may underestimate predation as it is negligible compared to predation on foliage (Dahlsten et al., 1985; Dahlsten and Copper, 1979).

The shelters have been used on two occasions to evaluate control programmes. A pheromone disruption programme was conducted near the El Dorado County plots by the USDA Forest Service in 1986 (L. L. Sower, personal communication). Cryptic shelters were used as one of the methods for evaluating efficacy. Females pupating in the cryptic shelters produced 80% fewer egg masses in the treated plots. The egg parasitoid Telenomus californicus was not adversely affected by the treatment. In another study by the Forest Service, the shelters were used in Plumas County, California to evaluate the effect of two concentrations of Bacillus thuringiensis. The shelters were placed as described above on host trees in treated and untreated areas. Preliminary results show no effects of *Bacillus* on natural enemies of the tussock moth, and a small treatment effect.

References

BROOKES, M. H., STARK, R. W. and CAMPBELL, R. W., (eds) (1978). *The Douglas-fir Tussock Moth: A Synthesis.* Washington, D C: United States Department of Agriculture, Forest Service. Technical Bulletin 1585. : 331 pp.

DAHLSTEN, D. L. and COPPER, W. A. (1979). The use of nesting boxes to study the biology of the mountain chickadee (*Parus gambeli*) and its impact on selected forest insects. In: *The Role of Insectivorous Birds in Forest Ecosystems,* pp. 217–260 (eds J. G. Dickson, R. W. Conner, R. R. Fleet, J. C. Kroll and J. A. Jackson). Proceedings of a symposium: The Role of Insectivorous Birds in Forest Ecosystems, 13–14 July, 1978, Nacogdoches, Texas. New York: Academic Press.

DAHLSTEN, D. L., LUCK, R. F., SCHLINGER, E. I., WENZ, J. M. and COPPER, W. A. (1977). Parasitoids and predators of the Douglas-fir tussock moth, *Orgyia pseudotsugata* (Lepidoptera: Lymantriidae), in low to moderate populations in central California. *Canadian Entomologist* **109**, 727–746.

DAHLSTEN, D. L., NORICK, N. X., WENZ, J. M., WILLIAMS, C. B. and ROWNEY, D. L. (1985). The dynamics of Douglas-fir tussock moth populations at low levels at chronically infested sites in California. In: *Site Characteristics and Population Dynamics of Lepidopteran and Hymenopteran Forest Pests,* pp. 132–139 (eds D. Bevan and J. T. Stokley). Proceedings of International Union of Forest Research Organizations Conference Subject Group S2.07.06, Dornoch, Scotland, 1–7 September, 1980. Edinburgh: Forestry Commission Research and Development Paper 135.

DAHLSTEN, D. L., COPPER, W. A, ROWNEY, D. L., and KLEINTJES, P. K. (1990). Quantifying bird predation of anthropods in forests. In: *Studies in Avian Biology* 13 (eds M. L. Morrison, C. J. Ralph and J. Verner). Proceedings of a symposium: Food Exploitation by Terrestrial Birds, 18 March, 1988, Asilomar, California. Cooper Ornithological Society.

MASON, R. R. (1970). Development of sampling methods for the Douglas-fir tussock moth, *Hemerocampa pseudotsugata* (Lepidoptera: Lymantriidae). *Canadian Entomologist* **102**, 836–845.

MASON, R. R. (1977). Sampling low density populations of the Douglas-fir tussock moth by frequency of occurrence in the lower tree crown. *USDA, Forest Service Research Paper* PNW–216, 8 pp. Pacific Northwest Forest and Range Experiment Station, Portland, Oregon.

MASON, R. R. (1987). Frequency sampling to predict densities in sparse populations of the Douglas-fir tussock moth. *Forest Science* **33**, 145–156.

MASON, R. R. and LUCK, R. F. (1978). Quantitative expression and distribution of populations. In: *The Douglas-fir Tussock Moth: A Synthesis,* pp. 39–41 (eds M. H. Brookes, R. W. Stark and R. W. Campbell). Washington, DC: United States Department of Agriculture, Forest Service Technical Bulletin 1585.

SHEPHERD, R. F. (1985). Pest management of Douglas-fir tussock moth: Estimating larval density by sequential sampling. *Canadian Entomologist* **117**, 1111–1115.

SHEPHERD, R. F., OTVOS, I. S. and CHORNEY, R. J. (1984). Pest management of Douglas-fir tussock moth (Lepidoptera: Lymantriidae): a sequential sampling method to determine egg mass density. *Canadian Entomologist* **116**, 1041–1049.

WICKMAN, B. E. (1978). Tree injury. In: *The Douglas-fir Tussock Moth: A Synthesis*, pp. 66–75 (eds M. H. Brookes, R. W. Stark and R. W. Campbell). Washington DC: United States Department of Agriculture, Forest Service Technical Bulletin 1585.

6

Species–area Relationships: Relevance to Pest Problems of British Trees?

M. F. CLARIDGE* AND H. F. EVANS**

*School of Pure and Applied Biology, University of Wales, PO Box 915, Cardiff CF1 3TL, UK
**Forestry Commission, Forest Research Station, Alice Holt Lodge, Wrecclesham, Farnham, Surrey GU10 4LH, UK

Introduction
Species–area relations: total insect faunas of British trees
Feeding guilds and species–area relationships
Quality of data: new estimates of areas for British trees
Complete herbivore fauna of British trees
Herbivore guilds on British trees
Tree area and pest status
Discussion and Conclusions

Introduction

Why should different plant species support often startlingly different numbers of associated insect herbivores? This apparently simple question has provided the stimulus for much research and even more discussion and argument by insect ecologists over the past 30 or so years. The relatively well known British flora, and in particular the tree flora, and its associated insects have been central to these discussions. Southwood (1961) was the first to gather from the literature data on insects associated with particular British trees and to attempt to find an explanation for his findings. His now classic study showed a significant relationship between the numbers of associated species and the abundance of host plant remains during recent geological time. Thus he concluded that both the present and past abundance of a plant largely determined the size of its associated insect fauna. Clearly, if such generalizations are valid, we may then have a basis for determining possible pest loads and even probability of invasion by additional herbivores on recently introduced tree species or on greatly expanded areas of forest plantations.

Population Dynamics of Forest Insects
© Intercept Ltd, PO Box 716, Andover, Hampshire, SP10 1YG, UK

Species–area relations: total insect faunas of British trees

Strong (1974a,b) was one of the first to follow the suggestion of Janzen (1968) that suitable host-plant species for phytophagous insects could be regarded as islands in a sea of other unsuitable habitats and thus could be interpreted in terms of island biogeography theory. Most importantly it led Strong to examine the insect data from Southwood's (1961) study and to analyse them as a species–area relationship. In order to obtain area data for trees in Britain, Strong counted the number of 10 x 10 km squares from which each was recorded in the Atlas of the British Flora (Perring and Walters, 1962). Thus, for each tree species or group of species a number of occupied squares was derived. Using the species–area function

$$S = k A^z$$

where S is number of associated insect species, A area occupied by host tree, and k and z are constants, he obtained a significant linear relationship using the log transformation of the function

$$\log S = \log k + z \log A$$

The relationship explained approximately 61% of the variation around the regression line ($r = 0.78$, $P < 0.001$). Further, using multiple regression analysis, Strong showed that, of Southwood's original data on present and past abundance of trees in the British flora, only present abundance contributed significantly to the relationship. Thus, Strong suggested that present area of a host tree overwhelmingly determined the species richness of associated herbivores in any region. Though other studies have been made, the British data are central to the validity of the species–area explanation of herbivore species richness on trees.

Very considerable arguments and controversy have raged and continue to do so over the detailed interpretation of regression data obtained from these and related studies (e.g. Connor and McCoy, 1979; Kuris, Blaustein and Alio, 1980; Rey, McCoy and Strong, 1981, 1982; Kennedy and Southwood, 1984). Although there is no agreement about the interpretation of k and z in these relationships, most authors agree that some form of species–area relationship is generally to be expected.

Among criticisms of the use of species–area relationships are the unreliability of faunal lists, and, to a lesser extent, also of the area data. Kennedy and Southwood (1984) attempted to address both of these criticisms, first by using more rigid criteria for selecting species for inclusion in the faunal list and second by refining the area data from the previously used 10 x 10 km square distribution data with limited 2 x 2 km square area (tetrad) information available for a few counties of Britain. There is no doubt that the species lists were greatly improved by using the new criteria, although it is questionable whether exclusion of polyphagous species is valid in terms of the interactions between the total herbivore load on a given tree. Of more significance are the limitations imposed by using presence or absence in a given quadrat as a criterion for plant area. The all or nothing nature of such a method will consistently overestimate the apparent abundance of trees that are widely distributed geographically but which are present in relatively low numbers. A further criticism, echoing the point made by Strong (1974) concerning the under-representation of data on recent distribution of trees, is the poor recording of widely planted trees, particularly conifers, in the distribution records in Perring and Walters (1962). Although Kennedy and Southwood (1984) acknowledge that 'the greatest source of error in the data arises from the common under-recording of widely planted (e.g. *Malus*, *Pinus* and *Picea*) and introduced species (*Juglans, Robinia, Larix* and *Quercus ilex* L.)', they did not allow for this in their own distribution estimates.

In their recent study with the improved data, Kennedy and Southwood (1984) were able to demonstrate a significant species–area relationship ($r = 0.76$, P <0.005) that explained 58% of the variance about the regression. Multiple regression analysis using a series of other variables increased to 74% the proportion of the variance explained. Significant other factors were 'time' and 'evergreenness'.

The present conclusion from studies on trees and other plants (review in Strong *et al.*, 1984) then must be that abundance (area) of host plants is generally likely to be the single most important determinant of associated herbivore species richness.

Feeding guilds and species–area relationships

One way to overcome the problem of inadequate species lists is to study particular ecological groupings or guilds as subsets of the total herbivore assemblages associated with plants. Here original data or at least more consistent and reliable literature data may be used.

Opler (1974) was the first to study a feeding guild of insect herbivores for area relationships. He worked on leafmining larvae of oaks in California and obtained a very significant relationship ($r = 0.95$, P <0.001) which accounted remarkably for 90% of the variation. Claridge and Wilson (1982b) studied the leafmining guild on 37 taxonomically diverse species of trees in Britain. A significant species–area relationship was obtained ($r = 0.43$, P <0.007) but the predictive value of area was only 19%. A study of mesophyll-feeding leafhoppers (Claridge and Wilson, 1981) on a similar array of British trees gave a similar relationship ($r = 0.4$, P <0.02) with the predictive value of area at 16%. It was concluded from these and other studies that in taxonomically diverse floras species–area relations were rarely helpful in explaining the variation in species richness, at least for ecological guilds (Claridge, 1987; Claridge and Wilson, 1982a, b). On the other hand, for groups of taxonomically related hosts, area may often provide much better levels of explanation (e.g. Opler, 1974; Leather, 1986).

Quality of data: new estimates of areas for British trees

Bearing in mind the above limitations, we have considered the question of improving one or both of the two criteria used in species–area relationships. Species lists continue to present major problems in that they will always be underestimates of the total number of species that are associated with particular plants. There is also the danger that some regions will have been more thoroughly studied than others. Owen (1987) has a been highly critical of published data and goes so far as to deny completely the value of such lists. This seems too extreme a view. We feel that the species lists produced by Kennedy and Southwood (1984) provide the best measures currently available. Additional data have been produced by Evans (1987) for Sitka spruce, *Picea sitchensis*. The leafhopper and leafminer data of Claridge and Wilson (1981, 1982) remain valid for specialist feeding guilds.

Estimates of area, on the other hand, can be improved considerably. Between 1979 and 1982 the Forestry Commission in Britain undertook a thorough Census of Woodlands and Trees, the results of which have recently been published (Locke, 1987). In this compendium details of estimated numbers of trees and areas covered by major species are given under four categories: high forest, coppice, scrub and non-woodland. We have abstracted the data for 18 major trees, comprising four conifers and 14 broadleaves (*Table 1*). These all represent different genera. Some include more than one related species, e.g. oak (*Quercus robur* and *Q. petraea*), birch (*Betula pendula* and *B. pubescens*), pine (*Pinus sylvestris* plus several exotic

Table 1. Areas occupied (hectares) by 18 trees in Great Britain (after Locke, 1987), total number of associated insect herbivore species (after Kennedy and Southwood, 1984; Evans, 1987), and total numbers of associated leafminer species (after Claridge and Wilson, 1982) and mesophyll-feeding leafhopper species (after Claridge and Wilson, 1981). Tree areas are broken down by woodland category.

Tree	Woodland categories					Herbivores		
	High forest	Coppice	Scrub	Non-woodland	Total area	Total species	Miners	Hoppers
Fraxinus excelsior	69 581	1747	7876	35 963	115 167	68	3	0
Quercus spp.	171 990	11 001	18 054	28 289	229 334	423	36	10
Fagus sylvatica	73 936		1070	10 451	85 457	98	6	6
Acer pseudo- platanus	49 426	2499	2366	21 476	75 767	43	7	7
Betula spp.	68 131		63 260	18 577	149 968	334	48	8
Castanea sativa	9871	19 091	264	1112	30 338	11	4	3
Ulmus spp.	9514			7220	16 734	124	12	9
Populus spp.	13 590			4563	18 153	189	23	2
Salix spp.			4964	8251	13 215	450	16	6
Corylus avellana		3095	8561		11 656	106	15	9
Carpinus betulus		3413	410		3823	51	10	7
Alnus glutinosa		8363		24 060	32 423	141	17	11
Aesculus hippo- castaneum				1857	1857	9	0	3
Tilia spp.				3264	3264	57	5	5
Pinus spp.	415 356			5665	421 021	172	6	1
Abies spp.	7944				7944	16	0	0
Picea spp.	642 748			4099	646 847	90	0	0
Larix spp.	151 763			2957	154 720	38	1	0

species) and spruce (*Picea abies* and *P. sitchensis*), others include only single species, e.g. ash (*Fraxinus excelsior*), beech (*Fagus sylvaticus*), hornbeam (*Carpinus betulus*) and sycamore, (*Acer pseudoplatanus*). For the first time, therefore, we have reliable estimates of areas covered by most of the major trees in Britain. There are immediately some surprises. For example, by far the biggest area (646 847 ha) is taken up by the two spruces. The biggest area of broadleaved trees is provided by oak (229 334 ha). The 10 x 10 km counts give willows (*Salix* species), birch and alder (*Alnus glutinosa*) bigger estimated areas than oak. The new area estimates give them much lower figures (*Table 1*). It seems, then, that the square counting method has led to gross errors of current area estimation. These new more accurate area data have been used to recalculate some of the previously published species–area relationships.

Complete herbivore fauna of British trees

We have calculated species–area relationships using the new area data for all 18 trees, and separately also for broadleaves and conifers. In contrast to Kennedy and Southwood (1984), the new analyses resulted in reduction in both levels of significance and amounts of variation explained by area. For all trees ($r = 0.29$, $P = 0.09$) area accounts for only 17% of the variation (*Figure 1*). For broadleaved trees only, 22% of the variation is explained ($r = 0.47$, $P = 0.09$). For conifers 80% of the variation was explained, but the relationship was not significant even at the 10% level ($r = 0.89$, $P = 0.11$). However, in his study of conifers alone, Evans

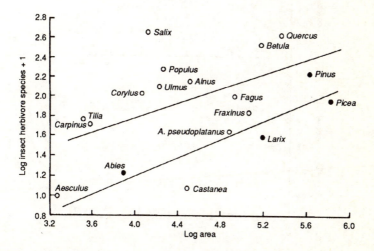

Figure 1. The relationship between the numbers of insect herbivore species associated with trees and the present area of each tree species in Britain. Open circles represent broadleaved species and closed circles conifers. Upper line represents calculated regression for broadleaves only and lower one for conifers only.

(1987) included separately figures for Norway and Sitka spruce from Forestry Commission sources. Thus with five points for his species–area relationship, he obtained a significant regression that explained 79% of the variance ($r = 0.89$, P <0.05).

Clearly the new data dramatically reduce the statistical significance of the species–area relationships for the complete herbivore fauna of British trees, but the analysis of Evans (1987) strongly suggests that broadleaves and conifers should be treated separately.

Herbivore guilds on British trees

The data from the studies on leafminers (Claridge and Wilson, 1982b) and mesophyll feeding leafhoppers (Claridge and Wilson, 1981; *Table 1*) were also used to recalculate species–area relationships with the new area data.

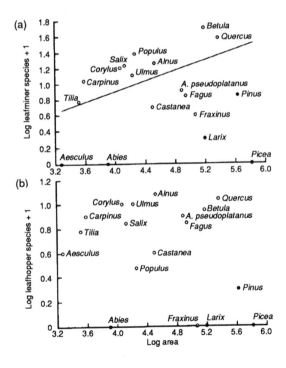

Figure 2. The relationships between the numbers of leafminer species (a) and mesophyll-feeding leafhoppers (b) associated with trees and the present area of each tree in Britain. Open circles represent broadleaved species and closed ones conifers. Calculated regression line given only for leafminers on broadleaved trees alone.

For leafminers (*Figure 2a*) no significant relationships were found for all trees ($r = 0.08$, $P = 0.75$). For broadleaved trees alone a poor relationship was obtained ($r = 0.49$, $P = 0.08$), 24% of the variation of which could be attributed to area. For leafhoppers, similarly no significant relationship was obtained either for all trees ($r = 0.29$, $P = 0.23$), or for broadleaved species alone ($r = 0$, $P = 0.99$; *Figure 2b*).

Thus no significant species–area relationships are now available for either guild when the better quality host area data are used.

Tree area and pest status

There is considerable interest in the use of area indices as tools for designing habitats for conservation. In addition, there is the danger that increasing availability of resources will increase the probability of pest outbreaks. This was discussed by Evans (1987) in relation to the increasing areas of conifers being planted in Britain and the fact that Sitka spruce, despite its prominent position in terms of area, has fewer insect species associated with it than would be predicted on the basis of a species–area relationship. He concluded that either the faunal lists were inaccurate as a result of undercollecting or that area has increased so quickly that it has outstripped the rate of accrual of new species.

The definition of a pest is clearly subjective since it reflects economic as well as ecological factors. In considering the possibility that area data may be used to

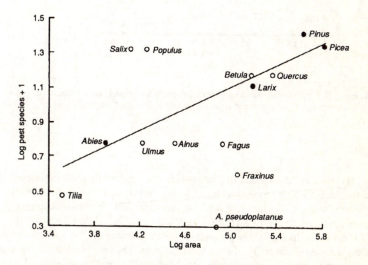

Figure 3. Relationship between numbers of associated insect pests associated with trees in Britain and the present area of each tree. Conventions as in *Figures 1* and *2*. Calculated regression line for conifers only.

predict pest outbreaks, we have made a retrospective analysis using the lists in Bevan (1987) as measures of the numbers of species that have reached pest status. In our analysis we have included all insects in Bevan's categories 'XX' to 'XXXXX' and those deemed to be secondary, 's', but which are still important.

We have plotted the logarithms of pest numbers against area and, using the convention developed elsewhere in this chapter, have distinguished between broadleaves and conifers (*Figure 3*). It is noteworthy that there is no clear separation between the conifers and broadleaves among the scatter of data points. However, when the regressions were calculated the variances explained by the two

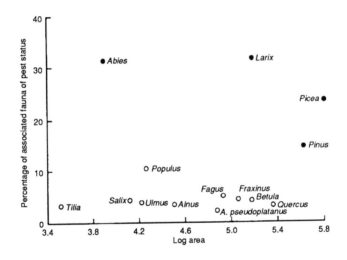

Figure 4. Relationship between the proportion of the associated fauna recorded as pests on each tree species and the present area of each tree in Britain. Conventions as in *Figures 1* and *2*.

data sets point to the numbers of pest species being more closely correlated with area for those feeding on conifers than for those on broadleaves. For the relationships for all trees, and broadleaves and conifers separately, the variances explained by area are respectively 18% ($P = 0.13$), 1% ($P = 0.77$) and 94% ($P = 0.03$).

We have also looked at the possibility that the number of pest species simply reflects the overall diversity of insects on the trees. Thus we plotted pest species as a proportion of the total species list for both conifers and broadleaves (*Figure 4*). The proportion of those that have become pests on broadleaved trees appears to be remarkably constant. For conifers, although total species diversity rises with increasing area, this is not reflected directly in an increased proportionate risk from pest attack.

Discussion and conclusions

The new area data for trees in Britain reduce without exception the significance of previously calculated species–area relationships. Indeed the present analysis suggests that, in contrast to the conclusions of Kennedy and Southwood (1984), host area may not generally be the best predictor of herbivore species richness. However, it is clear that for total species lists, our improved area data together with those of Evans (1987), have highlighted what appear to be different relationships for broadleaves and conifers.

Of the many other possible factors frequently cited as significant in the recruitment of new herbivores, probably one of the most important is the presence of taxonomically related trees in the flora (Connor *et al.*, 1980; Claridge and Wilson, 1981, 1982a, b; Neuvonen and Niemela, 1981). Indeed Claridge and Wilson (1982a, b) suggested that good species–area relationships were only likely to be demonstrated for herbivore guilds where hosts were taxonomically related. When new tree species are introduced into a region, whether by accident or design, the potential for colonization by local herbivores may be more dependent on the degree of taxonomic isolation in the new plant community than on area occupied. One well documented example of commercial plantings in Britain is of the southern beeches, *Nothofagus*, natives of southern South America and Australasia. At least two species have been planted as forest trees in Britain since about 1930 (Nimmo, 1971). Claridge and Wilson (1981a) found them to support six species of mesophyll-feeding leafhoppers, many more than would be predicted from species–area studies. However, the six species were found to be derived from related British Fagaceae.

Although our new analyses cast doubts on the use of area to explain the variability of insect herbivore richness on British trees, area, nevertheless, carries possible implications for the status of some pest species. Increasing area, particularly for conifers, carries with it an increased probability of insects reaching numbers sufficient to be regarded as pests (*Figure 3*). However, numbers of pest species do not rise at the same rate as total species diversity, thus resulting in a net gain of species richness over increased pest numbers (*Figure 4*). This is especially relevant to the enhancement and conservation of insects on trees. It would appear, for example, that increases in areas of trees planted may carry an overall benefit, despite the probability of some insects reaching pest status. Indeed, the relationship in *Figure 3* may simply reflect the fact that increased area is more likely to result in trees growing in sites that would normally be regarded as marginal. It is often on trees growing in such sites that insects find conditions suitable for rapid and destructive population growth.

The extent to which pest development may be predictable on the basis of areas planted is uncertain in the light of our new analyses. It will be no surprise to forest entomologists that we conclude that many diverse factors must contribute to the colonization of trees by insect herbivores. Area is only one and not necessarily the most important one.

References

BEVAN, D. (1987). *Forest Insects. A Guide to Insects Feeding on Trees in Britain.* Forestry Commission Handbook 1. London: Forestry Commission.

CLARIDGE, M . F. (1987). Insect assemblages — diversity, organization and evolution. In: *Organization of Communities Past and Present,* pp. 141–162 (eds. J.H.R. Gee and P.S. Giller). Oxford: Blackwell.

CLARIDGE, M. F. and WILSON, M. R. (1978). British insects and trees: a study in island biogeography or insect/plant coevolution. *American Naturalist* **112,** 451–456.

CLARIDGE, M. F. and WILSON, M. R. (1981). Host plant associations, diversity and species–area relationships of mesophyll-feeding leafhoppers of trees and shrubs in Britain. *Ecological Entomology* **6,** 217–238.

CLARIDGE, M .F. and WILSON, M. R. (1982a). Species–area effects for leafhoppers on British trees: comments on the paper by Rey *et al. American Naturalist* **119,** 573–575.

CLARIDGE, M. F. and WILSON, M. R. (1982b). Insect herbivore guilds and species–area relationships: leafminers on British trees. *Ecological Entomology* **7,** 19–30.

CONNOR, E. F. and McCOY, E. D. (1979). The statistics and biology of the species–area relationship. *American Naturalist* **113,** 791–833.

CONNOR, E. F., FAETH, S. H., SIMBERLOFF, D. S. and OPLER, P. A. (1980). Taxonomic isolation and the accumulation of herbivorous insects: a comparison of introduced and native trees. *Ecological Entomology* **5,** 205–211.

EVANS, H. F. (1987). Sitka spruce insects: past, present and future. *Proceedings of the Royal Society of Edinburgh* Sect. B **93,** 157–168.

JANZEN, D. (1968). Host plants as islands in evolutionary and contemporary time. *American Naturalist* **102,** 592–595.

KENNEDY, C. E. J. and SOUTHWOOD, T. R. E. (1984). The number of species of insects associated with British trees: a re-analysis. *Journal of Animal Ecology* **53,** 455–478.

KURIS, A. M., BLAUSTEIN, A. R. and ALIO, J. J. (1980). Hosts as islands. *American Naturalist* **116,** 570–586.

LAWTON, J. H. and CORNELL, H. (1981). Species as islands: comments on a paper by Kuris *et al. American Naturalist* **117,** 623–627.

LEATHER, S. R. (1986). Insect species richness of the British Rosaceae: the importance of host range, plant architecture, age of establishment, taxonomic isolation and species–area relationships. *Journal of Animal Ecology* **55,** 841–860.

LOCKE, G. M. (1987). Census of woodlands and trees 1979–82. *Forestry Commission Bulletin* **63.**

NEUVONEN, S. and NIEMELA, P. (1981). Species richness of Macrolepidoptera on Finnish deciduous trees and shrubs. *Oecologia* **51,** 364–370.

NIMMO, M. (1971). *Nothofagus* plantations in Great Britain. *Forestry Commission Forest Record* **79.**

OPLER, P. A. (1974). Oaks as evolutionary islands for leaf-mining insects. *American Scientist* **62,** 67–73.

OWEN, D. F. (1987). Insect species richness on the Rosaceae: are the primary data reliable? *Entomologist's Gazette* **38**, 209–213.

PERRING, F. H. and WALTERS, S. M. (1962). *Atlas of the British Flora*. London: Nelson.

REY, J. R., McCOY, E. D. and STRONG, D. R. (1981). Herbivore pests, habitat islands, and the species area relation. *American Naturalist* **117**, 611–622.

REY, J. R., STRONG, D. R. and McCOY, E. D. (1982). On overinterpretation of the species–area relationship. *American Naturalist* **119**, 741–743.

SOUTHWOOD, T. R .E. (1961). The number of species of insects associated with various trees. *Journal of Animal Ecology* **30**, 1–8.

STRONG, D. R. (1974a). Nonasymptotic species richness models and the insects of British trees. *Proceedings of the National Academy of Sciences* **71**, 2766–2769.

STRONG, D. R. (1974b). The insects of British trees: community equilibrium in ecological time. *Annals of the Missouri Botanic Gardens* **61**, 692–701.

STRONG, D. R., LAWTON, J. H. and SOUTHWOOD, R. (1984). *Insects on Plants. Community Patterns and Mechanisms*. Oxford: Blackwell.

Insect–Plant Interactions

7

Role of Site and Insect Variables in Forecasting Defoliation by the Gypsy Moth

MICHAEL E. MONTGOMERY

US Department of Agriculture, Forest Service, Northeastern Forest Experiment Station, Center for Biological Control of Northeastern Forest Insects & Diseases, Hamden, Connecticut 06514, USA

Introduction

Many forest defoliators, particularly those that attack deciduous trees, undergo eruptions in abundance followed by equally dramatic crashes. Density-dependent feedback mediated through host defences or natural enemies is usually responsible for population crashes when there is no shortage of food. A decision to apply pesticide based solely on pest density may be a wasteful and needless assault on the environment because the pest population may subside naturally without causing damaging defoliation. Feedback factors often are not considered in management decisions because these data are expensive to collect and/or must be collected for two or more years beforehand. Research models that focus on identifying fundamental relations between the defoliation response and predictor variables usually do not consider cost or if the data will be available to the user. My goal is not to produce a best-fit mathematical model, but to formulate low-cost, effective procedures for forecasting defoliation by the gypsy moth using multiple regression procedures. The gypsy moth exemplifies the 'messy data problems' of intercorrelation and non-normality inherent in constructing forest insect predictive models.

This analysis, hopefully, will provide insight not only for future experimentation on the gypsy moth, but other insects as well. It seems that theory regarding host-plant insect relationships has overreached corroborating data, particularly data from unmanipulated field observation. The old, neglected data set resurrected here supports not only modern ideas about tri-trophic interactions, but also shows the torment of both researcher and forest manager in predicting the outcome of these interactions in communities.

The gypsy moth

LIFE HISTORY

The gypsy moth, *Lymantria dispar* L., a native pest of north temperate deciduous forests throughout Eurasia (Giese and Schneider, 1979), was introduced to North America in 1869. Outbreaks of this insect occur periodically, during which several million hectares may be stripped of foliage. Species in the genera *Quercus*, *Betula*, *Salix*, *Populus*, *Sorbus* and *Larix* are most likely to be defoliated, but over 300 species in more than 40 genera are attacked (Lechowicz and Mauffette, 1986). Polyphagy and other life history characteristics give the gypsy moth a high potential for population increase, but also impose risks. For example, larvae hatch in early spring when foliage quality is good but weather may be unfavourable; they can grow to large size and lay as many as 1500 eggs, but the lengthy, eight-week growth period may result in high mortality by parasites and predators. The gypsy moth can be described as a 'risk taker' (Nothnagle and Schultz, 1987) that plays high odds of mortality against periodic, extraordinary increases in population.

EUROPEAN POPULATIONS

Fluctuations in the population density of the gypsy moth are very regular in areas of Europe. In Yugoslavia, population peaks were separated by 8.5 (1.8 s. d.) years (Montgomery and Wallner, 1988). The amplitude of the density peaks was less regular, however. At some cycle peaks, little or no defoliation occurred; while at others, extensive defoliation of host trees occurred over an entire region. An important factor in regulation of European gypsy moth populations is the delayed density-dependent response of oligophagous tachinid parasites (Sisojevic, 1975).

Mathematical models have been used successfully to predict changes in population density in Europe. An example is a model (Znamenski and Liamcev, 1983a) for mixed stands in the European USSR that predicts density of eggs in the following year from measurements of current egg density, fecundity, egg weight, proportion of oak in the stand and minimum May temperature. The latter is important because the region has spring frosts that kill newly hatched larvae. The model is relatively simple, but describes 88% of the variability in density. Models of North American populations lack this simplicity and accuracy (Sheehan, 1989).

NORTH AMERICAN POPULATIONS

Gypsy moth populations are chaotic in North America and regular cycles of population density have not been demonstrated. This instability may reflect a reduced gene pool due to the founder principle (see Wright, 1977), the absence of microsporidian disease (Montgomery and Wallner, 1988), as well as a less efficient parasite fauna (Montgomery and Wallner, 1988; Chapter 27, this volume). Campbell and Sloan (1978) proposed that North American populations are numerically bimodal. This postulate was based on populations that remained low in one region, populations that remained high in another region, and populations in a third region that, after several years at high densities, existed at very low density for several years. Miller, Mo and Wallner (1989) found that simple geometric expansion of the population accounted for 60% of the year-to-year variation in defoliation in the states of Connecticut and Massachusetts. Liebhold and Elkinton (1989) noted that regions within Massachusetts were different in timing and frequency of defoliation. The great amount of unexplained numerical variation within populations coupled with regional differences in population behaviour makes it difficult to characterize the dynamics of gypsy moth in North America.

Attempts to develop mathematical models of the relationships between defoliation and biotic and physical variables have had little success (Etter, 1981). A complex, empirical model based on 25 population, habitat and weather factors accounted for only 41% of the variability in defoliation (Campbell and Standaert, 1974). The model had the difficult task of estimating defoliation over a 20-year period in a three-state region. A model restricted to new infestations in central Pennsylvania that used only density of egg masses to predict defoliation had fairly good fit ($R^2 = 0.61$), but overestimates at high egg mass densities were a problem (Gansner, Herrick and Ticehurst, 1985). Populations lose vitality after a year or two at high densities and may collapse even if there is not a food shortage. Virus disease, oligophagous parasites and decline in food quality are the chief agents thought to be responsible (Campbell, 1963; Sisojevic, 1975; Rossiter, Schultz and Baldwin, 1988). Although these factors may be functional, they are difficult to measure and unlikely to be used in management models. Population trend, the ratio of density in year 1 to year 2, has been used to assess vitality (Campbell and Standaert, 1974), but managers have been reluctant to gather data for two years in order to make a defoliation estimate. What is needed are easily obtained indicator variables that characterize the susceptibility of a forest stand and the vitality of the gypsy moth population.

Procedures

SITE VARIABLES

Data for this analysis were collected from 32 sites located in Massachusetts, New York and New Jersey that were monitored yearly from 1972 to 1979. A site was defined as an area of 5–10 ha of fairly uniform forest and soil composition. Sampling units consisted of five 0.042-ha plots in a site. Plot means of each site were used in the analyses since some variables were collected on a site basis. Geological survey maps were used to classify the soil of each site from dry (=1) to wet (=4). The site index was based on the average height of the dominant species at age 50 years. Species, stem diameter (dbh), and percentage defoliation of each tree >5 cm dbh was measured at the end of the larval feeding period. Total defoliation for the plot was the sum of individual tree defoliation times its basal area divided by total basal area of all trees in the plot. Oak in the stand was recorded as percentage of the basal area of all species.

INSECT VARIABLES

Inspection for gypsy moth egg masses was made on all surfaces (tree boles, upper branches, rocks, debris on the ground and leaf litter) in each plot. Counts were stratified into <30 cm or >30 cm from the ground. Twenty egg masses from outside the plots in each site were taken to the laboratory where eggs/mass, proportion of eggs parasitized, and proportion that hatched were determined. Larvae hatching from the eggs (14 groups of 50 larvae per site) were placed on artificial diet and survival after 15 days was recorded. Each group was also scored if any dead larvae were found to have nucleopolyhedrosis virus (NPV); thus, each site received an NPV score of 0-14. Fifteen days was sufficient to observe mortality from infection at hatching but too brief for horizontal infection within a rearing group to be expressed. Only data for the years 1972, 1973 and 1974 were used since little or no defoliation occurred in other years. A site-year was excluded if insects other than gypsy moth caused defoliation and if one or more of the variables was missing. This left 81 site-years for analysis which are summarized in *Table 1*.

Table 1. Summary of variables evaluated.

Variable	Acronym	Mean	Range
Dependent			
Defoliation, %	*DEF*	28.2	0.1–95.7
Independent			
Egg masses/ha	*EM*	6705	79–52 410
Eggs/mass	*FEC*	366	108–846
Mass location, % up	*ELOC*	73.1	20.4–95.2
Egg parasitization, %	*EPAR*	25.4	5.3–50.1
Egg hatch, %	*EHAT*	40.5	0.9–80.9
Virus incidence	*NPV*	4.1	0–14
Larval survival, %	*SUR*	62.3	2.9–97.2
Basal area, m²/ha	*BA*	19.5	9.7–43.2
Oak as % of BA	*OAK*	67.8	2.1-95.7
Site index	*SI*	50.1	20.9–82.0
Soil moisture class	*SOIL*	1.7	1-4

Proportions, e.g. 0.278, rather than percentages, were used in the models.

STATISTICAL ANALYSIS

Statistical routines (Wilkinson, 1987) were used to compute correlations, to select and to test multivariate regression equations, and to plot data. Stepwise regression with alpha-to-enter and alpha-to-remove set at 0.15 was used for preliminary selection of variables for inclusion in the model. The stepwise procedure was satisfactory in eliminating variables of little importance and those that were redundant, but could not be relied on to make final selections. The final model was found by backing off the last one or two steps and plotting the residuals of the reduced model against unselected predictors and the cross-products of selected and unselected terms and then retesting the model with the term that showed greatest dependence on the residuals. To be included in the model, terms had to be significant at $P<0.05$ and the model had to have a higher adjusted R^2 than a model without the term. By examining residuals (which should be done anyway) rather than relying on an abstract statistical parameter generated from a canned program, outliers, skewness, serial effects and so forth can be observed and the scientist can visualize the logic of including a term in the equation. Transformations were done only where gross violations of equality of error variance occurred; only egg mass counts and percent defoliation were transformed. I do not know if soil moisture was originally intended to be a categorical or graded variable, but do know that soil classifications were made after placement of sites which was random within areas. Soil class was assumed to be a continuous variable in order to run stepwise regression; in the end, this was mainly a theoretical problem, since defoliation was found to be a linear function of soil class. A full detailing of the analysis and technical assumptions cannot be done here, but the reader may consult the guide used (Draper and Smith, 1981).

Results

THE PROBLEM

A density of 600 egg masses per hectare (*EM*/ha) is used by many forest managers as a threshold for applying pesticides. Only one case of defoliation >50% was observed below this density (*Figure 1*). Above this threshold, defoliation was sporadic; in one case, an egg mass density of 600 *EM*/ha resulted in 55% defoliation, while in some cases, densities above 10 000 *EM*/ha resulted in less than 20% defoliation. Neither effects associated with year nor population trend explained why high egg mass densities frequently resulted in little defoliation. There were 13 cases where densities >1000 *EM*/ha resulted in <50% defoliation in the first year and 18 cases where this occurred in the second year. There were 12 positive and 13 negative correlations between change in the amount of defoliation and change in population density from year 1 to year 2 of the study. In the third year of the outbreak, both egg mass density and defoliation were low. Exclusion of population trend from the model was thus not only based on lack of availability, but also because it was not useful. The inability to predict generational trend from the density at the start of the generation unless it is known if populations are in an

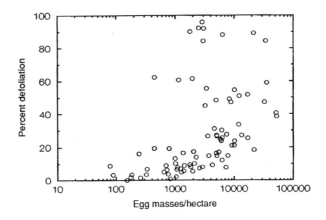

Figure 1.   Mean overwintering egg mass density and subsequent defoliation.

outbreak or collapse phase has been explained in more detail using this same data source (Campbell and Sloan, 1978).

Selection of the most appropriate type of model is difficult because of the data scatter. Theoretically, a logistic function seems most appropriate, but it was difficult to evaluate several variables simultaneously with nonlinear models. A double logarithmic transformation gave the best fit ($r^2 = 0.406$), but it deemphasized midrange observations where decisions are made. The model chosen was the arcsin square root (asn) of defoliation and the natural logarithm (log) of the egg mass density:

$$asnDEF = -0.346 + 0.110 \log EM \qquad (1)$$

Although this model accounted for only 25.2% of the variability in defoliation, it was a good starting point to test if other factors account for the unexplained variation in defoliation.

A MANAGEMENT MODEL

An important consideration for management models is the cost and effort to collect the variables. A manager who uses fixed or variable radius plots to count egg masses can also easily obtain the habitat variables and the location of egg masses. Obtaining fecundity, egg parasitism, egg hatch, disease and larval survival requires additional expenditures and a laboratory. The independent variables in *Table 1* most easily available for a management decision model are *EM*, *ELOC*, *BA*, *OAK*, *SI* and *SOIL*. A test model, which included these variables was evaluated by stepwise regression followed by further testing for non-linearity and interactions. The best-fit model was:

$$asnDEF = -0.583 + 0.246 \log EM - 0.123 \ ELOC * \log EM - 0.063 \ SOIL \qquad (2)$$

The model r^2 was 0.615 with all predictors significant at $P<0.009$. Decisions to apply pesticides are often based on whether or not defoliation above 60% is expected because above this level trees refoliate, which expends energy reserves and weakens them (Campbell and Valentine, 1972). The 95% confidence intervals for a prediction of 60% defoliation are 79% and 40%; hence, caution and judgement must be exercised if a prediction is within the confidence interval.

Figure 2 provides a less abstract method for a manager to decide if defoliation above 60% is likely to occur. It is a solution of equation (2) for a dry site (*SOIL* = 1) and 60% defoliation (asn*DEF* = 0.8861) rearranged so that *ELOC* is the *Y* variable and *EM* is the *X* variable. The plot of the result, $Y = 1.99 - 12.23/X$, on back-transformed scales is the isogram of 60% defoliation. This plot shows that 60% defoliation is expected at increasingly greater egg mass densities as the proportion of egg masses above the forest floor increases. For example, more than 60% defoliation on a dry site would be expected if there were 1000 *EM*/ha of which 25% were located above 30 cm; if 75% of the egg masses were above 30 cm, more than 60% defoliation would not be expected unless the *EM*/ha exceeded 12 000. The

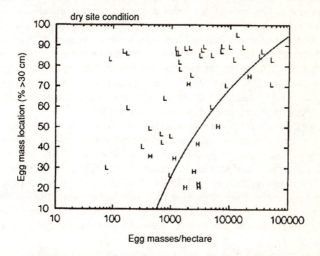

Figure 2. An isogram of expectation of 60% defoliation on sites with dry soil conditions. Actual observations of <60% and >60% defoliation on dry soil sites are designated by L and H, respectively.

model correctly classified 34 of 36 cases of observed defoliation <60% and 7 of 10 cases of observed defoliation >60%. The model needs improvement, but is better than using the traditional criterion of 600 *EM*/ha which would incorrectly class 28 of the 35 cases of low defoliation as high defoliation!

Nearly all potentially damaging defoliation occurred on dry sites. The 60% defoliation isograms for wetter sites would be to the right and almost parallel to the dry site isogram given in *Figure 2*, but were not constructed since this would be an extrapolation.

A RESEARCH MODEL

All the variables listed in *Table 1* — except *ELOC* — were evaluated in a model in order to determine mechanisms underlying the relationship between egg mass density and defoliation. Egg location is a behavioural variable and any direct effect it has on the number of larvae available to defoliate trees is reflected in egg hatch. The best fit model was:

$$asnDEF = 0.065 + 0.080 \log EM + 0.145\ EHAT*\log EM - 0.005\ NPV*\log EM$$
$$-0.439\ SUR - 0.008\ BA*OAK - 0.096\ SOIL \quad\quad (3)$$

The model r^2 was 0.702, and each predictor was significant at $P<0.002$ except $BA*OAK$, where $P = 0.039$.

Egg hatch and virus incidence were cross-products with egg mass density. This implies that the influence of egg hatch and virus on defoliation increased as the population density increased. This is reasonable behaviour for *NPV* since horizontal transmission of virus among larvae is greater at high larval densities. When examined across all sites, the incidence of NPV in the egg mass was negatively correlated ($r = -0.326$) with *EM*/ha. Virus contamination of egg masses was higher in the wetter sites and these had lower egg mass densities. The negative coefficient of *SUR*, survival of newly hatched larvae, is difficult to explain. Survival was positively correlated with defoliation within soil class one sites and negative in the others, but the *SUR*SOIL* interaction was not significant ($P = 0.182$). The interrelationships of *NPV* and *SUR* with population density and habitat were difficult to identify because densities were distributed unevenly among the habitats. Planned experiments are needed to clarify this.

The coefficient of *BA*OAK*, the basal area of oak on a site, was negative. This may seem puzzling since oak is a favourite food and percent defoliation was higher on sites that had a higher proportion of oak. Because of dry soil moisture conditions, the total basal area on these sites is small. Thus, the basal area of oak generally was less on the more defoliated sites than on the better stocked, more resistant sites with a lower proportion of oak. A higher percent defoliation at a given egg mass density occurs on dry sites not only because the percent of favoured host foliage is greater, but also because the total biomass of foliage is less than on wetter sites.

Some variables that may be useful predictors were excluded from the model because of co-linearity with included variables. Fecundity was entered first into the model by the stepwise procedure but removed in a later step. Fecundity was as good a predictor of defoliation as egg mass density in a single factor model. Fecundity was correlated with *EHAT*, *SUR* and *NPV*, $r = 0.683, 0.490$ and -0.466, respectively. A model with log *EM*, *FEC* and *SOIL* as predictors had an r^2 of 0.581. Site index was correlated with soil moisture class ($r = 0.740$) and % oak in the site ($r = -0.653$). Replacing *SOIL* with *SI* in equation (3) resulted in an r^2 of 0.645. Future efforts to develop models of defoliation response should consider fecundity and site index as predictors.

Discussion

These models forecast defoliation from observations made on one generation and, hence, do not include time lag effects since these data would not be available. The Campbell and Standaert (1974) model included population trend, but the poor fit of this model implies that trend may not be as important as other variables not identified in that analysis. Factors that define the vigour of a population, such as egg location, fecundity and incidence of disease, seem to be more useful predictors of gypsy moth defoliation than the rate of change of a population. Estimation of virus contamination in egg masses by hatching eggs and rearing larvae is not practical for management, but development of DNA hybridization (Keating, Burand and Elkinton, 1990) and other molecular assays may make it possible to include this factor in decision models. Future models may want to include weather conditions since they were identified as influential by Campbell and Standaert (1974) and Znamenski and Liamcev (1983a). Traditionally, gypsy moth egg mass densities are measured per unit of land surface area (*EM*/ha). Because the amount of food available depends on site conditions, future models may want to express egg mass density in units per host plant or foliage biomass. In the Soviet Union, the number of eggs per 100 shoots is commonly used (Znamenski, personal communication). Since the number of growing tips (=shoot) is estimated from basal area, egg masses per basal area may be a convenient measure.

Sites susceptible to gypsy moth defoliation in both Europe and North America have poor tree growth. As in this study, stands on these sites are thinly stocked with trees of small diameter for their age and have a high proportion of oak species adapted to dry soil conditions (Znamenski and Liamcev, 1983b; Valentine and Houston, 1984). A small amount of pine in the stand has also been associated with susceptibility (Montgomery, McManus and Berisford, 1989). Susceptibility of these stands is a consequence not only of the abundance of suitable food, but also the dynamics of natural enemies. An abundance of bark flaps and fissures that provide larval resting and pupation sites above the forest floor in susceptible stands reduces predation by small mammals (Bess, Spurr and Littlefield, 1947; Campbell, Hubbard and Sloan, 1975; Smith, 1989). Inadequate predation is thought to be important in the build-up of populations (Campbell and Sloan, 1977; Smith, 1989), whereas disease is important in the decline of outbreaks (Bess, Spurr and Littlefield, 1947; Semevsky, 1973). The percentage of larvae dying from virus at a given larval density was higher on wet sites than on dry sites (Campbell, 1963). Consequently, as observed in this analysis, egg masses would be fewer on wet sites, but would have more viral contamination than on dry sites. Tannin levels are higher in oak foliage from dry ridgetops than in oak foliage from mesic bottomlands (Kleiner, Montgomery and Schultz, 1989) and it has been shown that dietary tannins give the gypsy moth resistance to virus infection (Keating, Yendol and Schultz, 1988). Similarly, Rossiter (1987) found that larvae dosed with NPV survived longer when feeding on pine foliage than on oak foliage and that fewer larvae from eggs laid on pine rather than oak were infected with NPV. Effects of food quality on virus epizootics are a plausible explanation for the more prolonged and higher levels of defoliation that occur on dry sites.

Acknowledgement

I would like to thank Michael L. McManus, Northeastern Forest Experiment Station, for the laboratory measurements on gypsy moth egg masses and for the files of all the data.

References

BESS, H. A., SPURR, S. H. and LITTLEFIELD, E. W. (1947). Forest site conditions and the gypsy moth. *Harvard Forest Bulletin* **22**.

CAMPBELL, R. W. (1963). The role of disease and desiccation in the population dynamics of the gypsy moth, *Porthetria dispar* L. (Lepidoptera: Lymantriidae). *Canadian Entomologist* **95**, 426–435.

CAMPBELL, R. W. and SLOAN, R. J. (1977). Natural regulation of innocuous gypsy moth populations. *Environmental Entomology* **6**, 315-322.

CAMPBELL, R. W. and SLOAN, R. J. (1978). Numerical bimodality among North American gypsy moth populations. *Environmental Entomology* **7**, 641-646.

CAMPBELL, R. W. and STANDAERT, J.P. (1974). Forecasting defoliation by the gypsy moth in oak stands. *US Department of Agriculture, Forest Service, Research Note* **NE-193**.

CAMPBELL, R. W. and VALENTINE, H. T. (1972). Tree condition and mortality following defoliation by the gypsy moth. *US Department of Agriculture, Forest Service, Research Paper* **NE-236**.

CAMPBELL, R. W., HUBBARD, D. L. and SLOAN, R. J. (1975). Patterns of gypsy moth occurrence within a sparse and numerically stable population. *Environmental Entomology* **4**, 535-542.

DRAPER, N. R. and SMITH, H. (1981). *Applied Regression Analysis*. New York: John Wiley.

ETTER, D. O. (1981). Pest management systems development. In: *The Gypsy Moth: Research Toward Integrated Pest Management*, pp. 697-727 (eds C. C. Doane and M. L. McManus). *US Department of Agriculture Technical Bulletin* **1584**.

GANSNER, D. A., HERRICK, O. W. and TICEHURST, M. (1985). A method for predicting gypsy moth defoliation from egg mass counts. *Northern Journal of Applied Forestry* **2**, 78-79.

GIESE, R. M. and SCHNEIDER, M. L. (1979). Cartographic comparisons of Eurasian gypsy moth distribution (*Lymantria dispar* L.; Lepidoptera: Lymantriidae). *Entomological News* **90**, 1-16.

KEATING, S. T., BURAND, J. P. and ELKINTON, J. S. (1990). DNA hybridization assay for detection of gyspy moth nuclear polyhedrosis virus in infected gypsy moth (*Lymantria dispar* L.) larvae. *Applied and Environmental Microbiology* **55**, in press.

KEATING, S. T., YENDOL, W. G. and SCHULTZ, J. C. (1988). Relationship between susceptibility of gypsy moth larvae (Lepidoptera: Lymantriidae) to a baculovirus and host plant foliage constituents. *Environmental Entomology* **17**, 952-958.

KLEINER, K. W., MONTGOMERY, M. E. and SCHULTZ, J. C. (1989). Variation in leaf quality of two oak species: Implications for stand susceptibility to gypsy moth defoliation. *Canadian Journal of Forest Research* 19, 1445–1450.

LECHOWICZ, M.J. and MAUFFETTE, Y. (1986). Host preferences of the gypsy moth in eastern North American versus European forests. *Revue d'Entomologie du Quebec* 31, 43-51.

LIEBHOLD, A. M. and ELKINTON, J. S. (1989). Characterizing spatial patterns of gypsy moth regional defoliation. *Forest Science* 35, 557-568.

MILLER, D. R., MO, T. K. and WALLNER, W. E. (1989). Influence of climate on gypsy moth defoliation in Southern New England. *Environmental Entomology* 18, 646-650.

MONTGOMERY, M. E. and WALLNER, W. E. (1988). The gypsy moth: A westward migrant. In: *Dynamics of Forest Insect Populations*, pp. 353-375 (ed. by A.A. Berryman). New York: Plenum Publishing Corporation.

MONTGOMERY, M. E., MCMANUS, M. L. and BERISFORD, C. W. (1989). The gypsy moth in pitch pine-oak mixtures: Predictions for the South based on experiences in the North. In: *Proceedings of Pine-Hardwood Mixtures: A Symposium on Management and Ecology of the Type,* pp. 43-49 (ed. T. A. Waldrop). 18-19 April, 1989, Atlanta, Georgia. Southeastern Forest Experiment Station, General Technical Report SE-58.

NOTHNAGLE, P.J. and SCHULTZ, J.C. (1987). What is a forest pest? In: *Insect Outbreaks*, pp. 59-80 (eds. P. Barbosa and J.C. Schultz). New York: Academic Press.

ROSSITER, M. (1987). Use of a secondary host by non-outbreak populations of the gypsy moth. *Ecology* 68, 857-868.

ROSSITER, M., SCHULTZ, J.C. and BALDWIN, I.T. (1988). Relationships among defoliation, red oak phenolics, and gypsy moth growth and reproduction. *Ecology* 69, 267-277.

SEMEVSKY, F. N. (1973). Studies of the dynamics of gypsy moth, *Porthetria dispar* L. (Lepidoptera: Lymantriidae) at low population density levels. *Entomological Review* 52, 25-29.

SHEEHAN, K. A. (1989). Models for the population dynamics of *Lymantria dispar*. In: *Proceedings, Lymantriidae: A Comparison of Features of New and Old World Tussock Moths*, pp. 533-547 (co-ord. by W.E. Wallner and K.A. McManus). 26 June-1July, 1988, New Haven, Connecticut. Northeastern Forest Experiment Station, General Technical Report NE-123.

SISOJEVIC, P. (1975). Population dynamics of tachinid parasites of the gypsy moth (*Lymantria dispar* L.) during a gradation period (in Serbo-Croatian). *Zast. Bilja* 26, 97-170.

SMITH, H. R. (1989). Predation: its influence on population dynamics and adaptive changes in morphology and behavior of the Lymantriidae. In: *Proceedings, Lymantriidae: A comparison of features of New and Old World tussock moths*, pp. 469-488 (co-ord. by W. E. Wallner and K. A. McManus). 26 June-1 July, 1988, New Haven, Connecticut. Northeastern Forest Experiment Station, General Technical Report NE-123.

VALENTINE, H. T. and HOUSTON, D. R. (1984). Identifying mixed-oak stand susceptibility to gypsy moth defoliation: an update. *Forest Science* 30, 270-271.

WILKINSON, L. (1989). *SYSTAT: The System for Statistics*. Evanston, IL: SYSTAT.

84 MICHAEL E. MONTGOMERY

WRIGHT, S. (1977). *Evolution and the Genetics of Populations*, vol. 3, Experimental Results and Evolutionary Deduction. Chicago: University of Chicago Press.
ZNAMENSKI, V. S. and LIAMCEV, N. E. (1983a). Regression models for prognosis of gypsy moth density. *Lesnoie Khoszastvo* 9, 61-63.
ZNAMENSKI, V. S. and LIAMCEV, N. E. (1983b). Criterion for classification of forest stands-gypsy moth reservations (in Russian). *Lesnoie Khozastvo* 1, 60-61.

8

Water Stress in Sitka Spruce and its Effect on the Green Spruce Aphid *Elatobium abietinum*

E. J. MAJOR

Entomology Branch, Forestry Commission Research Station, Alice Holt Lodge, Farnham, Surrey, GU10 4LH, UK

Introduction

The green spruce aphid *Elatobium abietinum* (Walk). feeds exclusively on spruce. *Picea abies* L. and *Picea sitchensis* (Bong.) Carr. are the most susceptible to attack (Fox-Wilson, 1948). The aphid is anholocyclic in Britain. Two population peaks may occur, one in May/June and another, if conditions are mild, in October. Initial signs of damage are yellow chlorotic bands at the site of stylet insertion. Severe infestations rarely kill the tree but can result in reduced growth (Carter, 1977). Field observations suggest that periods of water shortage influence the severity of *E. abietinum* attack (Bevan and Carter, 1975).

    As virtually every plant process is affected during drought stress (Hsiao, 1973) it would seem likely that plant–insect relationships will be influenced. The physiological condition of the host is often quoted as a factor contributing to the success of forest insect herbivores although evidence is largely circumstantial (Mattson and Haack, 1987). Insect outbreaks which have been related to stress have representatives from three feeding classes; phloem feeders, leaf feeders and

Population Dynamics of Forest Insects
© Intercept Ltd, PO Box 716, Andover, Hampshire, SP10 1YG, UK

sap suckers. Climate plays an important role in the growth of pine beetle populations, both by affecting the ability of the host to resist attack and by influencing mortality and reproduction of the beetle population (Hicks, 1982). Pine sawflies are particularly responsive to host condition, a drought in Holland resulted in serious plantation damage by *Diprion pini* (L.) and *Neodiprion sertifer* (Geoff.) (Knerer and Atwood, 1973). White (1969) related two outbreaks of the psyllid *Cardiaspina densitexta* (Tayl.) feeding on leaves of *Eucalyptus fasiculosa* (F.V.M.) to stress caused by changes in the rainfall pattern. Theories to explain the success of insects on stressed plants include elevation of plant nutrient levels (Rhoades, 1983; White, 1984), compromised plant defences (Rhoades, 1983, 1985) and a more suitable physical environment (Rhoades, 1983). In this study aspects of host plant physiology and chemistry were investigated along with the performance of *E. abietinum* on water-stressed and control trees.

Materials and Methods

A clone of three year old *Picea sitchensis* was used in the experiments. The trees were potted, using a standard compost mix.

WATER REGIMES

Twenty-four trees were divided into three groups and randomly assigned one of the three treatments: control (CO), intermittent stress (IS) and a continuous stress (CS). The treatments were started during the growing season. Intermittent and continuous stress were included to determine the response of the trees to a long term stress as opposed to short periods of stress. The soil moisture characteristic curve of the compost was determined using the filter paper method of Fawcett and Collis-George (1967). From the moisture characteristic curve the moisture content at various soil water potentials can be derived. The CO trees were maintained near field capacity, the CS trees at a soil moisture content of 10–13%, and the IS trees were alternated between the two to prevent any adaptation to the stress. A gravimetric method involving destructive sampling of trees during the growing season was adopted to determine when the trees required watering. When the trees had become dormant they were transferred to control temperature rooms and left for two weeks to acclimatize prior to aphid infestation. The experimental trees were only used once and were in the control temperature rooms for a maximum of five weeks.

MEASUREMENT OF XYLEM WATER POTENTIAL

Leaf water potential was measured using a pressure bomb adapted to take individual needles based on that of Roberts and Fourt (1977). Ten one-year-old needles were taken from the third whorl of each experimental tree. Each needle was excised at the base and placed immediately into the top of the pressure bomb until the basal 0.5 mm was protruding. This was then cut flush with the surface. The cut surface was viewed through a microscope as pressure was applied until water was seen to exude from the xylem; when this occurred a pressure reading was taken.

APHID PERFORMANCE

Elatobium abietinum was collected from Dalby forest North Yorkshire and maintained on healthy dormant potted trees. All aphid performance experiments were carried out in constant temperature rooms at 15°C (15–17°C) with a 16-hour day length. A newly moulted adult of known weight was caged onto the one-year-old needles of the dormant experimental trees.

The first nymph produced was discarded as it is usually atypical (Murdie, 1969). The second nymph was removed and weighed, the next three nymphs were allowed to remain on the tree, the sixth nymph being removed and weighed and the adult discarded. The second and sixth nymphs were used to obtain initial weight. Removing and replacing nymphs significantly affects the growth rate and so was avoided (Nichols, 1984). The average weight of the nymphs removed was taken to be comparable to those left. The three remaining nymphs were left to become adults. On becoming adult the aphids were removed and weighed and replaced to allow the production of a nymph which was removed and weighed. The adult was then dissected and the number of embryos with pigmented eyes were recorded. This technique does not allow for determination of individual growth rates. The mean relative growth rate (*MRGR*) was obtained for each tree. This was calculated using the mean weight of newly born nymphs (*N*), newly moulted adults (*A*) and the overall development time (*T*) (i.e. birth of nymph to moult to adult (van Emden, 1969).

$$MRGR = \frac{\ln (A - N)}{T}$$

POPULATION GROWTH

Five IS,CO,CS trees were used to assess the growth of the aphid population. Ten newly moulted adults of similar weight were caged on each tree and counts made daily. Only aphids present on the foliage were included in the counts. The population was left until space and feeding material became limiting and it suffered a 'natural decline'.

Chemical Analysis

Foliar terpenes were analysed using a Hewlett Packard 5008A gas chromatogram with a computerized integrator. 5 μl of sample was injected manually. A slow run time of 90 min ensured good separation of the monoterpenes.

Results

COMPARISON OF NEEDLE XYLEM WATER POTENTIAL BETWEEN THE THREE TREATMENTS

Analysis of variance showed that the xylem water potential values obtained were significantly different ($P<0.001$) (see *Table 1*).The xylem water potential readings were lower in the CS and IS trees than the controls (Duncan's multiple range test $\alpha = 0.01$). When the IS trees were watered xylem water potentials recovered to the same level as the controls within 2–3 days.

Table 1.   The mean xylem water potential readings in megapascals (MPa) of a *P. sitchensis* clone subject to different water regimes. Mean ± SE, $n = 8$.

Treatment	Xylem water potential
CO	0.862 ± 0.01
IS	1.347 ± 0.02
CS	1.406 ± 0.01

POPULATION GROWTH

Aphid numbers at weekly intervals are shown in *Figure 1*. An analysis of variance on aphid numbers at the initial population peak (for the IS and CO treatments this was at 21 days for the CS treatment the peak was at 14 days) indicated a significant

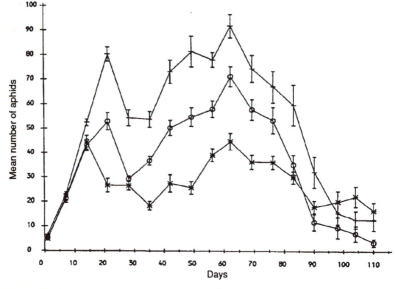

Figure 1.   The mean number of aphids per tree (±SE) on control (o), intermittently stressed (+), and continuously stressed (*) *P. sitchensis*.

difference between treatments ($P<0.001$). All treatments were significantly different from each other (Duncan's multiple range test $\alpha = 0.05$), with the IS trees supporting the highest number of aphids. If the number of nymphs produced per adult per day is plotted against time in days (*Figure 2*) it can be seen that the production of nymphs is sustained on IS trees whereas it declines after approximately 13 days on both CO and CS trees.

Restless behaviour (wandering and short interrupted periods of feeding) which resulted in loss of aphids was more pronounced on CO and CS trees where there was a 50% drop in numbers over a seven- day period compared to 30% on IS trees (*Figure 1*). The loss in numbers shown in all treatments does not appear again as it is compensated for by a continuous production of nymphs. Total numbers of aphids produced was significant between the three treatments ($P<0.015$). After 70 days crowding and lack of suitable needles for feeding resulted in the final decline.

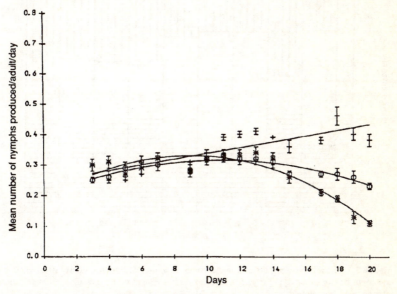

Figure 2. The mean number of nymphs produced per adult per day ($\pm$SE) on control (o), intermittently stressed (+), and continuously (*) *P. sitchensis*.

Control	$Y = 0.20 + 0.02x - 0.001x^2$
Intermittently stressed	$Y = 0.25 + 0.009x$
Continuously stressed	$Y = 0.19 + 0.03x - 0.002x^2$

INDIVIDUAL PERFORMANCE PARAMETERS

There were no significant differences in the performance parameters of initial nymphal weight, adult weight, overall development time, pre-reproductive period, nymphal weight or embryo count between treatments.

MONOTERPENES

Analysis of variance on monoterpenoid compounds revealed nine compounds which were significantly different between the IS trees and the CO and CS trees. Five of these compounds were identified as volatile monoterpenes. *Figure 3* shows that generally the volatile monoterpenes increased in the IS trees and the others decreased when compared with CO and CS trees. The ratio of the volatile monoterpenes to those of higher molecular weight was different between the CO trees and IS trees. The ratio of volatiles to others was 1.9:1 in IS trees and 1.6:1 in the CO trees.

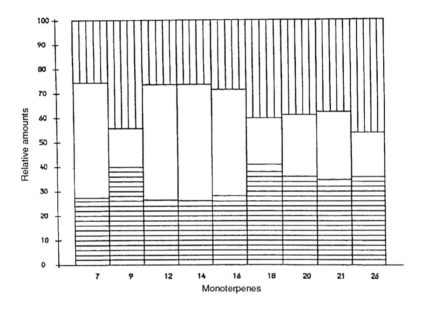

Figure 3. The relative amounts of monoterpenes in control ▭ intermittently stressed, ▢ and continuously stressed ▥ *P. sitchensis*. Each column represents the relative amount of the terpene in each treatment.

Discussion

The results indicate that intermittent water stress may play a role in promoting the success of an aphid population. Aphids achieved higher numbers on intermittently stressed than on control or continuously stressed trees. This effect may be due to changes in host physiology, such as turgor, nutrient levels and secondary metabolites which have been associated with stress (Kramer,1983; Mattson and Haack,1987). The vast majority of work on plant water stress and aphid performance has been done on herbacous plants. The literature suggests that there is great variation in aphid response to stressed host plants both in the laboratory and the field situation (Kennedy, 1958). Lower plant turgor and an increase in sap soluble nitrogen have been shown to be the two main factors determining aphid perform-ance on water stressed plants (Kennedy, Lamb and Booth, 1959). Loss of turgor in stressed raddish plants decreased the fecundity of *Lipaphis erysimi* (Kalt.) (Chadda and Arora, 1982). Wearing and van Emden, (1967) showed that *Brevico-ryne brassicae* (L.) and *Myzus persicae* (Sulz.) varied in their dependence on turgor pressure for food uptake. In this study turgor was not thought to be a limiting factor because it would not explain why the intermittently stressed trees which had xylem water potentials as high as the controls and as low as the continuously stressed trees should have higher overall aphid population densities.

Observations made on the water-stressed trees, suggest that water stress promoted the rate at which chlorotic bands appeared as a result of aphid feeding. Fisher (1987) found that chlorosis induced in spruce needles by aphid feeding improved aphid growth by providing a better quality food source. To explain the poor performance of the aphids on the continuously stressed trees there must be other factors preventing them taking advantage of an improved food source.

In this study monoterpenoid compounds were found to be affected by the water deficits. The intermittently stressed trees differed from the control and continu-ously stressed trees. Although monoterpenes are under strong genetic control in conifers developmental, seasonal, and environmental factors can cause modifica-tions (Gilmore, 1977; Gershenzon, 1984). Secondary metabolites may increase during a period of drought stress due to the increased levels of available carbon and nitrogen. Their role at the plant level is not fully understood. In the intermittently stressed trees improved aphid population growth was associated with changes in the monoterpenes. Changes in terpenes associated with stressed trees have been linked with the success of other insects, for example, the spruce budworm *Choristoneura occidentalis* Freem. where significant increases in weight and survivorship were attributed to a change in the terpene defence chemistry (Cates, Redak and Henderson,1982). Changes in the volatile monoterpenes can be used by insects to locate a stressed host, α-pinene in particular has been recognized as a common bark beetle attractant (Mattson and Haack, 1987). There is less informa-tion available regarding aphid response to terpenoid compounds. DeHayes (1981) suggested that the Balsam twig aphid *Mindarus abietinus* Koch. was attracted or repelled by certain monoterpenes within Balsam fir foliage. Alexander (1987) found that the adelgid *Pineus floccus* (Patch.) infested trees that had relatively low amounts of camphor and santene. The role of volatile monoterpenes as attractants could explain the observation that intermittently stressed trees were preferentially colonized when left to acquire natural infestations.

A particular ratio of monoterpenes may have an inhibitory effect on the aphid. This effect may be seen in behavioural changes, short interrupted periods of feeding and restlessness. An increase in restless behaviour was observed on the control and continuously stressed trees. The similarity in the monoterpenes of continuously stressed trees to the controls may suggest an adaptation to the stress, with that particular complement of monoterpenoid compounds being the most cost effective defence strategy for the plant.

Although significant differences were found between final population size on the three treatments there was no significant difference in individual aphid performance. This may reflect a methodological problem, where the experiments needed to be extended to cover the next generation of aphids. The effect on the population only became apparent after more than 13 days.

The ability of aphids to respond rapidly to changes in host quality will enable them to take advantage of short periods of stress experienced by their host. During a prolonged period of water deficit any increases in free amino acids, sugars, etc. beneficial to the aphid may be unavailable due to an unpalatable imbalance of nutritional components or changes in secondary metabolites.

Acknowledgements

I thank Sara Smith for much useful discussion, staff at the Forestry Commission Research Station, technical staff within the Forestry Department at Bangor University for their assistance during this project, and Drs N. Straw and S. R. Leather for commenting on the manuscript. This work was carried out while the author was in receipt of a NERC CASE studentship.

References

ALEXANDER, H. J. (1987). Inhibition of *Pineus floccus* colonization by volatile compounds found in leaf tissue of Red Spruce. *Virginia Journal of Science* **38**, 27–34.

BEVAN, D. and CARTER, C. I.(1975). Host plant susceptibility. In: *Report on Forest Research*. 37 pp. London: HMSO.

CARTER, C. I.(1977). Impact of the green spruce aphid on growth.Forestry *Commission Research and Development Paper* **116.**

CATES, R. G., REDAK, R. A. and HENDERSON,C.B. (1982). Patterns in defensive natural product chemistry. Douglas Fir and Western Spruce Budworm interactions. In: *Plant Resistance to Insects*, pp. 1–19 (ed. P.A. Hedin). ACS Symposium Series **208.**

CHADDA, I. C. and ARORA, R. (1982). Influence of water stress in the host plant on the mustard aphid *Lipaphis erysimi*. *Entomon* **7**, 75–78.

DEHAYES, D. H. (1981). Genetic variation in susceptibility of *Abies balsamea* to *Mindarus abietinus*. *Canadian Journal of Forest Research* **11**, 30–35.

FAWCETT, R. G. and COLLIS-GEORGE, N. (1967) . A filter paper method for determining the moisture characteristics of soil. *Australian Journal of Experimental Agriculture and Animal Husbandry* **7**, 162–167 .

FISHER, M. (1987). The effect of previously infested spruce needles on the growth of the green spruce aphid *Elatobium abietinum* and the effect of the aphid on the amino acid balance of the host plant. *Annals of Applied Biology* **111**, 33–41

FOX-WILSON, G. (1948). Two injurious pests of Conifers. *Journal of the Royal Horticultural Society* **73**, 73–78 .

GERSHENZON, J . (1984) . Changes in the levels of plant secondary metabolites under water and nutrient stress. In: *Recent Advances in Phytochemistry*.

Phytochemical Adaptation to Stress, pp. 273–320 (eds B. N. Timmuermann,   C. Steelink and F.A. Loewus). vol. 18, New York: Plenum Press.

GILMORE, A. R. (1977). Effect of soil moisture stress on monoterpenes in lobolly pine. *Journal of Chemical Ecology* 3, 667–676 .

HSIAO, T.C. (1973). Plant responses to water stress. *Annual Review of Plant Physiology* 24, 519–570.

HICKS, R. R. (1982). Climatic, site and stand factors. *USDA Forest Service Science Education Administrative Technical Bulletin* 1631, 55–68.

KENNEDY, J. S. (1958). Physiological condition of the host plant and susceptibility to aphid attack. *Entomologia Experimentalis et Applicata* 1, 50–65.

KENNEDY, J. S. LAMB, K. P. and BOOTH, C. O. (1959). Responses of Aphis fabae to water shortage in host plants in pots. *Entomologia Experimentalis et Applicata* 2, 274–291.

KNERER, G. and ATWOOD, C. E. (1973). Diprionid sawflies polymorphism and speciation. *Science* 179, 1090–1099.

KRAMER, P. J. (1983). *Water Relations of Plants.* New York: Academic Press.

MATTSON,W.J. and HAACK, R. A. (1987). Role of drought stress in provoking. outbreaks. In: *Insect Outbreaks,* pp. 365–407 (eds P. Barbosa and J. C. Schultz). New York: Academic Press.

MURDIE, G. (1969). The biological consequences of decreased size caused by crowding or rearing temperature in apterae of the pea aphid *Acyrthosiphon pisum* (Harris). *Transactions of the Royal Entomological Society London* 121, 443–455.

NICHOLS, J. (1984). The performance of the green spruce aphid on various spruce species and the effect of foliar amino acids and secondary compounds. Unpublished M. Phil. Thesis, Reading University

RHOADES, D. F. (1983). Herbivore population dynamics and plant chemistry. In: *Herbivore Population Dynamics and Plant Chemistry,* pp. 155–200 (eds R.F. Denno, and M. S. McClure). New York: Academic Press.

RHOADES, D. F. (1985). Offensive–defensive interactions between herbivores and plants. Their relevance in herbivore population dynamics and ecological theory. *The American Naturalist* 125, 205–238.

ROBERTS, J. and FOURT, D. F. (1977). A small pressure chamber for use with plant leaves of small size. *Plant and Soil* 48, 545–546.

VAN EMDEN, H. F. (1969). Plant resistance to *Myzus persicae* induced by a plant regulator and measured by aphid relative growth rate. *Entomologia Experimentalis et Applicata* 12, 125–131.

WEARING, C. H. and VAN EMDEN, H. F. (1967). Effects of water stress in host plants on infestations by *Aphis fabae, Myzus persicae* and *Brevicoryne brassicae. Nature* 213, 1052–1053.

WHITE, T. C. R. (1969). An index to measure weather induced stress of trees associated with outbreaks of Psyllids in Australia. *Ecology* 50, 905–909.

WHITE, T. C .R. (1984). The abundance of invertebrate herbivores in relation to the availability of nitrogen in stressed food plants. *Oecologia* 63, 90–105.

9

Insect Outbreaks in Forests of Western Australia

IAN ABBOTT

Department of Conservation and Land Management PO Box 104, Como WA 6152, Australia

Introduction

The native hardwood forest estate of SW Australia covers 1.65 M ha and is dominated by Jarrah *Eucalyptus marginata* (85%) and Karri *E. diversicolor* (10%) (*Figure 1*). Over much of their range both species occur mixed with Marri *E. calophylla*. These stands have been commercially exploited for 80–100 years and have been managed under an Act of Parliament since 1918. Plantations of two introduced species of pine *(Pinus radiata, P. pinaster)* account for a further 62 000 ha of commercial forest. The first softwood plantations date from the 1920s, though most were established after 1956.

The original inhabitants of these hardwood forests (the Australian Aborigines) used fire but did not extract timber. European settlement in 1829 led to changes in use of fire (particularly in incidence of wildfires) and eventually commercial logging. Rainfall records from 1880 to the present show that both long periods of drought and above average rainfall have been recorded (Sadler, Mauger and Stokes, 1988). Yet these decades of climatic fluctuations and human disturbance of the hardwood forests did not result in any recorded outbreaks of forest insects.

No economic problems with insects in Jarrah forest were reported until the early 1960s, when populations of a native leafminer *Perthida glyphopha* Common (Lepidoptera: Incurvariidae) erupted. By 1987 some 750 000 ha of forest were infested. A second native insect species, gumleaf skeletonizer *Uraba lugens*

Population Dynamics of Forest Insects
© Intercept Ltd, PO Box 716, Andover, Hampshire, SP10 1YG, UK

Walker (Lep., Nolidae) erupted in 1982 and by 1985 infested about 300 000 ha of forest.

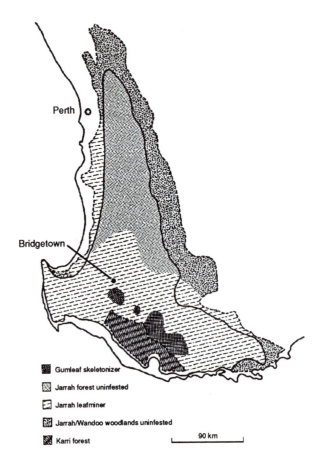

Figure 1.   Map of SW Australia, showing the distribution of Jarrah and Karri forests and maximal extent of outbreaks of Jarrah leafminer and gumleaf skeletonizer.

One-fifth of the Karri forest has been clearfelled, resulting in extensive even-aged regrowth. Significant damage in these stands by bullseye borer *Tryphocaria acanthocera* (Macleay) (Coleoptera, Cerambycidae) was first reported in 1980 when thinning of regrowth stands began.

Departmental records indicate that the American bark beetle *Ips grandicollis* (Eichhoff) (Col., Scolytidae) had been present in plantations of introduced pine for 20–30 years before first erupting in 1970, and again in 1973, 1980 and since 1986.

Biology and Life History

Jarrah leafminer and gumleaf skeletonizer eat the tissue of senescing Jarrah leaves (*c.* 24 weeks old). Bullseye borer eats sapwood and heartwood in the bole of Karri, and *I. grandicollis* is a cambial feeder of pine, mainly of fresh slash and recently felled logs. The larvae of leafminer and skeletonizer feed for *c.* 20 (May – October) and 40 (April – January) weeks respectively, and the larvae of bullseye borer and *I. grandicollis* feed for *c.* 68 weeks and 4–5 weeks respectively. Leafminer, skeletonizer and bullseye borer have one generation per year, and *I. grandicollis* has 6 generations per year. Further details can be obtained from Abbott (in press), Mazanec (1988), Morgan (1967) and Rimes (1959).

The Outbreaks

The first map of the leafminer outbreak showed that 150 000 ha of Jarrah forest were infested by 1964. *Figure 2* shows the infested area during 1964 – 83. Research was initiated in 1962 (Wallace, 1970). The outbreaks spread west and then north. Currently about 750 000 ha, half of the area of Jarrah forest, is affected (*Figure 1*), although the severity of infestation is not uniform spatially or from year to year.

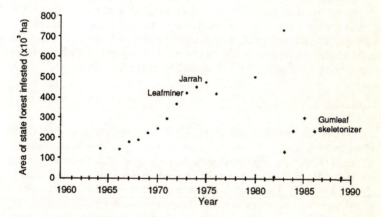

Figure 2. Area of Jarrah forest under outbreak by Jarrah leafminer since spring 1964, and by gumleaf skeletonizer since January 1983, evidenced by moderate-severe browning of Jarrah crowns.

The first infestation of skeletonizer in State Forest was recorded in January 1983, when a total of 95 000 ha were mapped (Strelein, 1988). The infestation (evidenced by browning of forest canopy) has fluctuated from year to year but disappeared by January 1989 (*Figure 2*). Currently, remote sensing of leafminer and skeletonizer outbreaks from aircraft and satellite is being developed by the Commonwealth Scientific and Industrial Research Organization under contract to the Department of Conservation and Land Management.

Damage to even-aged Karri regeneration by bullseye borer was first noted in 1967 and began to cause concern in 1980 when thinnings were first utilized commercially. Stands are not infested until age 14 years. The borer has been recorded within the entire range of Karri forest, some 170 000 ha.

I. grandicollis was accidentally introduced to Western Australia before 1952 and within 10 years was present in nearly all plantations (Abbott, unpubl.). The first outbreak was not recorded until 1970; others followed in 1973,1980, and 1986–88.

Impact of the Outbreaks

None of these outbreaks has yet caused significant tree mortality (Mazanec, 1988; Abbott, pers. obs.). Economic significance has only been quantified for leafminer — it results in a dramatic decline in quantity of foliage carried by the crown (Abbott, 1990) and (depending on site and dominance status) a 40–70% loss of diameter increment (Mazanec 1974; Abbott, in press). Annual loss in sawlog royalty to Government is estimated at $A1–2M. Skeletonizer infestations by analogy with leafminer should also result in crown decline and wood decrement.

Although bullseye borer causes defect, its major impact is on the smaller sections of wood cut from regrowth stands. Mechanical damage to pine timber by *I. grandicollis* is not serious; however they introduce the wood-staining fungus *Ophiostoma ips* ('blue stain') which limits the use of affected timber to structural purposes. Ips is usually a secondary pest of pines in Western Australia, though in recent years 'feeding attacks' (Morgan, 1967) have been observed.

Causes of the Outbreaks

For none of the four outbreaks does a single factor explain why outbreaks began when they did and why they persisted for as long as they have. Research by Mazanec (1988) has implicated weather, logging and fire in causing leafminer outbreaks. My research has implicated weather, fire (but not consistently) and logging not at all. The discrepancy may reflect the different approaches used, though none has been experimental. An experiment underway is manipulating fire intensity and logging. Although years with wet winters are followed by increased damage to foliage (*Figure 3*), this factor appears insufficient to explain the initiation and continuation of the outbreak.

Although the area of forest logged and burned by prescribed low intensity fire in spring increased rapidly during the 1950s and 1960s, a spatial and temporal connection between logging, fire and spread of outbreak is unproven. Furthermore, the outbreak spread widely outside State Forest in which there was no logging or burning. These issues are currently being addressed using a CALM computerized data base containing maps of all fires, logging activities and outbreak areas in State Forest.

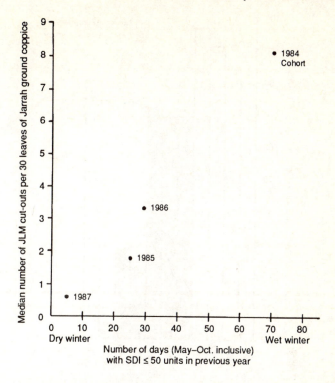

Figure 3. Abundance of mature Jarrah leafminer larvae in spring of four years in relation to the number of days from May to October inclusive in the previous year with Soil Dryness Index (Burrows, 1987) ≤50. This index is a measure of seasonal droughting over a large area.

Dry winters are associated with initiation and continuation of skeletonizer outbreaks (Strelein, 1988), and a wet winter with the remission of the outbreak (*Figure 4*). Biomass of caterpillars of skeletonizer relative to biomass of Jarrah foliage varies inversely with wetness of the winter. However, dry winters have occurred in the past and no skeletonizer outbreak followed them, so clearly other factors are involved. The role of fire and logging is under investigation, both experimentally and using the computerized information system described above.

A survey of 30 Karri regrowth stands (aged 11–108 y) in which 12 tree/stand factors were measured or estimated, along with more elaborate site data including 14 soil properties and 17 climatic variables, revealed several significant correlations with bullseye borer damage. These were proximity to mature Karri stands, sites where Karri was a minor component before felling, small coupe size and site susceptibility to drought (Abbott *et al.*, unpubl.).

Outbreaks of *I. grandicollis* are associated with low winter rainfall followed by summer drought (*Figure 5*). The 1969–70 drought coincided with canopy closure in many plantations on sites highly susceptible to drought. The combination of

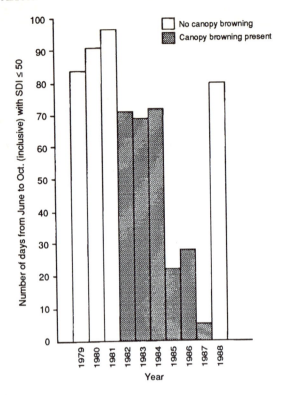

Figure 4. Occurrence of outbreaks from 1979 to 1988 of gumleaf skeletonizer (evidenced in January of each year by moderate–severe browning of Jarrah crowns) in relation to the number of days from June to October inclusive in which Soil Dryness Index (Burrows, 1987) ≤50. Modified from Strelein (1988) and updated.

drought with recent prunings/thinnings on the plantation floor or with overstocked plantations has resulted in *I. grandicollis* outbreaks.

Control Options and Implications for Forest Management

Application of chemical insecticides to forests in Western Australia has not been seriously considered for several reasons: outbreaks began relatively recently and we learned from mistakes evident in spraying forests in Canada and USA; tree mortality has been negligible, so a quick fix solution has not been needed; and much of the potable water supply of SW Australia comes from the hardwood forests.

Research to date on the hardwood forest insect pests has yielded few proven recommendations that forest managers could adopt to ameliorate outbreaks. This

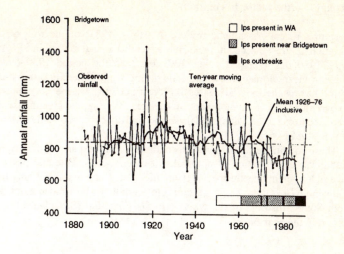

Figure 5. Outbreaks of *I. grandicollis* in 15 000 ha of *P. radiata* plantations in relation to rainfall at Bridgetown (rainfall data based on Borg, Stoneman and Ward, 1987).

should not be taken to signify that the research has been useless. A long-term detailed study of the ecology and biology of the 10 species of wasp parasitic on leafminer has shown that all are ineffective in controlling outbreaks (Mazanec, 1988). This has led to the current thrust by CALM to devise appropriate silvicultural methods, using fire intensity and logging intensity to disfavour leafminer and skeletonizer. The first hypothesis under test is that a single, moderate intensity (500–3000 kW/m) fire in autumn would disrupt the life cycle of both species by reducing the availability of suitable oviposition sites for at least one year.

Recent research on bullseye borer (Abbott *et al.*, unpubl.) is still at a preliminary level. It is therefore premature to formulate specific management options.

Silvicultural methods of disfavouring Ips outbreaks have long been known (Rimes, 1959; Morgan, 1967). When pine plantations are established, unsuitable sites should be avoided and subsequent thinning must be carried out to schedule and the thinnings properly disposed of. Recent studies of deaths of P. *radiata* in 15 000 ha of plantations near Bridgetown have implicated (in order of importance): overstocking, shallow soils, NE facing stands and landscape position as critical site factors (J. McGrath, pers. comm.).

Two predator species *Thanasimus dubius* (Fabricius) (Col., Cleridae) and *Temnochila virescens* (Fabricius) (Col., Ostomidae) and two parasitoid species *Roptrocerus xylophagorum* (Ratzeburg) (Hym., Torymidae) and *Dendrosoter sulcatus* (Muesbeck) (Hym., Braconidae) have been released in Western Australian pine plantations. To date only *R. xylophagorum* has established. It is now widespread.

Drought — the factor in common

Despite a recent pessimistic review (Martinat, 1987) of the role of climatic variation in forest insect outbreaks, it is a fact that annual rainfall in SW Australia since 1960 has averaged about 10% less than the period 1930–59. In this Mediterranean climate, most rain falls during winter; therefore, this recent dry period has resulted in water stress in forest trees and warmer days presumably favourable to insect development. In the next 50 years, the greenhouse effect is predicted to exacerbate this trend in warmer and drier winters (Sadler, Mauger and Stokes, 1988).

These conditions should result in continuation and expansion of current insect outbreaks, and initiation of outbreaks of other species. Potential pests include several leaf-eating beetle species (Curculionidae, Chrysomelidae) and the larvae of the moth *Doratifera quadriguttata* (Limacodidae). The accidental introduction from South Australia of the woodwasp *Sirex noctilio* (Siricidae) also appears certain in the near future.

References

ABBOTT, I. (1990). Ecological implications of insect pests in jarrah and karri forest. *Department of Conservation and Land Management Western Australia. Occasional Paper*, in press.

BORG, H., STONEMAN, G. and WARD, C. G. (1987). Stream and ground water response to logging and subsequent regeneration in the southern forest of Western Australia: Results from four catchments. *Department of Conservation and Land Management Western Australia. Technical Report* 16 (113 pp.).

BURROWS, N. D. (1987). The soil dryness index for use in fire control in the south-west of Western Australia. *Department of Conservation and Land Management Western Australia. Technical Report* 17 (37 pp.).

MARTINAT, P. J. (1987). The role of climatic variation and weather in forest insect outbreaks. In: *Insect Outbreaks*, pp. 241–268 (eds P. Barbosa and J.C. Schultz). San Diego: Academic Press.

MAZANEC, Z. (1974). Influence of Jarrah leaf miner on the growth of Jarrah. *Australian Forestry* 37, 32–42.

MAZANEC, Z. (1988). Jarrah leafminer, an insect pest of jarrah. In: *The Jarrah Forest: a Complex Mediterranean Ecosystem,* pp. 123–131 (eds B. Dell, J. J. Havel and N. Malajczuk). Dordrecht: Kluwer.

MORGAN, F. D. (1967). *Ips grandicollis* in South Australia. *Australian Forestry* 31, 137–155.

RIMES, G. D. (1959). The bark beetle in West Australian pine forests. *Journal of Agriculture of Western Australia* 8 (3rd series), 353–355.

SADLER, B. S., MAUGER, G. W. and STOKES, R. A. (1988). The water resource implications of a drying climate in south-west Western Australia. In: *Greenhouse: Planning for Climatic Change*, pp. 296–311 (ed. G.I. Pearman). Leiden: Brill.

STRELEIN, G. J. (1988). Gum leaf Skeletoniser moth, *Uraba lugens,* in the forests of
   Western Australia. *Australian Forestry* **51**, 197–204.
WALLACE, M.M.H. (1970). The biology of the Jarrah Leaf Miner, *Perthida glyphopa*
   Common (Lepidoptera:Incurvariidae). *Australian Journal of Zoology* **18**, 91–104.

10

Effect of *Neomycta pulicaris* Pascoe (Coleoptera: Curculionidae) on Hard Beech *Nothofagus truncata*

G. P. HOSKING, R. M. J. MacKENZIE, R. ZONDAG
AND J. A. HUTCHESON

Forest Research Institute, Rotorua, New Zealand

Introduction

For more than 100 years areas of extensive dieback have been recorded in New Zealand beech forests (Cockayne, 1926; Conway, 1949; Skipworth, 1981; Cunningham and Stribling 1978; Wardle, 1984). While catastrophic events such as wind and snow have been shown to be important in the dynamics of mountain beech (*Nothofagus solandri* var. *cliffortioides* (Hook. f.) Poole) (Wardle and Allen 1983) much beech decline is not obviously associated with such events (Hosking and Hutcheson, 1986). Concern in recent years, by both forest managers and researchers, at the extent of dieback led to major research by forest health specialists at the Forest Research Institute. An examination of the role of insects in the process of stand collapse formed an important part of this programme.

Many insect species live in association with *Nothofagus* spp. but only a few reach epidemic levels or show population increases before or during the course of tree mortality. The lepidopterous oecophorid and tortricid defoliators, *Proteodes carnifex* (Butler) and *Epichorista emphanes* (Meyrick) periodically cause extensive damage to mountain beech (Milligan, 1974) but this rarely results in death. The beech scale *Inglisia fagi* Maskell has been known to precede extensive mortality in red beech (*Nothofagus fusca* (Hook. f.) Oerst.) but like the beech buprestid *Nascioides enysi* (Sharp) only affects trees already under serious stress (Milligan, 1974; Hosking and Kershaw, 1985). The pinhole borers, *Platypus* spp., can transmit a pathogenic fungus capable of killing beech trees (Faulds, 1977).

Population Dynamics of Forest Insects
© Intercept Ltd, PO Box 716, Andover, Hampshire, SP10 1YG, UK

A recent study of the decline of isolated stands of hard beech (*Nothofagus truncata* Col. (Ckn.) on the Mamaku Plateau east of Rotorua (Hosking and Hutcheson, 1986) revealed very high populations of a leaf-mining weevil *Neomycta pulicaris* Pascoe (Curculioninae) associated with debilitated stands. *N. pulicaris* was described in 1877 from a specimen collected at Tairua on the Coromandel Peninsula. The other two species in the genus are *N. rubida* Broun and *N. seticeps* Broun. No information has been published on the biology of any of the species, although Horak-Kaenel (1970) briefly mentions *N. pulicaris* in her unpublished thesis. *N. rubida* is believed to be confined to *Metrosideros* spp. and the other two species to *Nothofagus* spp. *N. pulicaris* has been recorded from all four New Zealand species of *Nothofagus*. There are several other New Zealand species of leaf-mining weevils in the subfamily Curculioninae, three of which have been recorded from *Nothofagus*. This paper summarizes and discusses recently completed and published work (Hosking and Hutcheson, 1986) and adds to knowledge of the insect's biology.

Biology of *Neomycta pulicaris*

Adult *N. pulicaris* overwinter beneath mosses and liverworts on the trunks and branches of various trees and shrubs including those of their hosts. The weevils are not gregarious, are sexually immature and show no evidence of winter activity such as feeding or flight. Adults become active just before *Nothofagus* bud burst in spring and can be collected from adjacent vegetation before they migrate onto their host. Maturation feeding occurs on expanding leaves and results in a shot-hole pattern. Almost all foliage is affected on heavily infested trees and feeding continues until the leaves harden 2–3 weeks after flushing.

Oviposition only begins when substantial leaf expansion has occurred. The eggs are inserted from the lower surface of the leaf usually into the midrib close to the petiole, and within a few days the leaf falls from the tree. There is clearly a close relationship between oviposition and formation of an abscission zone.

While up to six eggs have been recorded on a single leaf, 75% of all infested leaves have only one. Adults are long-lived and probably continue to oviposit as long as suitable foliage is available. Fallen infested leaves appear to remain green longer than uninfested ones which wilt and curl within a day or two of having fallen.

Eggs laid early in the spring take up to four weeks to hatch, while those laid in late spring take only about eight days. The larvae feed within the leaf blade for approximately three weeks producing a vermiform mine which eventually comes to occupy most of the blade. Development can be successfully completed in leaves as small as 6 mm x 5 mm.

The very active pre-pupal larva emerges from the leaf and moves into the litter layer in search of a suitable pupation site, which is usually just below the soil surface. Pupation takes 6–9 days under laboratory conditions and after a further 2–3 days the teneral adult emerges from the litter. Two generations occur annually in the field where a secondary flush in late summer provides suitable oviposition sites.

Impact of *Neomycta pulicaris* on host

During the flushing period large numbers of adult *N. pulicaris* can be beaten from individual branches. Recognition of the close relationship between the number of *N. pulicaris* in the canopy of hard beech and number of leaves shed (*Figure 1*) is based on cage studies carried out by Hosking and Hutcheson (1986). These showed that 98% of the loss of newly flushed foliage was attributable to oviposition by *N. pulicaris*. It was estimated that moderately attacked hard beech lost more than one-third of their newly flushed foliage. In this Mamaku study unhealthy stands of hard beech shed almost 3000 newly flushed leaves per m² of ground area below the canopy over an 11-week period, which was more than 10 times that lost by healthy stands (*Figure 2*). New leaf shedding peaked in mid November, with 48% occurring in a two week period, then declined rapidly.

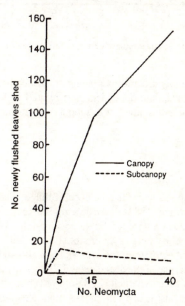

Figure 1. Relationship between newly flushed leaves shed and number of *N. pulicaris* caged on canopy and subcanopy foliage.

Heavily defoliated hard beech produce a limited secondary flush in late summer. This is invariably heavily attacked by *N. pulicaris*, most leaves are shed, and overwintering populations of the insect are boosted. Repeated attack on hard beech over successive years leads to crown thinning and contraction, and ultimately death (Hosking and Hutcheson, 1986). The impact of *N. pulicaris* on other New Zealand *Nothofagus* species is unknown.

On the Mamaku Plateau Hosking and Hutcheson (1986) compared healthy and unhealthy stands and found high *N. pulicaris* populations only in the latter, suggesting an element of predisposition to damage. They found a strong correlation

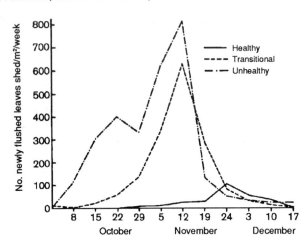

Figure 2.   A comparison of three sites for newly flushed foliage loss in hard beech.

between decreasing soil moisture and increasing stand debility, and concluded hard beech decline here was site-dependent and related to low soil moisture retention capability. *N. pulicaris* attack was seen as a contributing rather than a causative factor.

Discussion

Herbivory associated with the new flush of beech foliage is not unique to New Zealand. Selman and Lowman (1983) describe a similar situation for the Australian chrysomelid *Novacastria nothofagi* Selman occupying a very similar niche. The larvae of this chrysomelid emerge synchronously with bud burst and feed only on the expanding foliage of *Nothofagus moorei* Maiden, although leaf shedding is not involved. In Europe, larvae of the beech weevil *Rhynchaenus fagi* L. mine the newly flushed foliage of *Fagus sylvatica* L. (Bale, 1984) but, although showing a strong preference for that host it is by no means monophagous (Bale and Luff, 1978), nor is it associated with leaf shedding, the unique feature of the hard beech–*N. pulicaris* association.

Increasing evidence suggests food quality, in particular available nitrogen, has a major influence on the population dynamics of many herbivorous invertebrates (White 1984, Hain 1987). According to Larcher (1980) highest concentrations of N, P and K usually occur in young rapidly expanding leaves, with concentrations declining with leaf maturity.

Mooney, Gulman and Johnson (1983) discuss this 'window of vulnerability' associated with the production of unsclerotized, nitrogen rich, new leaves and suggest its potential for reduction by the plant is fairly limited. An obvious opportunity exists for niche exploitation by many unrelated herbivores with similar lifestyles.

Overlying this seasonal opportunity are longer term changes in tree physiology associated with ageing and periodic environmental stress. A link between herbivore populations and environmental stress is convincingly made by White (1986) and supported by Landsberg and Wylie's (1983) findings of higher foliar nutrients in stressed eucalypts. The coincidence of newly flushed foliage with severe environmental stress sets the stage for a very high herbivore load on affected trees.

The particularly close association between *N. pulicaris* and hard beech appears very sensitive to increased environmental stress. Under normal conditions the 'window of vulnerability' is exploited but populations of the weevil remain low and only a small proportion of newly flushed foliage is lost to the tree. At times of severe stress, such as that caused by drought, populations increase dramatically with resultant heavy foliage loss. On good sites tree recovery is swift once the stressful period is passed whilst on marginal sites the insect outbreak may precipitate the irreversible decline of the stand.

We suggest the production of a secondary flush by trees under extreme pressure may drive the system into a positive feedback loop similar to that postulated by Landsberg and Wylie (1983) for eucalypts. The secondary flush boosts the overwintering weevil population which puts increased defoliation pressure on the tree the following spring.

Perhaps the most interesting feature of the biology of *N. pulicaris* is the associated abscission of the host leaf. The evolution of this leaf shedding can be considered as either insect or tree initiated. It might be argued that weevil initiated shedding benefits the insect by preventing the loss of soluble amino acids (Mooney, Gulman and Johnson, 1983; White, 1984) and thus preserving higher food quality, with resultant higher weevil survival. Oviposition at the base of the leaf blade, if associated with injury site ethylene production leading to leaf abscision, would provide a character on which selection could act.

Alternatively, if the ancestors of *N. pulicaris* normally completed their larval growth whilst leaves were still attached, then a tendency by the host to shed infested leaves may have led to increased weevil mortality and reduced crown herbivore pressure, giving a selective advantage to the tree. Such an occurrence would place strong selective pressure back on the weevil for individuals which could complete their development in the shed leaf, thus striking the present balance. As greater photosynthetic area is lost by leaf shedding than if leaves were retained, we favour the first scenario. Although all species of *Nothofagus* are attacked by *N. pulicaris*, mountain and black beech (*Nothofagus solandri* var. *solandri* (Hook. f.) Oerst.) have not been observed supporting high populations, and it seems unlikely leaf shedding would be so universal through New Zealand *Nothofagus* populations if initiated by selection pressure on the tree.

Whatever mechanism led to the present association, it is a striking example of a sensitive balance between herbivore and host, which responds rapidly to environmental stress with the potential to precipitate major changes in forest structure.

Acknowledgements

The authors would like to express their gratitude to colleagues Mike Nuttall, John Dugdale and John Herbert for their very constructive review of the manuscript.

References

BALE, J. S. (1984). Bud burst and success of the beech weevil, *Rhynchaenus fagi*: feeding and oviposition. *Ecological Entomology* **9**, 139–148.

BALE, J. S. and Luff, M. L. (1978). The food plants and feeding preferences of the beech leaf mining weevil, *Rhynchaenus fagi* L. *Ecological Entomology* **3**, 245–249.

COCKAYNE, L. (1926). Monograph on the New Zealand beech forests. Part I. The ecology of the forests and taxonomy of the beeches. *New Zealand State Forest Service Bulletin* **4** 71 pp..

CONWAY, M. J. (1949). Beetle damage to beech forest. *New Zealand Journal of Forestry* **6**: 66–67.

CUNNINGHAM, A. and STRIBLING, P.W. (1978). The Ruahine Range. *New Zealand Water and Soil Technical Publication* **13**, 60 pp.

FAULDS, W. (1977). A pathogenic fungus associated with *Platypus* attack on New Zealand *Nothofagus* species. *New Zealand Journal of Forestry Science* **7**, 384–396.

HAIN, F. P. (1987). Interactions of insects, trees and air pollutants. *Tree Physiology* **3**, 93–102.

HORAK-KAENEL, M. (1970). Systematische und Oekologische Untersuchungen an Lepidopteren auf *Nothofagus cliffortiodes* in Neuseeland. Diplomarbeit (Abt. XA) ausgeführt am Entomologischen Institut der ETH.

HOSKING, G. P. and HUTCHESON, J. A. (1986). Hard beech decline on the Mamaku Plateau, North Island, New Zealand. *New Zealand Journal of Botany* **24**, 263–269.

HOSKING, G. P. and KERSHAW, D. J. (1985). Red beech death in the Maruia Valley South Island, New Zealand. *New Zealand Journal of Botany* **23**, 201–211.

LANDSBERG, J. and WYLIE, F. R. (1983). Water stress, leaf nutrients and defoliation: a model of dieback of rural eucalypts. *Australian Journal of Ecology* **8**, 27–41.

LARCHER, W. (1980). *Physiological Plant Ecology*. New York: Springer-Verlag.

MILLIGAN, R. H. (1974). Insects damaging beech (*Nothofagus*) forests. *Proceedings of the New Zealand Entomological Society* **21**, 32–40.

MOONEY, H. A., GULMAN, S. L. and JOHNSON, N. D. (1983). Physiological constraints on plant chemical defences. In: *Plant Resistance to Insects* pp. 21–36 (ed. P.A. Hedin) Washington DC: American Chemical Society.

SELMAN, B. J. and LOWMAN, M. D. (1983). The biology and herbivory rates of *Novacastria nothofagi*, a new genus and species on *Nothofagus moorei* in Australian temperate rainforests. *Australian Journal of Zoology* **31**, 79–91.

SKIPWORTH, J. P. (1981). Mountain beech mortality in the West Ruapehu Forests. *Wellington Botanical Society Bulletin* **41**, 26–34.

WARDLE, J. A. (1984). *The New Zealand Beeches*. Wellington: New Zealand Forest Service, 447 pp.

WARDLE, J. A. and ALLEN, R. B. (1983). Dieback in New Zealand *Nothofagus* forests. *Pacific Science* **37**, 397–404.

WHITE, T. C. R. (1984). The abundance of invertebrate herbivores in relation to the availability of nitrogen in stressed food plants. *Oecologia* **63**, 90–105.

WHITE, T. C. R. (1986). Weather, *Eucalyptus* dieback in New England, and a general hypothesis of the cause of dieback. *Pacific Science* **40**, 58–78.

11
Positive and Negative Feedbacks in Insect–Plant Interactions

ERKKI HAUKIOJA

Laboratory of Ecological Zoology, Department of Biology; and Kevo Subarctic Research Station; University of Turku, SF-20500 Turku, Finland

Introduction
Effects on insects
The plant's point of view
Discussion
Acknowledgements
References

Introduction

There are a growing number of studies showing that previous artificial damage to foliage induces changes in current leaves with adverse effects on herbivores (see Haukioja and Neuvonen, 1987; Coleman and Jones, 1990), and that natural damage may affect insect performance even more strongly (Haukioja and Neuvonen, 1985; Hartley and Lawton, 1990; Neuvonen, Haukioja and Molarius, 1987). Such induced responses have been called plant defences and have aroused interest especially among zoologists (Benz, 1974; Haukioja and Hakala, 1975; Haukioja *et al.*, 1983; Rhoades, 1985; Haukioja and Neuvonen, 1987; Schultz, 1988). For insect populations the interesting point is that the herbivore-triggered responses in plants depend on herbivore density and may regulate their populations.

Not surprisingly, variable results have been found when effects of herbivore-induced responses in plants have been tested. Some studies have found no effects or that only young trees responded, (e.g. Fowler and Lawton, 1985; Haukioja and Neuvonen, 1987; Leather, Watt and Forrest, 1987; Karban, 1987). Finally there are some tests in which conspecific insects have grown better on trees which had been previously damaged (Niemela *et al.*, 1984; Roland and Myers, 1987; Craig, Price and Itami, 1986; Price, Roininen and Tahvanainen, 1987; Haukioja *et al.*, 1990a). These positive effects have not elicited much interest, perhaps because no theory has predicted them. Such theories might also seem counter-intuitive because it may be hard to understand why plants would behave in ways making them more vulnerable to herbivory. An easy way out of the potential dilemma is that there are methodological flaws in many studies. This is especially true where naturally

Population Dynamics of Forest Insects
© Intercept Ltd, PO Box 716, Andover, Hampshire, SP10 1YG, UK

defoliated trees have been tested for leaf quality either chemically and/or by biotests (Benz, 1974; Schultz and Baldwin, 1982; Niemelä *et al.*, 1984; Williams and Myers, 1984; Myers and Williams, 1984; Roland and Myers, 1987; Haukioja *et al.*, 1990a): insects, instead of researchers, had chosen the plants which were defoliated. Therefore, defoliated trees may represent a biased sample of the tree population and may have been better from the start compared to undefoliated trees which were used as controls (Neuvonen and Haukioja, 1985). Significance testing has often been performed with the wrong error variance which has easily exaggerated differences between defoliated and control trees (Hurlbert, 1984; Fowler and Lawton, 1985; Haukioja and Neuvonen, 1987).

But such methodological sources of error may not explain all the variance in the results. In this paper it is emphasized that, in addition to induced resistance, there exists another type of herbivore-induced response, induced amelioration in plant quality (Haukioja *et al.*, 1990a). These two kinds of induced alterations in forage quality will be first considered from an entomological and then from a botanical point of view. In both cases both positive and negative feedbacks from the insects point of view will be discussed. From the botanical viewpoint the functional organization and regulation of plants will be emphasized.

Effects on insects

For population biologists the important question is whether induced plant responses have any effect on insect populations. This possibility triggered the original interest in induced defences (Benz, 1974; Haukioja and Hakala, 1975). However, the role of herbivore-induced alterations in food quality to population dynamics of insects is extremely difficult to study directly. In practice, we can try to evaluate the effects of plant responses by calculating probable outcomes, preferentially on a per generation basis (Haukioja and Neuvonen, 1987), and try to infer whether potential effects are large enough to be important for population dynamics. Insect growth rate is usually the first parameter to change when food quality changes; it may or may not indicate change in final size (which often determines fecundity of the insect). Typical outcomes are that growth rates of herbivorous insects decline by 5–10% when trees show induced resistance (Haukioja and Neuvonen, 1987). Differences in calculated per generation fecundity between insects on control and on induced foliage of several species of plants vary from 5–10% to over 70% (for references, see Haukioja and Neuvonen, 1987). Even small changes in growth rates and weights may be important for insect performance (Wratten, Edwards and Winder, 1988) but density dependent regulation in populations may modify the effects of decreased (or increased) fecundity per capita caused by induced plant reactions.

Effects of induced resistance on insects are confounded by the two types of induced resistance: short-term and delayed. These make opposite predictions of the stability of insect population dynamics (Haukioja, 1982; Haukioja and Neuvonen, 1987). If resistance is rapidly induced, has a strong effect on insects, and decays in a less time than the generation time of the herbivore, it can potentially stabilize insect densities. By contrast, resistance due to a delayed induction response by the plant may contribute to population cycles (see Benz, 1974; Barbosa and Schultz, 1987; Haukioja *et al.*, 1988): delayed defensive responses may cause or contribute to decline of herbivore populations, and the effects may last at least up to three or four generations (Benz, 1974; Haukioja *et al.*, 1988).

The adverse effects of previous consumption may also be experienced by other species later consuming the plant (Faeth, 1986; 1987; Karban, Adamchak and Schnathorst, 1987; Neuvonen *et al.*, 1988) and the effects may be carried over the next growth season (Neuvonen *et al.*, 1988). For delayed induced resistance, different species of birch herbivores seem to be affected equally negatively (Neuvonen et al., 1988; Hanhimäki, 1989), indicating that the chemical alterations in plant quality are non-specific. On the other hand, the rapid induced resistance may affect herbivore species differently, some positive, some negative (Hartley and Lawton, 1987).

A positive insect-induced feedback has been postulated for insects defoliating *Eucalyptus* trees (Landsberg, Morse and Khanna, 1989); previous damage makes trees more suitable for defoliating beetles and may lead to a circle which finally kills the tree. Perhaps surprisingly, this kind of response does not exclude induced resistance. Birch displays induced resistance after foliar damage (e.g. Haukioja, Suomela and Neuvonen, 1985), but Danell and Huss-Danell (1985) showed that simulated or real moose browsing made trees better for some herbivores and such trees harboured higher densities of insects, especially aphids. Neuvonen and Danell (1987) tested effects of previous browsing on *Epirrita* larvae but found no significant difference in performance of larvae although there was a tendency for larvae to grow better on previously browsed birches. Experiments by K. Danell and E. Haukioja (unpublished) indicated that, after simulated moose browsing (removing apical tips of twigs in winter), leaves of mountain birch were significantly richer in nutrients, poorer in phenols and were significantly favoured by a generalist snail. Haukioja *et al.* (1990a) repeated the experiment and found a positive, although not quite significant, effect in growth of *Epirrita* larvae. They also showed that removal of apical buds alone, or clipping of apical parts of twigs with buds, made the tree respond. Ramets from which the same number of basal buds was removed did not differ from controls. Haukioja *et al.* (1990a) interpreted that it is hormonal disturbance in the tree that was responsible for the ameliorated foliage quality. These studies referred to changes after simulated or natural browsing, therefore, they may not be directly relevant to damage made by larvae. However, the feeding behaviour of *Epirrita* satisfies two important conditions: larvae consumed buds and preferred apical buds (Haukioja *et al.*, 1990a).

The plant's point of view

Whether induced resistance is an evolutionary response to herbivores has not been solved. Much of the discussion is at least partially semantic and may have masked more fundamental problems: how are plants constructed and integrated and how do these configurations modify their responses to herbivory? This is important because without an explicit model of plant function it is hard to understand which options are available for plants (Haukioja, 1990).

Unlike unitary animals, regulation and control of the behaviour of an individual plant is not based on any centralized organ but is a product of modular interactions (White 1979). Trees, for instance, are not completely integrated holistic organisms in which resources are equally available for the whole tree (Haukioja, 1990). On the contrary, trees are made up of compartmentalized units (Shigo, 1984), and resources may be transported only at relatively short distances within the tree (Kozlowski and Winget, 1964). Accordingly, it is not surprising that individual branches, not whole trees seem to respond to defoliation (Morrow and LaMarche, 1978; Watson and Casper, 1984: Haukioja and Neuvonen, 1985; Tuomi et al., 1988; Haukioja et al., 1990a).

To demonstrate how regulation of the integrity of trees affects some problems in plant–herbivore interactions, the possible benefits of induced resistance for plants are discussed first. The most obvious proximate benefit is reduced consumption. However, this is no self-evident consequence because, theoretically, herbivores might compensate for low quality diets by eating for a longer period and therefore consuming more (Moran and Hamilton, 1980; but see Haukioja et al., 1990b).

For delayed induced resistance in birch, Neuvonen and Haukioja (1984) calculated that trees showing the response presumably lost less foliage than control trees, largely because of higher larval mortality and perhaps also due to feeding inhibition. Rapidly induced resistance may also increase insect mortality, although there is little proof of this (see Wratten, Edwards and Winder, 1988). Accordingly, it has been proposed that the benefits of rapid induced responses are due to repelling herbivores from the most valuable young leaves, often in developing shoots which are important for the plant's competitive ability (Janzen, 1979; Edwards and Wratten, 1983; 1987). Herbivores seem to avoid previously damaged leaves or parts of them (e.g. Edwards, Wratten and Gibberd, 1990) but total consumption in the tree need not decrease (Fowler and MacGarvin, 1986). A fundamental problem is how the plant benefits if defensive reactions just redistribute the damage. Two reasons have been proposed. First, dispersed damage may be easier to compensate (see Watson and Casper, 1984), especially if resource intake is sink-regulated and modules compete locally for resources (see Haukioja, 1990). Locally increased photosynthesis after damage is in accordance with this hypothesis. Secondly, reduced consumption of young leaves in developing shoots may help the shoots to grow and the plant to compete better with other plants (Edwards and Wratten, 1987).

Relevance of the latter explanation depends on the role of leaves in the plant's competitive ability. Leaves are carbon sources although young leaves first function as sinks (Coleman and Jones, 1990). In addition, leaves affect development of the modules in which they belong (Haukioja, 1990; Haukioja et al., 1990a). Birches have two types of shoots, short shoots which produce leaves only in the spring, and long shoots which form new leaves as long as the shoot grows. Haukioja et al. (1990a) showed that removal of young leaves either in developing long shoots or in short shoots in birch did not affect shoot or tree growth that season. The reason is that shoot growth in birch happens via resources sequestered in the previous year. However, damage to individual long shoot leaves affects the shoot which potentially develops the next year from the axillary bud of the leaf.

Therefore the importance of rapidly induced resistance in young long shoot leaves may not be to protect shoot growth (which seems to be insensitive to foliar damage in the current year) but to protect long shoot leaves until they have completed subtending the axillary bud (Haukioja *et al.*, 1990a).

It is imperative to understand plant functional structure and regulation before plant responses to herbivory can be understood and predicted. It is true that the theory of natural selection can predict what happens to different genetically determined traits and basically to individuals provided with those traits. Still general evolutionary theories — largely built by using non-modular organisms as models — may not be very efficient in unravelling the plant's internal rules or constraints caused by the basic modular design. A plant simply may not behave in ways which are optimal for a holistic, totally integrated organism. An adaptationist explanation is that compartmentalization is beneficial for the plant. For instance, compartmentalization can help to isolate infected parts from the rest of the plant, and the plant can easily reorganize hierarchies of meristems after herbivory and therefore be able to tolerate and to recover from many types of damage, including herbivory (Haukioja, 1990). A non-adaptationist explanation is that a modular organism simply cannot build a totally integrated system. If it cannot, there exists no such heritable variation on which natural selection could operate.

The manner in which plants are constructed and integrated may also be important for the old debate on whether plants benefit from herbivory. Owen and Wiegert (1976, 1987) claimed that plants may require herbivory to develop a canopy with an optimal form. Janzen (1979) was concerned about the conditions which would cause a plant to relinquish control of its own physiology to an external factor. Haukioja (1990) contended that consequences of herbivory simply depend on the rules of how the plant is constructed and regulated, not on the hypothetical benefits of herbivory. Integration of individual plants is based on local hormonal control by active meristems whose function makes growth possible and keeps nearby meristems subordinated or dormant. Because herbivores can directly affect the control points by consumption, they also can alter the growth and influence the form and other characteristics of plants. But, contrary to Owen and Wiegert (1976), there is no reason why this would happen for the benefit of the plant.

Discussion

An explicit separation between the two possible herbivore-induced responses — leading either to higher or lower plant suitability — may erase some controversies in studies where inducible plant defences have been studied. First, positive changes in foliage quality are possible. Even the same plant may be capable of both induced amelioration and resistance, depending upon the type of damage (Haukioja *et al.*, 1990a). When apical buds were removed either by clipping tips of birch twigs or by removing buds only, foliage quality for herbivores improved. In experiments in which induced resistance was elicited (Haukioja *et al.*, 1985, 1988) leaves were damaged by tearing laminas. Buds were not destroyed.

For insect population dynamics the hormonal disturbance hypothesis by Haukioja *et al.* (1990a) proposes a new causal agent for how plant–herbivore interactions might promote eruptive outbreaks (see Berryman, 1988). Not only the decline but also the initiation of an outbreak may depend on induced plant responses. Although a direct proof of the importance of the hormonal disturbance hypothesis under natural conditions is lacking, it is worth discussing how the hypothesis might modify a number of opinions of herbivore-induced effects on insect populations.

If insect-induced amelioration in plant quality happens because of the disturbed hormonal control, probability of amelioration in plant quality should be highest in cases in which plants are particularly sensitive to manipulation in their hormonal balance and/or herbivores are capable of making plants to respond. The most extensive outbreaks of forest pests seem to occur on poor quality soils, on mature stands and in species-poor open-canopy forests (see Mattson *et al.*, 1987; Barbosa and Schultz, 1987). There are several possible explanations. Predation may be inefficient in controlling insect populations in such areas because species richness is low and only generalist predators can survive troughs in prey density. In addition, due to poor resources, tolerance to damage may be low and plants have to defend themselves but not necessarily after their peak reproductive value (see Mattson *et al.*, 1987). But trees in these areas tend to grow slowly, especially when they mature. In such a case minor changes in dominance relations among shoots might diminish the weak apical dominance and — for unknown physiological reasons — lead to higher quality foliage for herbivores (Haukioja *et al.*, 1990a).

The insect-induced amelioration in foliage quality may lead to a delayed positive feedback which might automatically follow after the herbivore density reaches a high level. Above the threshold density, insects may alter foliage quality in a favourable way which makes possible high densities which last for years and which may move from the epicenter. Alternatively, the ability of herbivores to alter plants may vary. Haukioja *et al.* (1990a) proposed that insect species with feeding patterns which easily reshuffle module dominances in trees may be the most probable to experience outbreaks. Colonial feeding with preference for tipmost leaves in the distal parts of shoots, and active use of buds, represent such characteristics. It may not be just chance that some of the most serious pests of mature tree stands are species which consume buds (like budmoths of the genera *Choristoneura* and *Zeiraphera*).

The hormonal disturbance hypothesis indicates some unsuspected methodological pitfalls in previous tests where researchers have tried to test the importance of induced plant responses for insect outbreaks. Plant–herbivore theories have been defence-dominated and, therefore only defensive responses were looked for. Taking induced resistance and amelioration into account, it is possible to make the following predictions about herbivore-induced alterations in tree quality: (i) effective induced 'defences' do not occur while insect populations are expanding; (ii) delayed induced 'defences' are most likely found during or after pest population declines; (iii) induced amelioration is more common than induced resistance in plants which experience irregular insect outbreaks; (iv) *both* induced amelioration *and* induced resistance are most likely found in plant species whose pests show regular cyclic outbreaks; (v) induced amelioration is most common in mature trees whose growth is slowing down, while delayed induced resistance is most common in younger plants.

Acknowledgements

I thank Sinikka Hanhimäki, Simon Leather, Josef Senn, Janne Suomela, Mari Walls and an anonymous reviewer for comments on an earlier version. This work was financed by the Academy of Finland.

References

BARBOSA, P. and SCHULTZ, J. (1987). *Insect Outbreaks*. San Diego: Academic Press.

BENZ, G. (1974). Negative Rückkoppelung durch Raum- und Nahrungskonkurrenz sowie zyklische Veränderung der Nahrungsgrundlage als Regelprinzip in Populationsdynamik des Grauen Lärchenwicklers, *Zeiraphera diniana* (Guenée) (Lep., Tortricidae). *Zeitschrift für Angewandte Entomologie* **76**, 196–228.

BERRYMAN, A. A. (1988). The theory and classification of outbreaks. In: *Insect Outbreaks*, pp.3–30 (eds P. Barbosa and J. Schultz). San Diego: Academic Press.

COLEMAN, J. S. and JONES, C. G. (1990). In: *Phytochemical Induction by Herbivores* (eds M. Raupp and D. Tallamy). New York: John Wiley, in press.

CRAIG, T. P., PRICE, P. W. and ITAMI, J .K. (1986). Resource regulation by a stem–galling sawfly on the arroyo willow. *Ecology* **67**, 419–425.

DANELL, K. and HUSS-DANELL, K. (1985). Feeding by insects and hares on birches earlier affected by moose browsing. *Oikos* **44**, 75–81.

EDWARDS, P. J. and WRATTEN, S. D. (1983). Wound induced defences in plants and their consequences for patterns of insect grazing. *Oecologia (Berlin)* **59**, 88–93.

EDWARDS, P. J. and WRATTEN, S. D. (1987). Ecological significance of wound-induced changes in plant chemistry. In *Insect-Plant Relationships*, pp 213–218 (eds V. Labeyrie, G. Farbres and D. Lachaise). Dordrecht: Dr W Junk.

EDWARDS, P.J., WRATTEN, S.D. and GIBBERD, R. (1990). The impact of inducible phytochemicals on food selection by insect herbivores and its consequences for the distribution of grazing damage. In: *Phytochemical Induction by Herbivores* (eds D.W. Tallamy and M. J. Raupp): New York: John Wiley, in press.

FAETH, S . H . (1986). Indirect interactions between temporally separated herbivores mediated by the host plant . *Ecology* **67**, 479–494

FAETH, S. H. (1987). Community structure and folivorous insect outbreaks: the role of vertical and horizontal interactions. In: *Insect Outbreak*, pp 135–171. (eds P. Barbosa and J. C. Schultz). San Diego: Academic Press.

FOWLER, S.V . and LAWTON, J. H. (1985). Rapidly induced defences and talking trees: the devil's advocate position. *The American Naturalist* **126**, 181–195

FOWLER, S. V. and MacGARVIN, M. (1986). The effects of leaf damage on the performance of insect herbivores on birch, *Betula pubescens. Journal of Animal Ecology* **55**, 565–573 .

HANHIMÄKI, S. (1989). Induced resistance in mountain birch: defence against leaf-chewing insect guild and herbivore competition. *Oecologia (Berlin)* **81**, 242–248.

HARTLEY, S. E. and LAWTON, J. H. (1987). Effects of different types of damage on the chemistry of birch foliage, and the responses of birch feeding insects. *Oecologia (Berlin)* **74**, 432–437.

HARTLEY, S. E. and LAWTON, J. H. (1990). Biochemical aspects and significance of the rapidly induced accumulation of phenolics in birch foliage. In: *Phytochemical Induction by Herbivores* (eds M. Raupp and D. Tallamy). New York: John Wiley, in press.

HAUKIOJA, E. (1982). Inducible defences of white birch to a geometrid defoliator, *Epirrita autumnata*. Proc. 5th Int. Symp. *Insect-Plant Relationships*, pp. 199–203. Wageningen: Pudoc.

HAUKIOJA, E. (1990) Toxic and nutritive substances as plant defence mechanisms against invertebrate pests. In: *Pest Pathogens and Plant Communities*, pp. 219–231 (eds J. Burdon and S. Leather). Oxford: Blackwell Scientific Publications.

HAUKIOJA, E. and HAKALA, T. (1975). Herbivore cycles and periodic outbreaks. Formulation of a general hypothesis. *Reports from the Kevo Subarctic Research Station* **12**, 1–9.

HAUKIOJA, E. and NEUVONEN, S. (1985) Induced long-term resistance of birch foliage against defoliators, defensive or incidental? *Ecology* **66**,1303–1308

HAUKIOJA, E. and NEUVONEN, S. (1987). Insect population dynamics and induction of plant resistance: the testing of hypotheses. In: *Insect Outbreaks*, pp. 411–432 (eds P. Barbosa and J. Schultz). New York: Academic Press.

HAUKIOJÄ, E., KAPIAINEN, K., NIEMELÄ, P. and TUOMI, J. (1983). Plant availability hypothesis and other explanations of herbivore cycles: complementary or exclusive alternatives. *Oikos* **40**, 419–432.

HAUKIOJA, E., SUOMELA, J. and NEUVONEN, S. (1985). Long-term inducible resistance in birch foliage, triggering cues and efficacy on a defoliator. *Oecologia (Berlin)* **65**, 363–369.

HAUKIOJÄ, E., NEUVONEN, S., NIEMELÄ, P. and HANHIMÄKI, S. (1988). The autumnal moth *Epirrita autumnata* in Fennoscandia. In: *Dynamics of Forest Insect Populations*, pp. 167–178 (ed. A.A. Berryman). New York: Plenum Press.

HAUKIOJA, E., RUOHOMÄKI, K., SENN, J., SUOMELA, J. and WALLS, M. (1990a). Consequences of herbivory in the mountain birch (*Betula pubescens* ssp *tortuosa*): importance of the functional organization of the tree. *Oecologia (Berlin)*, **82**, 238–247.

HAUKIOJA, E., RUOHOMÄKI, K., SUOMELA, J. and WORISALO, T. (1990b). Nutritive quality as a defense against herbivores. *Forest Ecology and Management*, in press.

HURLBERT, S. H. (1984). Pseudoreplication and the design of ecological field experiments. *Ecological Monographs* **54**,187–211.

JANZEN, D. H. (1979). New horizons in the biology of plant defenses. In: *Herbivores. Their Interaction with Secondary Plant Metabolites*, pp. 331–350 (eds G.A. Rosenthal and D.H. Janzen). New York: Academic Press.

KARBAN, R. (1987). Herbivory dependent on plant age: hypothesis based on acquired resistance. *Oikos* **48**, 336–337.

KARBAN, R., ADAMCHAK, R. and SCHNATHORST, W.C. (1987). Induced resistance and interspecific competition between spider mites and a vascular wilt fungus . *Science* **235**, 678–680.

KOZLOWSKI, T. T. and WINGET, C. H. (1964). The role of reserves in leaves, branches, stems, and roots on shoot growth of red pine. *American Journal of Botany* 51, 522–529 .

LANDSBERG, J., MORSE, J. and KHANNA, P. (1989) . Tree dieback and insect dynamics in remnants of native woodlands on farms. *Proceedinqs of the Ecological Society of Australia* 16 (Australian Ecosystems: 200 years of Utilization, Degradation & Reconstruction), in press.

LEATHER, S. R., WATT, A .D. and FORREST, G. I. (1987) . Insect-induced chemical changes in young lodgepole pine (*Pinus contorta*): the effect of previous defoliation on oviposition, growth and survival of the pine beauty moth, *Panolis flammea. Ecological Entomology* 12, 275–281.

MATTSON, W. J., LAWRENCE, R. K., HAACK, R. A., HERMS, D. A. and CHARLES, P. -J. (1987) . Defensive strategies of woody plants against different insect feeding guilds in relation to plant ecological strategies and intimacy of association with insects. In: *Mechanisms of Woody Plant Defenses Against Insects: Search for Pattern*, pp. 1–36 (eds W. J. Mattson, J. Levieux and C. Bernard–Dagan). New York: Springer-Verlag.

MORAN, N. and HAMILTON, W. D. (1980) . Low nutritive quality as defence against herbivores . *Journal of Theoretical Biology* 86, 247– 254.

MORROW, P. A. and LAMARCHE, V. C., Jr. (1978) . Tree ring evidence for chronic suppression of productivity in subalpine *Eucalyptus. Science* 201, 1244–1246.

MYERS, J. H . and WILLIAMS, K. S. (1984). Does tent caterpillar attack reduce the food quality of red alder foliage? *Oecologia (Berlin)* 62, 74–79 .

NEUVONEN, S. and DANELL, K. (1987). Does browsing modify the quality of birch foliage for *Epirrita autumanta* larvae? *Oikos* 49, 156–160 .

NEUVONEN, S., HANHIMÄKI, S., SUOMELA, J. and HAUKIOJA, E. (1988). Early season damage to birch foliage affects the performance of a late season herbivore. *Journal of Applied Entomology* 105, 182–189 .

NEUVONEN, S. and HAUKIOJA, E. (1984). Low nutritive quality as defence against herbivores: induced responses in birch. *Oecologia (Berlin)* 63, 71–74.

NEUVONEN, S. and HAUKIOJA, E. (1985). How to study induced plant resistance? *Oecologia (Berlin)* 66,456–457

NEUVONEN, S., HAUKIOJA, E. and MOLARIUS, A. (1987). Delayed induced resistance against a leaf-chewing insect in four deciduous tree species. *Oecologia (Berlin)* 74, 363–369 .

NIEMELÄ, P., TUOMI, J., MANNILA, R. and OJALA, P. (1984) . The effect of previous damage on the quality of Scots pine foliage as food for Diprionid sawflies. *Zeitschrift für angewandte Entomologie* 98, 33–43 .

OWEN, D. F. and WIEGERT, R. G. (1976). Do consumers maximize plant fitness? *Oikos* 27, 488–492 .

OWEN, D. F. and WIEGERT, R. G. (1987). Leaf eating as mutualism. In: *Insect Outbreaks,* pp. 81–95 (eds P. Barbosa and J. C. Schultz). San Diego: Academic Press.

PRICE, P. W., ROININEN, H. and TAHVANAINEN, J. (1987) . Why does the bud-galling sawfly, *Euura mucronata,* attack long shoots? *Oecologia (Berlin)* 74, 1–6.

RHOADES, D. F. (1985). Offensive-defensive interactions between herbivores and plants, their relevance in herbivore population dynamics and ecological theory. *The American Naturalist* **125**, 205–238.

ROLAND, J. and MYERS, J. H. (1987). Improved insect performance from host-plant defoliation: winter moth on oak and apple. *Ecological Entomology* **12**, 409–414

SCHULTZ, J. C. (1988). Plant responses induced by herbivores. *Trends in Ecology and Evolution* **3**, 45–49.

SCHULTZ, J. C. and BALDWIN, I.T. (1982). Oak leaf quality declines in response to defoliation by Gypsy moth larvae. *Science* **217**, 149–151.

SHIGO, A. L. (1984). Compartmentalization: a conceptual framework for understanding how trees grow and defend themselves. *Annual Review of Phytopathology* **22**, 189–214.

TUOMI, J., NIEMELÄ, P., HAUKIOJA, E., SIRÉN, S. and NEUVONEN, S. (1984). Nutrient stress: An explanation for plant anti-herbivore responses to defoliation. *Oecologia (Berlin)* **61**, 208–210.

TUOMI, J., NIEMELÄ, P., ROUSI, M., SIRÉN, S. and VUORISALO, T. (1988). Induced accumulation of foliage phenols in mountain birch: branch response to defoliation? *The American Naturalist* **132**, 602–608.

WATSON, M. A. and CASPER, B. B. (1984). Morphogenetic constraints on patterns of carbon distribution in plants. *Annual Review of Ecology and Systematics* **15**, 233–258.

WHITE, J. (1979). The plant as a metapopulation. *Annual Review of Ecology and Systematics* **10**, 109–145.

WILLIAMS, K. S. and MYERS, J. H. (1984). Previous herbivore attack of red alder may improve food quality for fall webworm larvae. *Oecologia (Berlin)* **63**, 166–170.

WRATTEN, S. D., EDWARDS, P. J. and WINDER, L. (1988). Insect herbivory in relation to dynamic changes in host plant quality. *Biological Journal of the Linnean Society* **35**, 339–350.

12
Variation in the Effects of Spring Defoliation on the Late Season Phytophagous Insects of *Quercus robur*

MARK D. HUNTER* AND CHRIS WEST

University of Oxford, Department of Zoology, South Parks Road, Oxford OX1 3PS UK
*Current address: Pennsylvania State University, Pesticide Research Laboratory, University Park, PA 16802, USA

Introduction

There is now a considerable body of data which describes the effects of defoliation on changes in leaf quality for arthropod herbivores (Green and Ryan, 1972; Haukioja and Niemela, 1977, 1979; Edwards and Wratten, 1982, 1983, 1985; Baldwin and Schultz, 1983; Raupp and Denno, 1984; Bergelson, Fowler and Hartley, 1986; Hartley and Firn, 1989). The effect of wound-induced changes in plants on arthropod populations can be negative (Croft and Hoying, 1977; Karban and Carey, 1984; West, 1985a; Harrison and Karban, 1986), positive (Hiechel and Turner, 1983; Niemela *et al.*, 1984; Williams and Myers, 1984; Kidd, Lewis and Howell, 1985; Rhoades, 1985) or a dynamic balance between positive and negative forces (Faeth, 1986; Hunter, 1987a; Roland and Myers, 1987; Silva-Bohorquez, 1987). Variation in herbivore response to leaf damage has important ramifications for our understanding of the population dynamics of plant pests, and we propose to examine some of that variability here.

Our paper differs in two important ways from others in this volume which consider the consequences of wound-induced changes in plant quality. First, we explore the effects of early season defoliation on late-season phytophagous insect species — we concentrate on between-species rather than within-species interactions mediated by changes in the host plant. Secondly, we argue that the nutritional quality of foliage is only one of several leaf traits influenced by defoliation. Plants respond both physically and chemically to defoliation and we would stress, as others have previously (Heinrich, 1976; Heinrich and Collins, 1983; Schultz, 1983; Southwood, 1985), that leaves represent a place to live for phytophagous arthropods as well as a food source. Like many other 'parasites', insect herbivores live on what they consume, and we would argue that to understand fully the role defoliation plays in the population dynamics of insects on plants, we must consider all aspects of wound-induced change in habitat quality. These will include the nutritional quality of the foliage, variation in the susceptibility of herbivores to their natural enemies, and the ability of phytophagous insects to utilize defoliated plants as refuges from adverse climatic conditions.

This paper describes some of our research on the insect fauna of the pedunculate oak, *Quercus robur* (L.) in Oxfordshire, England (West, 1985a, 1985b; Hunter, 1987a, 1987b, 1988; Hunter and Willmer, 1989) which has led us to three main conclusions about the effects of wound-induced changes in plants on insect populations:

1. Different insect guilds may respond in different ways to equivalent levels of defoliation of their host plant.
2. Insect herbivore species within one guild can exhibit positive, neutral and negative responses to defoliation depending on the level of leaf damage — there is not a linear relationship between defoliation by one herbivore and the response of a second.
3. Since defoliation by one insect species can have a positive or negative effect on a second depending on degree, plant-mediated interactions between organisms differ from most direct interactions because there can be competition and commensalism simultaneously between two species; variation in defoliation levels between individual plants is sufficient that competition and commensalism can occur side by side.

The system

The host plant, *Q. robur*, was studied in Wytham Woods, Oxfordshire, where trees are attacked in spring (April to early June) by two major defoliators: *Tortrix viridana* (L.) (Lepidoptera: Tortricidae) and *Operophtera brumata* (L.) (Lepidoptera: Geometridae). These two species can completely defoliate trees and, in an average year, they remove around 40% of *Q. robur* leaf area in Wytham (West, 1985a; Hunter, 1987a; Hunter and Willmer, 1989). This is higher than some other parts of the country (Crawley, 1985; Crawley, 1987) and higher than for many other deciduous tree species (Bray, 1964; Leigh and Smythe, 1978; Nielsen, 1978; Schowalter *et al.*, 1986). In Wytham, defoliation levels are greatest in the upper canopy of mature trees (West, 1985a; Hunter, 1987a). The dynamics of *O. brumata* and *T. viridana* populations in Europe are primarily determined by the degree to which caterpillar eclosion coincides with oak bud burst in spring (Schutte, 1957; Satchell, 1962; Varley and Gradwell, 1968) — broadly speaking, in years when eclosion precedes bud burst, caterpillars starve and/or disperse, and populations crash. When bud burst precedes eclosion, colonization rates and larval densities are high. Although very early bud burst ought to inhibit larval perform-ance due to increasing tannin levels, increasing toughness and decreasing water content in leaves (Feeny, 1970), we have always found that both mature trees and saplings which burst their buds first carry the highest spring defoliator populations and suffer the greatest leaf damage (Hunter, 1988).

Q. robur responds to spring defoliation by producing regrowth leaves in late June/July in direct proportion to the level of leaf damage by *O. brumata* and *T. viridana* (Crawley, 1983). Late-season insect herbivores (July to October) are therefore exposed to three different types of foliage; undamaged primary leaves, damaged primary leaves and regrowth leaves. The effect of feeding on these foliage types on the populations of late season insect herbivores has been assessed for the lepidopteran leaf-mining guild (West, 1985a), a leafsucker (the aphid *Tuberculoides annulatus* L. (Silva-Bohorquez, 1987)) and three species of oak leaf skeletonizer (Hunter, 1987a). For the purpose of this discussion, we want to compare the responses of leaf-miners and leaf skeletonizers to spring defoliation.

Lepidopteran leafminers

There are thirty-five species of lepidopteran leafminer feeding on *Q. robur* in Britain (Emmet, 1979), all but one of which feed after mid-June, and the most abundant of which are in the genus *Phyllonorycter* (Lepidoptera: Gracillariidae). We have recorded five *Phyllonorycter* species in Wytham, over 50% of which are *P. harrisella* L. *Phyllonorycter* species are bivoltine — the first generation (about 40 days) commences mining in the middle of June, after *T. viridana* and *O. brumata* have pupated, but before regrowth foliage has emerged. The first generation, therefore, is restricted to undamaged and damaged primary foliage. The second generation (about 60 days) feeds on undamaged primary, damaged primary and regrowth foliage in August and September (West, 1985a).

In both generations, there are strong negative relationships between the mean area of primary leaf removed from a *Q. robur* branch by spring defoliators and the proportion of miners reaching adulthood on that branch (*Figure 1*).

These relationships hold for both natural and artificial (scissor) damage and are reflected in female oviposition choice — fewer eggs are laid on branches with high levels of leaf damage and on defoliated trees in general (West, 1985b). However, even at moderate levels of spring defoliation, there are few undamaged leaves left on trees in Wytham Woods (Hunter, 1987a, 1988), and the combination of low oviposition and low larval survival on damaged leaves results in asymmetric competition between early season leaf-chewers and late season leafminers (West, 1985a).

Life table analysis of *Phyllonorycter* species has shown that predation, parasit-ism and leaf abscission are unrelated to leaf damage levels, and we assume that some unidentified change in plant chemistry is responsible for the high levels of *Phyllonorycter* mortality at high levels of leaf damage. This contrasts with other

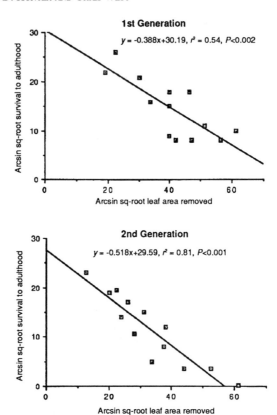

Figure 1. The relationship between survival to adulthood of first and second generation *Phyllonorycter* leafminers, and the mean leaf area removed by spring defoliators from the primary foliage of their host plant, *Quercus robur*.

studies of leaf-mining insects on oak which have shown that wound-induced responses operate through increasing levels of larval parasitism (Faeth, 1985, 1986; Faeth and Bultman, 1986).

There may be some degree of compensation for the second generation of miners on defoliated trees. As we have previously described, the production of regrowth foliage is directly related to spring defoliation levels, and leafminer survival is 10% higher on regrowth foliage than on undamaged primary foliage (West, 1985a). This may be related to the higher levels of nitrogen in regrowth foliage since late-season leafminers are thought to be feeding very close to their basal nutritional threshold on *Q. robur* (West, 1985a).

Lepidopteran leaf skeletonizers

There is considerable contrast between the patterns just described for leafminers, and those observed among free-living lepidopteran larvae. Although most species feed in spring when *Q. robur* leaves have their highest protein and water content and lowest toughness (Feeny, 1970), we have recorded 21 species during late season sampling. Three of these are relatively abundant on *Q. robur: Diurnea fagella* (D. & S.) (Lepidoptera: Oecophoridae), *Teleiodes luculella* (Hubn.) (Lepidoptera: Gelechiidae) and *Gypsonoma dealbana* (Frol.) (Lepidoptera: Tortricidae). All three are leaf-rolling skeletonizers, and larvae hatch either synchronously with (*D. fagella*) or after regrowth flush (*T. luculella* and *G. dealbana*). They therefore feed on the three leaf types (undamaged primary, damaged primary and regrowth) available to the second generation of *Phyllonorycter* species.

Unlike leafminers, the leaf skeletonizers perform poorly on regrowth foliage (Hunter, 1987a). Larvae are almost absent from trees which have been completely defoliated in spring and are refoliated with new growth. In experiments with *D. fagella*, larval survival and female pupal weights are depressed by 48% and 23% respectively compared with undamaged primary foliage (Hunter, 1987a). This is despite the higher nitrogen and water content of regrowth foliage (West, 1985a). On *Q. robur* trees which are highly susceptible to *T. viridana* and *O. brumata* in spring, therefore, there is host-plant mediated asymmetric competition between early and late season free-living Lepidoptera.

In further contrast with leaf-miners, sampling data show that, when defoliation levels are moderate, all species of skeletonizer are most abundant at the tops of *Q. robur* trees (higher levels of spring defoliation) and on individual trees with increasing levels of primary leaf damage (*Figure 2*).

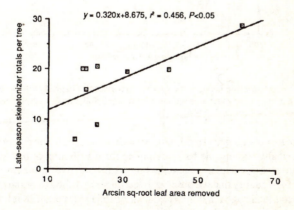

Figure 2.   The relationship between late-season skeletonizer totals from *Quercus robur* trees and the mean primary leaf area removed by spring defoliators from those trees.

Moreover, skeletonizers choose to build refuges and feed on individual leaves in higher damage classes than the background population of leaves. Distributions are therefore skewed in three ways (trees, canopy and leaves) towards primary leaves with moderately high (up to 75% leaf area removed) levels of spring defoliation (Hunter, 1987a). On most trees, therefore, there is an apparent commensalism between spring defoliators and late-season skeletonizers.

There is strong evidence to suggest that the skew of skeletonizer distribution towards spring leaf damage is habitat and not nutritionally based. We conclude this because the natural distribution of skeletonizers described above contrasts with the results of our bagging trials. Experimental enclosure of *D. fagella* larvae on artificially damaged *Q. robur* leaves causes a 26% reduction in larval survival compared with undamaged leaves (Hunter, 1987a). There is also a negative relationship between the mean spring damage level of an individual tree and the survival of *D. fagella* larvae bagged on primary leaves (exhibiting the natural distribution of damage around the mean) on that tree (*Figure 3*).

This drop in skeletonizer survival with increasing leaf damage is in direct contrast with natural skeletonizer distribution patterns, but in agreement with the experimental data for leafminers (*Figure 1*).

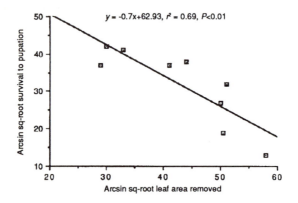

Figure 3.   The relationship between the survival to pupation of bagged *Diurnea fagella* larvae and the mean primary leaf area removed by spring defoliators from their host plant, *Quercus robur*.

We argue that bagging the free-living larvae has removed some powerful factor which is normally responsible for determining the distribution of oak leaf skeletonizers and, as a consequence, revealed the same negative influence of spring damage on food quality that we observe in leafminers. Although organza bags do not influence the nitrogen and water content of foliage (West, 1985a), they will increase the relative humidity, decrease air movement, and exclude predators and parasites from bagged leaves (Hunter and Willmer, 1989).

Bagging leaves may interact with spring leaf damage because *D. fagella*, *T. luculella* and *G. dealbana* are all leaf-rollers. Experiments with *D. fagella* have demonstrated that leafrolling is much more rapid on damaged leaves than undamaged leaves (*Figure 4*), probably because old, damaged leaves curl at the edges. The distributions of late-season skeletonizers may therefore be skewed towards damaged leaves because leaf-rolling rate (in the absence of bags) influences larval survival.

Leaf rolls are certainly of considerable importance to *D. fagella*. They defend their refuges from conspecifics by wrestling and regurgitation, and have evolved modified third thoracic legs which they scrape on skeletonised leaves, apparently 'singing' their territorial rights (Hunter, 1987b). Leaf rolls may help protect larvae from natural enemies and the adverse effects of weather — factors which lose their

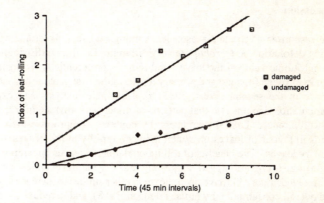

Figure 4.   A comparison between the mean leaf-rolling rate of *Diurnea fagella* larvae on damaged and undamaged foliage of their host plant, *Quercus robur*. Each point is the mean leaf-rolling score of 20 larvae.

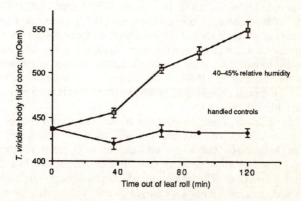

Figure 5.   Changes in the body fluid concentration (mOsm) of fourth instar *T. viridana* larvae after being removed from their leaf rolls (handled controls were reintroduced). Leaf rolls are critical for water balance in this species.

importance when leaves are bagged. We assume that some combination of enemies and weather results in the distribution of larvae on the best resource for building refuges (damaged leaves) rather than best resource for nutrition (undamaged leaves). The relative contribution of climate and enemies is unclear in the present system. However one of the spring defoliators of *Q. robur*, *T. viridana*, has been shown to rely almost entirely on leaf rolls for osmoregulation (*Figure 5*), and many other phytophagous insects rely on their within-plant distribution to fulfil their microclimatic requirements (Tahvanainen, 1972; Casey, 1976; Willmer, 1980; 1982).

Discussion

The responses of the late-season leaf-mining and leaf-skeletonizing guilds to spring defoliation on *Q. robur* are very different. Leafminer distribution reflects the nutritional value of the leaves, and females preferentially oviposit on leaves high in nitrogen (undamaged primary leaves in the first generation, and regrowth foliage in the second (West, 1985a)). Life table analysis and experimental manipulation demonstrate that patterns of miner performance mirror patterns of host preference. Leaf skeletonizers, by contrast, compromise their nutritional needs in favour of leaves on which they can rapidly build refuges — damaged primary leaves. The benefits of rolling on damaged leaves apparently disappear when those leaves are bagged, suggesting to us that natural enemies or climate may be involved. Leaf rolls have been shown in other studies to increase ingestion rate and overall consumption by larvae (Henson, 1958), and to reduce the deleterious effects of some plant secondary compounds (Berenbaum, 1978). Whatever the precise mechanism, it is clear that leaf structure is more important to late-season skeletonizers than leaf chemistry.

It is not unusual to find the distribution of phytophagous insect species to be determined in part by the structure and growth form of their host(s) foliage (Southwood, 1973; Wellso, 1973; Turnipseed, 1977; Lamb, 1980; Willmer, 1982; Kennedy, 1986; Schoener, 1988). Damman (1987) describes a group-living pyralid leaf tier which feeds on nutritionally poor mature foliage because young, nutritious leaves wilt when bound together. Aphids, too, may choose to feed on nutritionally inferior host-plant species if the leaf surface is more suitable for colonization (Keenlyside, 1989). Compromises of this kind may be a regular feature of the foraging patterns of herbivorous insects (Hassell and Southwood, 1978; Schultz, 1983).

We should like to stress, however, that the structural properties of host plants are as susceptible to wound-induced change as the chemical and nutritional properties. In the case of late-season skeletonizers on *Q. robur*, structural changes in leaves caused by chewing herbivores are more important than chemical or nutritional properties. The degree to which phytophages are influenced by physical and chemical/nutritional changes caused by leaf damage may depend upon their foraging strategy. The microclimate that larvae in the late-season leaf-mining guild on *Q. robur* are exposed to, for example, will probably be independent of spring leaf damage; feeding within leaf tissue will buffer miners from microclimatic changes. In contrast, leaf damage critically affects leaf-rolling and consequently osmoregulation by the oak leaf-roller *T. viridana* (Hunter and Willmer, 1989), and may influence the distribution patterns of the late-season skeletonizers described here (Hunter, 1987a).

Interactions between species of insect herbivore mediated by wound-induced responses in all aspects of habitat quality will vary from competitive through neutral to commensal. Even the response of an individual species will vary along this continuum because the responses of herbivores to leaf damage are sometimes a dynamic balance between positive and negative forces (Faeth, 1986; Hunter, 1987a). Our data demonstrate that leaf-skeletonizer populations are 'juggled' between the positive influence of leaf damage on refuge quality and the negative influence of leaf damage on nutritional quality — the interaction between spring leaf-chewers and late-season skeletonizers is commensal at low and moderate defoliation levels and competitive at high defoliation levels (Hunter, 1987a).

Other studies have found similar results. A range of responses is shown by the aphid *Tuberculoides annulatus* L. on *Q. robur* depending upon the degree of spring leaf damage (Silva-Bohorquez, 1987). At low defoliation levels, there is an asymmetric competitive interaction between spring leaf chewers and *T. annulatus*. Aphid reproduction is reduced on damaged primary foliage compared with undamaged foliage, reflecting a similar drop in nutritive quality mirrored in the survival of leafminers described here. At higher defoliation levels, however, the production of nutritious regrowth foliage on *Q. robur* greatly increases reproduction by and populations of aphids — the interaction between spring chewers and aphids has become commensal.

The production of damage-induced regrowth foliage on plants may often generate thresholds around which the form of species interactions can pivot. One reason for this is the variable responses of herbivorous insects to regrowth, from strongly positive (Rockwood, 1974) through neutral (Faeth, 1987) to strongly negative (Baldwin and Schultz, 1983; Hunter, 1987a). We know, for example, that spring defoliation approaching 100% on *Q. robur* (a regular, but patchy phenomenon) decreases rather than increases populations of late-season skeletonizers: the positive habitat component is removed and replaced with the negative nutritional (regrowth) component (Hunter, 1987a).

What are the consequences of plant-mediated interactions among herbivore species for forest insect population dynamics? We have shown that changes in the population of one species (an early season feeder) need not lead to a linear response in the population of a second (a late-season feeder). There is considerable natural variation in the densities of *T. viridana* and *O. brumata* larvae among years and among individual *Q. robur* trees in spring (Hunter and Willmer, 1989). The early-season feeders will enhance populations of late feeders on some trees, and reduce them on others. The overall balance will vary from year to year.

An attempt to incorporate wound responses of plants into population models, for example, must be structured to allow not only the magnitude, but also the direction of the effect of some species A on a second species B to vary with the population size of A. Understanding wound-induced responses in plants can increase our understanding of (and ability to predict) outbreaks of insects on plants. We require more field studies which examine the balance between the positive and negative impacts of defoliation on herbivore species, and which consider how variation in levels of defoliation, and other biotic and abiotic factors, influence that balance.

Acknowledgements

We should like to thank Dick Southwood, Willy Wint, Cathy Kennedy, Steve Greenwood and Irma Silva-Bohorquez for their support during our research on oak insects. Two anonymous reviewers made helpful comments on an earlier draft of this manuscript.

References

BALDWIN, I. T. and SCHULTZ, J .C. (1983). Rapid changes in tree leaf chemistry induced by damage: evidence for communication between plants. *Science* **221**, 277-279.

BERENBAUM, M. (1978). Toxicity of a Furanocoumarin to armyworms: a case of biosynthetic escape from insect herbivores. *Science* **201**, 532-534.

BERGELSON, J., FOWLER, S. and HARTLEY, S. (1986). The effects of foliage damage on casebearing moth larvae, *Coleophora serratella*, feeding on birch. *Ecological Entomology* **11**, 241-250.

BRAY, J. R. (1964). Primary consumption in three forest canopies. *Ecology* **45**, 165-167.

CASEY, T. M. (1976). Activity patterns, body temperature and thermal ecology in two desert caterpillars (Lepidoptera: Sphingidae). *Ecology* **57**, 485-497.

CRAWLEY, M. J. (1983). Herbivory, the dynamics of animal–plant interactions. Oxford: Blackwell Scientific.

CRAWLEY, M .J. (1985). Reduction in oak fecundity by low-density herbivore populations. *Nature* **314**, 163–164.

CRAWLEY, M. J. (1987). The effects of insect herbivores on the growth and reproductive performance of English oak. Proceedings of the 6th International Symposium on Insect/Plant Relationships, Pau, France, 1986. Dordrecht: W. Junk.

CROFT, B. A. and HOYING, S. A. (1977). Competitive displacement of *Panonychus ulmi* (Acarina: Tetranychidae) by *Aculus schlechtendali* (Acarina: Eriophyidae) in apple orchards. *Canadian Entomologist* **109**, 1025–1034.

DAMMAN, H. (1987). Leaf quality and enemy avoidance by the larvae of a pyralid moth. *Ecology* **68**, 88–97.

EDWARDS, P. J. and WRATTEN, S .D. (1982). Wound-induced changes in palatibility in birch (*Betula pubescens* Ehrh spp *pubescens*). *American Naturalist* **120**, 816–818.

EDWARDS, P. J. and WRATTEN, S.D. (1983). Wound-induced defenses in plants and their consequences for patterns in insect grazing. *Oecologia* **59**, 88–93.

EDWARDS, P. J. and WRATTEN, S. D. (1985). Induced plant defenses against insect grazing: fact or artefact? *Oikos* **44**, 70–74.

EMMET, A. M. (1979). *A Field Guide to the Smaller British Lepidoptera.* London: The British Entomological and Natural History Society.

FAETH, S. H. (1985). Host leaf selection by leaf miners: interactions among three trophic levels. *Ecology* **66**, 870–875.

FAETH, S. H. (1986). Indirect interactions between temporally separated herbivores mediated by the host plant. *Ecology* **67**, 479–494.

FAETH, S. H. (1987). Indirect interactions between seasonal herbivores via leaf chemistry and structure. *Chemical Mediation of Coevolution* (ed. K. Spencer). AIBS Symposium.

FAETH, S. H. and BULTMAN, T. L. (1986). Interacting effects of increased tannin levels on leaf mining insects. *Entomologia Experimentalis et Applicata* **40**, 297–301.

FEENY, P. (1970). Seasonal changes in oak leaf tannins and nutrients as a cause of spring feeding by winter moth caterpillars. *Ecology* **51**, 565–581.

GREEN, T. R. and RYAN, C. A. (1972). Wound-induced proteinase inhibitor in plant leaves: a possible defence mechanism against insects. *Science* **175**, 776–777.

HARRISON, S. and KARBAN, R. (1986). Effects of an early-season folivorous moth on the success of a later-season species, mediated by a change in the quality of the shared host, *Lupinus arboreus. Oecologia* **69**, 354–359.

HARTLEY, S. E. and FIRN, R. D. (1989). Phenolic biosynthesis, leaf damage, and insect herbivory in birch (*Betula pendula*). *Journal of Chemical Ecology* **15**, 275–283.

HASSELL, M. P. and SOUTHWOOD, T. R. E. (1978). Foraging strategies of insects. *Annual Review of Ecology and Systematics* **9**, 75–98.

HAUKIOJA, E. and NIEMELÄ, P. (1977). Retarded growth of a geometrid larva after mechanical damage to leaves of its host tree. *Annales Zoologici Fennici* **14**, 48–52.

HAUKIOJA, E. and NIEMELÄ, P. (1979). Birch leaves as a resource for herbivores: seasonal occurrence of increased resistance in foliage after mechanical damage of adjacent leaves. *Oecologia* **39**, 151–159.

HEICHEL, G. H. and TURNER, N. C. (1983). CO_2 assimilation of primary and regrowth foliage of red maple (*Acer rubrum* L.) and red oak (*Quercus rubra* L.): response to defoliation. *Oecologia* **57**, 14–19.

HEINRICH, B. (1976). Foraging strategies of caterpillars. Leaf damage and possible predator avoidance strategies. *Oecologia* **42**, 325–337.

HEINRICH, B. and COLLINS, S. L. (1983). Caterpillar leaf damage, and the game of hide and seek with birds. *Ecology* **64**, 592–602.

HENSON, W. R. (1958). Some ecological implications of the leaf-rolling habit in *Campsolechia niveopulvella* Chamb. *Canadian Journal of Zoology* **36**, 809–818.

HUNTER M. D. (1987a). Opposing effects of spring defoliation on late-season oak caterpillars. *Ecological Entomology* **12**, 373–382.

HUNTER, M. D. (1987b). Sound production in the larvae of *Diurnea fagella* (Lepidoptera: Oecophoridae). *Ecological Entomology* **12**, 355–357.

HUNTER, M. D. (1988). Interactions Between Phytophagous Insects on the Pedunculate Oak. PhD thesis, University of Oxford.

HUNTER, M. D. and WILLMER, P.G. (1989). The potential for interspecific competition between two abundant defoliators on oak: Leaf damage and habitat quality. *Ecological Entomology* 14, 267–277.

KARBAN, R. and CAREY, J. R. (1984). Induced resistence of cotton seedlings to mites. *Science* 225, 53–54.

KEENLYSIDE, J. (1989).   Host Choice in Aphids.   PhD thesis, University of Oxford.

KENNEDY, C. E. J. (1986). Attachment may be a basis for specialization in oak aphids. *Ecological Entomology* 11, 291–300.

KIDD, N. A. C., LEWIS, G. B. and HOWELL, C. A. (1985). An association between two species of pine aphid, *Schizolachnus pineti* and *Eulachnus agilis*. *Ecological Entomology* 10, 427–432.

LAMB, R. J. (1980). Hairs protect pods of mustard (*Brassica hista* Gisilba) from flea beetle feeding damage. *Canadian Journal of Plant Science* 60, 1439–1440.

LEIGH, E. G. JR. and SMYTHE, N. (1978). Leaf production, leaf consumption and the regulation of folivory on Barro, Colorado Island. In: *The Ecology of Arboreal Folivores* (ed. G. G. Montgomery). Washington DC: Smithsonian Institute Press,

NIELSEN, B. O. (1978). Above-ground food resources and herbivory in a beech forest ecosystem. *Oikos* 31, 273–279.

NIEMELA, P., TOUMI, J., MANNILA, J. and OJALA, P. (1984). The effect of previous damage on the quality of Scots pine foliage as food for Diprionid sawflies. *Zeitschrift für Angewante Entomologie* 98, 33–43.

RAUPP, M. J. and DENNO, R. F. (1984). The suitability of damaged willow leaves as food for the leaf beetle, *Plagiodera versicolora*. *Ecological Entomology* 9, 443–448.

RHOADES, D. F. (1985). Offensive-defensive interactions between herbivores and plants: their relevance to herbivore population dynamics and community theory. *American Naturalist* 125, 205–238.

ROCKWOOD, L. L. (1974). Seasonal changes in the susceptibility of *Crescentia alata* leaves to the flea beetle, *Oedionychus* sp. *Ecology* 55, 142–148.

ROLAND, J. and MYERS, J. H. (1987). Improved insect performance from host-plant defoliation: winter moth on oak and apple. *Ecological Entomology* 12, 409–414.

SATCHELL, J. E. (1962). Resistance in oak (*Quercus* spp.) to defoliation by *Tortrix viridana* L. in Roudsea Wood National Nature Reserve. *Annals of Applied Biology* 50, 431–442.

SCHOENER, T. W. (1988). Leaf damage in island buttonwood, *Conocarpus erectus*: correlations with pubescence, island area, isolation and the distribution of major carnivores. *Oikos* 53, 253–266.

SCHOWALTER, T. D., HARGROVE, W. W. and CROSSLEY, D. A., JR. (1986). Herbivory in forested ecosystems. *Annual Review of Entomology* 31, 177–196.

SCHULTZ, J. C. (1983). Habitat selection and foraging tactics of caterpillars in heterogeneous trees. In: *Variable Plants and Herbivores in Natural and Managed Systems* (eds. R.F. Denno and M.S. McClure). New York: Academic Press.

SCHUTTE, F. (1957). Untersuchungen uber die populationdynamik des Eichenwicklers *Tortrix viridana* L. *Zeitschrift für Angewante Entomologie* 40, 1–36.

SILVA-BOHORQUEZ, I. (1987). Interspecific Interactions Between Insects on Oak

Trees, with Special Reference to Defoliators and the Oak Aphid. PhD. thesis, University of Oxford.

SOUTHWOOD, T. R. E. (1973). The insect plant relationship — an evolutionary perspective. *Symposium of the Royal Entomological Society of London* **6**, 143–155.

SOUTHWOOD, T. R. E. (1985). Interactions of plants and animals: patterns and processes. *Oikos* **44**, 949–965.

TAHVANAINEN, J. O. (1972). Phenology and microhabitat selection of some flea beetles (Coleoptera: Chrysomelidae) on wild and cultivated crucifers in central New York. *Entomologica Scandinavica* **3**, 120–138.

TURNIPSEED, S. G. (1977). Influence of trichome variations on populations of small phytophagous insects on soybean. *Environmental Entomology* **6**, 815–817.

VARLEY, G. C. and GRADWELL, G .R. (1968) Population models for the winter moth. In: *Insect Abundance* (ed. T. R. E. Southwood). *Symposium of the Royal Entomological Society of London* **4**, 132–142.

WELLSO, S. G. (1973). Cereal leaf beetles: Feeding, orientation, development and survival on four small-grain cultivars in the laboratory. *Annals of the Entomological Society of America* **66**, 1201–1208.

WEST, C. (1985a). Factors underlying the late seasonal appearance of the lepidopterous leaf mining guild on oak. *Ecological Entomology* **10**, 111–120.

WEST, C. (1985b). The Effects on Phytophagous Insects of Variation in Defense Mechanisms Within a Plant. PhD. thesis, University of Oxford.

WILLIAMS, K. S. and MYERS, J. H. (1984). Previous herbivore attack of red alder may improve food quality for fall webworm larvae. *Oecologia* **63**, 166–170.

WILLMER, P. G. (1980). The effects of a fluctuating environment on the water relations of larval Lepidoptera. *Ecological Entomology* **5**, 271–292.

WILLMER, P. G. (1982). Microclimate and the environmental physiology of insects. *Advances in Insect Physiology* **16**, 1–17.

13

Consequences of Rapid Feeding-induced Changes in Trees for the Plant and the Insect: Individuals and Populations

STEPHEN D. WRATTEN, PETER J. EDWARDS AND
ALISON M. BARKER

Department of Biology, Building 44, University of Southampton, Southampton SO9 5NH, UK

Introduction

There is no doubt that chemical changes do occur in herbaceous and woody plants following both experimental damage and that caused by feeding insects (Smith, 1988; Edwards, Wratten and Gibberd, 1990; Wratten, Edwards and Barker, 1990). Field and laboratory bioassays involving leaf-chewing insects have commonly been used by biologists to detect the consequences of wound-induced chemical change for herbivores. These bioassays have shown that biological effects are often detectable within hours (Wratten, Edwards and Dunn, 1984) or sometimes a day or two of damage (Edwards, Wratten and Cox, 1985), and that these effects may last for days or weeks (Haukioja and Neuvonen, 1987). Many of these bioassays have been limited in their scope however, so this paper will review the evidence that feeding-induced changes may have a significant biological effect on herbivorous insects, concentrating mainly on tree-feeding species. The initial review of the factual evidence in the literature will be brief, however, because the papers mentioned above have dealt with this more thoroughly. The main aim of this paper is to present a critical review of published evidence in the light of a dozen or more frequently asked questions concerning the methodology of the experimen-

Population Dynamics of Forest Insects
© Intercept Ltd, PO Box 716, Andover, Hampshire, SP10 1YG, UK

tal approaches used and the ecological interpretation of the results as they concern the individual plant, the individual insect, and populations of both.

Evidence for biological effects of wound-induced changes in woody plants on herbivorous insects

It should be borne in mind that, despite a wide range of positive published evidence, some experimental results are more equivocal than others. While some authors have produced evidence of wound-induced changes affecting the insect, other workers on the same insect–plant system have produced negative results (Lawton, 1987).

This disparity of results is not altogether surprising given the range of bioassays which have been used, and the range of spatial scales over which a putative effect has been looked for. Important experimental variables include the range of intervals between damage and the beginning of the assay, the duration of the assay itself, and the types of effects measured (see Edwards, Wratten and Gibberd, 1990). For example, bioassays have ranged from two detached leaves in a container on the laboratory bench to which generalist molluscs or Lepidoptera larvae have been added as bioassay agents (e.g. Croxford, Edwards and Wratten 1989), to increasingly realistic assays involving cut shoots, growing seedlings (Fowler, 1984) or stands of trees in the field (Fowler and Lawton, 1985). The spatial scales for assessment of results have ranged from the vicinity of the wound itself to parts of leaves (Gibberd, Edwards and Wratten 1988), whole leaves and/ or, adjacent leaves (Wratten, Edwards and Dunn, 1984), shoots and whole plants (Fowler 1984, Fowler and Lawton 1985) or even between plants (Baldwin and Schultz, 1983). The effects measured have sometimes been derived from the confinement of insects on damaged tissue with subsequent measurements of their growth rate, survival and fitness. In other cases the work has involved simple choice experiments in Petri dishes or experimental designs in which freely foraging larvae have been allowed to explore heterogeneous shoots or plants, parts of which have received small amounts of damage. Smith (1988) set out to review the effects of mechanical damage to plants on insect populations but, in fact, evidence for true population effects brought about by wound-induced changes is still scanty. Making ecological deductions from a simple Petri dish bioassay to insect population dynamics represents a large extrapolation from the experimental situation to population effects in the field. The question now to be addressed in this paper is: what do the research results so far mean for the fitness and survival of plants and insects and for the population dynamics of both? The questions being asked by behaviourists and ecologists in this area, in the early 1990s, are segregated into important methodological queries as well as interpretative questions which probe the ecological meaning, if any, of the bioassay results. These questions will be analysed in the next section.

Wound-induced changes in plants and their effects on insects: methodology and interpretation.

METHODOLOGY

Chemical changes are not often measured in parallel with behavioural effects. When chemical analysis is undertaken, the chemical assays chosen are often those which are tractable and easily carried out by non-experts in plant biochemistry. However, Hartley and Firn (1989) analysed phenylalanine ammonia-lyases in birch rather than the phenols whose pathway this enzyme system is a part. Another example in which the possible chemical pathway of wound-induced changes was investigated was that of phenol glucosides (Clausen *et al.*, 1989). Phenolic compounds are indeed one of the commonest chemical groups measured in trees, but the conclusions reached about levels of these compounds in the leaves or the temporal pattern of their synthesis often do not mirror changes in leaf acceptability, demonstrated through bioassays (e.g. Wratten, Edwards and Dunn, 1984; this volume, Chapter 14). These inconsistencies are not surprising given the difficulty of identifying precisely the compound or compounds involved, although occasionally the link has been made (e.g. Green and Ryan, 1972: Solanaceae; Tallamy, 1985: cucurbitacins/Coccinellidae).

Damage type: natural or artificial?

Although there is evidence that insect damage may bring about a stronger response than artificial damage (this volume, Chapter 14), there is no evidence for the opposite effect. This means results based on mechanical damage are likely to give, at worst, minimal effects. There are some disadvantages associated with studying natural grazers: the date on which field-collected leaves were grazed is usually unknown, while, in the laboratory, it is difficult to control and standardize levels of grazing.

Detached leaves or whole plants?

If the control leaves decline in palatability following removal from the plant this is likely to make the detection of a difference between them and the experimental leaves more difficult, so that any significant effects that are detected can be assumed to be real. Most recent work uses cut shoots or growing plants rather than detached leaves (e.g. Fowler and MacGarvin, 1986; Croxford, Edwards and Wratten, 1989). Wratten, Edwards and Barker (1990) compared results from shoots damaged while still on the tree with those damaged after collection; there were significant wound-induced changes in the former but not in the latter.

Assesment of grazing

This is carried out by visually estimating the proportion of a leaf's area which is removed (e.g. Edwards, Wratten and Cox, 1985) or by computer-based image analysis (e.g. Croxford, Edwards and Wratten, 1989). Waller and Jones (1989) warned against methods which use only leaf area as a measurement of herbivory and suggested that because leaves differ in thickness and/or density, biomass removed would be a better measure.

The insect species used in bioassays

Generalist and highly mobile species have often been used. These have been chosen because: (1) specialist feeders may have evolved adaptations by which they can tolerate wound-induced changes in their host plant; and, (2) the use of mobile species allows insects to forage widely and to sample leaf quality. Despite this, there are many examples of complex trenching and severing behaviour by specialist insects which appear to be adaptations in those species to avoiding wound-induced secondary substances. Secondly, the work by Bergelson and Lawton (1988) on the specialist birch leaf miner *Coleophora* showed that larvae of this insect moved away from pin-pricks made artificially in the lamina around its mine. Thirdly the ideas of Schultz (1983) point to the inherent heterogeneity of plant canopies; we might therefore expect specialist herbivores to have evolved behaviour to forage for high quality sites, including those not showing a wound-induced reduction in acceptability.

The timing of the bioassay

Recent evidence shows (*Table 1*) that many small meals are taken by Lepidoptera larvae and leaves are usually abandoned with only a small proportion of the lamina consumed, even on previously undamaged leaves (Silkstone 1987; Wratten, Edwards and Barker, 1990). The time spent by a caterpillar on one leaf may be minutes or an hour or two, so bioassay intervals of 24 h are not unrealistic. Some bioassay intervals are of only a few hours' duration (see Smith, 1988). If an effect is short-lived but a bioassay is begun 24 h after damage is imposed, effects could be missed but this would not give spurious positive results.

Table 1. The numbers and sizes of meals taken by larvae of *Orthosia stabilis gothica* on previously damaged and undamaged shoots of birch (*B. pubescens* Ehrh.); 10% of each leaf lamina was removed in the field using a hole punch; 24 h later, shoots were collected and two third instar larvae were allowed to forage on each shoot for 24 h; $\chi^2 = 19.9$; $P<0.001$

	Control	Damaged
No. small meals (< 1% of leaf area)	21	55
No. large meals (>5% of leaf area)	90	61

INTERPRETATION

Interpretation of damage avoidance

It has been suggested that herbivores may show an evolved behaviour to avoid visually hunting vertebrate predators by moving away from small amounts of damage (Heinrich, 1979). There may also be an evolved behaviour to avoid detection by predators or parasitoids which respond to the plant chemicals associated with damage, leading again to the herbivores' moving away from such damage (e.g. Faeth, 1985). Background heterogeneity within and between leaves may lead to foraging insects continually abandoning low quality sites, as suggested by Schultz (1983), while diurnal rhythms in the herbivores' behaviour may be an evolved adaptation to avoid foraging predators or parasitoids (Lance, Elkinton and Schwalbe, 1986). Interrupted meals brought about by successful or unsuccessful predator or parasitoid attacks, may also lead to apparent avoidance of small amounts of damage.

In spite of these possibilities, which may be important effects in some species, several lines of evidence argue against the interpretation that insects are merely avoiding physical damage and point towards the animals' avoiding chemically induced leaves or parts of leaves:

1　Wound-induced responses differ between plant species in the same community (Edwards, Wratten and Greenwood, 1986).
2　There is a clear seasonal effect in that tree leaves damaged in the summer or later show a much reduced effect on the behaviour of laboratory-reared insects compared with leaves damaged in the spring (Wratten, Edwards and Dunn, 1984).
3　The age of the leaf influences the strength of the wound-induced change; young leaves appear to show the effect much more strongly than do old leaves on the same shoot (Edwards, Wratten and Gibberd, 1990).
4　The effect of the damage on foraging larvae declines with time following damage (Gibberd, Edwards and Wratten, 1988).
5　The effect of damage on foraging behaviour is often transmitted in some way to adjacent, undamaged leaves which are also relatively protected (Wratten, Edwards and Dunn, 1984; Edwards, Wratten and Cox, 1985).
6　The changes in insect behaviour sometimes correlate with measured chemical changes in the leaf (Tallamy, 1985; Chiang *et al.*, 1987).
7　Many of the documented cases of herbivores' moving following small meals concern conspicuous aposematic species; predator avoidance would appear to be less necessary in these groups, yet they still move away from damage (e.g. Edwards and Wanjura, 1989).

Reasons for avoidance of the wound edge

Holes may be avoided because the edge of a wound dries out and that area of the leaf, or the leaf as a whole, contains less water than the control leaves. The list of seven pieces of evidence supporting a chemical explanation (see above) also makes this an unlikely explanation for wound-avoidance by insects. In addition, leaves bearing small amounts of damage do not differ measurably, in their fresh or dry weights, from control leaves (Gibberd, 1987).

Individuals of a plant species give differing results

Given the range in bioassay design mentioned earlier variable results are not surprising. For instance, the tree species may have a range of genotypes (Gill and Davy, 1983) with differing strengths of wound-induction; the interval between leaf damage and assessment, and the duration of the assessment itself, may vary, as do the types of grazing measurements recorded (see Waller and Jones, 1989).

Ecological significance of wound-induced changes

Fowler and Lawton (1985) suggested that three important criteria must be satisfied before it is possible to conclude that wound-induced changes in plants are ecologically significant. First, that the individual herbivore must suffer in its performance. Secondly, that herbivore populations must be reduced as a result. Thirdly, that the plant must receive less overall damage than would a plant not showing these wound-induced changes. These questions and an 'alternative model' for the ecological rôle of wound-induced changes are discussed in detail in Edwards, Wratten and Gibberd (1990). The contrasts between the two models however are summarized in *Table 2*.

Table 2.   Contrasting models for the defensive role of wound-induced chemical changes in plants

Consequences of grazing	Model 1*	Model 2**
For foliage	Effective induction of chemical changes occurs after high levels of insect feeding (unlikely to occur at or below mean natural levels i.e. 5–10%).	Any level of grazing induces at least local changes.
For individual herbivore	Growth, survival and fecundity suffer.	Main effect is movement away from site of feeding; performance effects secondary.
For herbivore population dynamics	Significant reduction of population levels.	Reduction in population may or may not occur
For plant	Plant protected: less overall damage from subsequent grazing.	Grazing damage dispersed, especially away from young leaves: overall effects secondary.

*(Fowler and Lawton, 1985)
**(Edwards, Wratten and Gibberd, 1990)

Are the changes evolved defences?

A major area of contention is whether the changes in plants which are frequently recorded following damage are an adaptation to insect feeding (i.e. are they a defence) or are they merely part of wound repair processes in the plant? Edwards and Wratten (1985) listed four main criteria by which such a question can be judged: (1) the induced chemical substances are not always directly involved in the repair process; (2) chemical changes may be induced in tissues remote from the site of damage; (3) speed and magnitude of the chemical changes are sufficient to affect the behaviour or fitness of insects; (4) plant species producing induced defences show other general ecological similarities.

From the point of view of the ecology of the insects and the plant, however, whether or not the effects are evolved defences is not necessarily important. If wound-induced changes improve the plant's fitness by driving herbivores away from the most vulnerable regions, then the plant's competitive ability may be enhanced. In addition, the insect's populations may be reduced as a result (see Wratten, Edwards and Winder, 1988). However, evidence for population effects is still scanty, although growing (Hunter, 1987) and more field and laboratory evidence is needed.

References

BALDWIN, I.T. and SCHULTZ, J.C. (1983). Rapid changes in tree leaf chemistry induced by damage; evidence for communication between plants. *Science* **221**, 277–279.

BERGELSON, J. M. and LAWTON, J. H. (1988). Does foliage damage influence predation on the insect herbivores of birch? *Ecology* **69**, 434–445.

CHIANG, H., NORRIS, D. M., CIEPIELA, A., SHAPIRO, P. and OOSTERWYK, A. (1987). Inducible versus constitutive soybean resistance to Mexican bean beetle. *Journal of Chemical Ecology* **13**, 741–749.

CLAUSEN, T. P., REICHARDT, P. B., BRYANT, J. P., WERNER, R. A., POST, K. and FRISBY, K. (1989). Chemical model for short-term induction in quaking aspen (*Populus tremuloides*) foliage against herbivores. *Journal of Chemical Ecology* **15**, 2335–2346.

CROXFORD, A. C., EDWARDS, P. J. and WRATTEN, S. D. (1989). Temporal and spatial variation in palatability of soybean and cotton leaves following wounding. *Oecologia* **79**, 520–525.

EDWARDS, P. B. and WANJURA, W. J. (1989). Eucalypt-feeding insects bite off more than they can chew: sabotage of induced defences. *Oikos* **54**, 246–248.

EDWARDS, P. J. and WRATTEN, S. D. (1985). Induced plant defences against insect grazing: fact or artefact? *Oikos*, **44**, 70–74.

EDWARDS, P. J., WRATTEN, S. D. and COX, H. (1985). Wound-induced changes in the acceptability of tomato to larvae of *Spodoptera littoralis*: a laboratory bioassay. *Ecological Entomology* **10**, 155–158.

EDWARDS, P. J., WRATTEN, S. D. and GIBBERD, R. (1990). The impact of inducible phytochemicals on food selection by insect herbivores and its consequences for the distribution of grazing damage. In: *Phytochemical Induction by Herbivores* (eds M. J. Raupp and D. Tallamy). New York: John Wiley, (in press).

EDWARDS, P. J., WRATTEN, S. D. and GREENWOOD, S. (1986). Palatability of British trees to insects: constitutive and induced defences. *Oecologia* **69**, 316–319.

FAETH, S. H. (1985). Host leaf selection by leaf miners: interactions among three trophic levels. *Ecology* **66**, 870–875.

FOWLER, S. V. (1984). Foliage value, apparency and defence investment in birch seedlings and trees. *Oecologia* **62**, 387–392.

FOWLER, S. V. and LAWTON, J. H. (1985). Rapidly induced defenses and talking trees: the devil's advocate position. *American Naturalist* **126**, 181–195.

FOWLER, S. V. and MacGARVIN, M. (1986). The effects of leaf damage on the performance of insect herbivores on birch, *Betula pubescens. Journal of Animal Ecology* **55**, 565–573.

GIBBERD, R. (1987). Wound-induced Plant Responses and their Consequences for Insect Grazing. PhD thesis, University of Southampton.

GIBBERD, R., EDWARDS, P. J. and WRATTEN, S. D. (1988). Wound-induced changes in the acceptability of tree-foliage to Lepidoptera: within-leaf effects. *Oikos* **51**, 43–47.

GILL, J. A. and DAVY, A.J. (1983). Variation and polyploidy within lowland populations of the *Betula pendula/B. pubescens* complex. *New Phytologist* **94** 433–451.

GREEN, T. R. and RYAN, C. A. (1972). Wound-induced proteinase inhibitor in plant leaves: a possible defense mechanism against insects. *Science* **175**, 776–777.

HARTLEY, S. E. and FIRN, R. D. (1989). Phenolic biosynthesis, leaf damage and insect herbivory in birch (*Betula pendula*). *Journal of Chemical Ecology* **15**, 275–283.

HAUKIOJA, E. and NEUVONEN, S. (1987). Insect population dynamics and the induction of plant resistance: the testing of hypotheses. In: *Insect Outbreaks* (eds P. Barbosa, and J.C. Schultz). New York: Academic Press.

HEINRICH, B. (1979). Foraging studies of caterpillars. *Oecologia* **42**, 325–337.

HUNTER, M. D (1987). Opposing effects of spring defoliation on late season oak caterpillars. *Ecological Entomology* **12**, 373–382.

LANCE, D. R., ELKINTON, J. S. and SCHWALBE, C. P. (1986). Techniques for monitoring feeding of large larval Lepidoptera, with notes on feeding rhythms of late-instar gypsy months (Lepidoptera: Lymantriidae). *Annals of the Entomological Society of America* **79**, 390–394.

LAWTON, J. H. (1987). Food shortage in the midst of apparent plenty? The case for birch-feeding insects. In: *Proceedings of the Third European Congress of Entomology*, pp. 219–228. Amsterdam: Nederlandse Entomologische Vereniging.

SCHULTZ, J. (1983). Impact of variable plant defensive chemistry on susceptibility of insects to natural enemies. In: *Plant Resistance to Insects* (ed. P. A. Hedin). *American Chemical Society Symposium* No. **208**, 37–55.

SILKSTONE, B. E. (1987). Consequences of leaf damage for subsequent insect grazing of birch (*Betula* spp.): a field experiment. *Oecologia* **74**, 149–152.

SMITH, C. M. (1988). Effects of mechanical damage to plants on insect populations.

In: *Plant Stress-Insect Interactions*, pp. 321–340 (ed. E. A. Heinrichs). New York: John Wiley.

TALLAMY, D. (1985). Squash beetle trenching behaviour: an adaptation against induced cucurbit defenses. *Ecology* **66**, 1574–1579.

WALLER, D. A. and JONES, C. G. (1989). Measuring herbivory. *Ecological Entomology* **14**, 479–481.

WRATTEN, S. D., EDWARDS, P. J. and BARKER, A. (1990). Rapid wound-induced changes in plant chemistry: their ecological significance. In: *Insect–Plant Relationships*, (ed. T. Jermy). Budapest, Proceedings of the Seventh International Symposium: Insect-Plant Relationships, 1989, (in press).

WRATTEN, S. D., EDWARDS, P. J. and DUNN, I. (1984). Wound-induced changes in the palatability of *Betula pubescens* and *B. pendula*. *Oecologia* **61**, 372–375.

WRATTEN, S. D., EDWARDS, P. J. and WINDER, L. (1988). Insect herbivory in relation to dynamic changes in host plant quality. *Biological Journal of the Linnean Society*. **35**, 339–350.

14
Damage-induced Changes in Birch Foliage: Mechanisms and Effects on Insect Herbivores

S. E. HARTLEY AND J. H. LAWTON*

Department of Biology, University of York, Heslington, York YO1 5DD, UK
*Centre for Population Biology, Department of Pure and Applied Biology, Imperial College, Silwood Park, Ascot SL5 7PY, UK

Introduction

Biochemical changes are detectable in the foliage of birch trees (*Betula pendula* and *Betula pubescens*) within hours or days of the leaves being damaged, experimentally or by insect herbivores. This paper summarizes work on the impact of these rapidly induced responses (RIRs; Haukioja *et al.*, 1988) on insects, carried out at York (Hartley and Lawton, 1990). Important related studies are being carried out in Southampton (e.g. Wratten, Edwards and Dunn, 1984) and Finland (Haukioja and Neuvonen, 1987).

Contrary to the results of the Southampton group (Edwards and Wratten, 1987; Edwards, Wratten and Gibberd, 1990) we believe it is difficult to extrapolate from RIRs in foliage to predict effects on insect behaviour, and, even more difficult to predict effects on insect dynamics or plant fitness (Fowler and Lawton, 1985). However, some of the problems of interpretation that exist in this field may disappear if the more rigorous and detailed chemical techniques already used by plant pathologists (e.g. the spatial and temporal induction of secondary metabolites and enzyme induction at the molecular level) are employed (Bailey, Rowel and Arnold, 1980; Esnault *et al.*, 1987). We report some preliminary steps in that direction. Finally, we view RIRs in the plant as an integrated response to a variety of external threats; insects may not be a plant's biggest problem!

Population Dynamics of Forest Insects
© Intercept Ltd, PO Box 716, Andover, Hampshire, SP10 1YG, UK

Chemical changes in birch foliage

Wounding triggers a complex series of reactions in both *B. pendula* and *B. pubescens* leaves (Niemela, Aro and Haukioja, 1979; Wratten, Edwards and Dunn, 1984; Bergelson, Fowler and Hartley, 1986). For example, within a few hours phenolic levels (as measured by the Folin Denis method) increase in artificially damaged (hole-punched) leaves; this increase is largest in the youngest leaves (Hartley and Lawton, 1987). Insect-grazing triggers an even larger response and, unlike mechanical damage, insect damage causes an elevation of phenolics in immediately adjacent but undamaged leaves (Hartley and Firn, 1989). The undamaged portions of leaves mined by eriocranids (Lepidoptera: Eriocranidae) show an increase in phenolics which is larger than that caused by artificial damage, but smaller than that following insect grazing (Hartley and Lawton, 1987).

Phenolics and tannins can also be measured using their ability to precipitate proteins (Martin and Martin, 1983) and, paradoxically, this method gives different results, with RIRs in the sequence: artificial damage – mined damage – grazed leaves. Possible reasons for this puzzling result are discussed by Hartley and Lawton (1990, see also Martin and Martin, 1982; Mole and Waterman, 1987).

One way around these problems is to avoid measuring the end-products, and to concentrate instead on characterizing the induction of phenolic biosynthetic enzymes. Although this does not provide information about the phenolics themselves, or their ecological effects, it reveals a lot about the nature of the plant's response to wounding. For example, why do phenolics increase at sites remote from damage; is this due to transport of phenolics from damaged leaves or to *de novo* biosynthesis in the undamaged leaves? The answers to these and related questions help us to better understand the nature of RIRs, and hence to interpret their possible significance.

We have studied the enzyme Phenylalanine Ammonia Lyase (PAL), which catalyses the first committed step in the synthesis of phenolics. We used an intact cell assay, i.e. one which uses fresh tissue and so measures the actual rate of phenolic production in the leaf (Amrhein, Goedeke and Gerhardt, 1976). This produced a most intriguing result: caterpillar grazing triggered a far larger induction of PAL activity than an equivalent amount and distribution of artificial damage, and the largest increase in enzyme activity was in the undamaged leaves on grazed branches (Hartley and Firn, 1989). This confirmed the Folin Denis results, and shows that the observed increase in phenolics in undamaged leaves is due to on-site synthesis and not transport from the wounded area. Even more interestingly, mechanically damaged leaves 'painted' with caterpillar saliva also showed much larger responses than mechanically damaged leaves alone (Hartley and Lawton, 1990).

The fact that insects are more effective than artificial damage at increasing phenolic levels, and that they do so in adjacent undamaged foliage whilst hole-punching does not, could be taken as support for a defensive role for phenolics against insect herbivores. Unfortunately, a major problem exists with this interpretation, namely chemical changes following damage are very small compared to both the within- and between-tree variation in constituent phenolic levels in undamaged leaves. Between-tree differences of up to 50% make induced changes in phenolics of about 10% seem insignificant, as does the fact that in several experiments the undamaged control trees actually had the highest phenolics, even after induction in the damaged trees (Lawton, 1986; Hartley, 1987).

While we await more detailed information on the nature of the phenolic compounds induced and the mechanisms of induction in the undamaged foliage, one way to assess the likelihood of a defensive role for RIRs is to test for adverse effects on birch-feeding herbivores. An important additional reason for doing such experiments, rather than using solely chemical measurements, is that insects' responses to leaf damage may not be due to phenolics, or at least not due to phenolics alone. In the next section we briefly describe some of the experiments we have carried out in a search for such adverse effects. Again, the results are far from simple.

Responses of herbivores

One way RIRs could affect insect herbivores is via their feeding behaviour. Our first set of experiments involved laboratory preference tests with several species of birch-feeding caterpillars and three types of damaged leaves from *B. pendula*: artificially damaged (hole-punched), insect-grazed, and mined (by eriocranid spp.). The results (Hartley and Lawton, 1987) were rather surprising: most caterpillars were indifferent to artificial or grazing damage, or even preferred the damaged leaves, but mined leaves (retaining at least 50% undamaged tissue) were generally avoided, despite having lower phenolic levels than the grazed leaves. Overall feeding behaviour seemed to reflect damage type and herbivore species, rather than being related in any simple way to changes in chemistry.

However, the Folin Denis assay is crude, and may not be a good estimate of the chemical signals to which the insects are responding; specific phenolic compounds or other types of compounds altogether may be much more important in determining the effect of damage on preference. One way to test this is to separate the physical presence of damage from the induction of phenolic compounds. This can be achieved by blocking phenolic synthesis using an inhibitor of PAL, (aminooxy)acetic acid (AOA) (Amrhein, 1979). When sprayed onto birch trees it causes an 80% inhibition of PAL, and a marked reduction in leaf phenolics. *B. pendula* trees were damaged both artificially and by insect grazing, and half the damaged trees were sprayed with AOA. Feeding preferences of a birch-feeding and a non-birch-feeding caterpillar (*Apocheima pilosaria* and *Spodoptera littoralis* respectively) were then tested in laboratory trials, by comparing damaged and undamaged leaves with and without phenolic production inhibited (Hartley, 1988). Both species were indifferent to damage and their preferences were identical, whether or not damage-induced phenolic synthesis was blocked. Thus, in this experiment, neither damage *per se* nor phenolic levels appear to influence herbivore behaviour.

Overall, laboratory feeding trials with several species of herbivores, three types of damaged leaves and varying amounts of wound-induced chemicals (including leaves with phenolic production inhibited), generally failed to provide convincing evidence that RIRs act defensively, by reducing further insect attack on damaged leaves. These results stand in marked contrast to those of the Southampton group (e.g. Wratten, Edwards and Dunn, 1984; Edwards, Wratten and Greenwood, 1986; Silkstone, 1987). We have no idea why. However, this largely negative picture could possibly be due to the nature of the tests: laboratory trials with excised leaves are difficult to interpret (Risch, 1985), and a negative result does not mean that RIRs in damaged foliage necessarily have no adverse effect on these herbivores in the field. It could simply mean they do not respond behaviourally to those changes under laboratory conditions.

We have therefore conducted a series of much more detailed experiments in the laboratory and the field, using eriocranid leaf-mined leaves. We selected eriocranid-damaged leaves because these produced the biggest response in the earlier laboratory feeding trials, and showed levels of chemical change in the undamaged part of the leaves intermediate between mechanical damage and chewing damage. Similar experiments ought to be carried out using chewed leaves, but we have not yet been able to do them.

Responses to eriocranid leafminer damage

We tested the responses of three species of free-living caterpillars and the case-bearer *Coleophora serratella* to *B. pubescens* leaves damaged by eriocranid leafminers (mines occupied less than 50% of leaf area at the start of each experiment). The experiments again emphasise that different species of caterpillars respond differently to leaf damage.

1. In the laboratory, we used single third to fifth instar caterpillars presented with a pair of excised leaves (mined vs undamaged control leaf matched for age and size) in petri-dishes, i.e. experimental conditions identical to those in our earlier experiments.

    In the field we tested third to fifth instars:

2. Presented with a pair of leaves still attached to the tree, employing petri-dish sized foam-lipped clip-cages to hold the leaves and the caterpillar.
3. Confined singly in mesh bags on small branches with 6–12 leaves, of which a known number (about a third) were mined.
4. Bagged in groups of five on branches containing 50–115 leaves, 20–35% of which were mined (the figure was known exactly for each bag). Leaves damaged in other ways were removed prior to the experiment (never more than five leaves per bag).

Details of the experimental design and statistical analyses will be reported elsewhere (G. Valledares and J. H. Lawton, in preparation). Results are summarized in *Table 1*. November moth caterpillars were indifferent to leaf damage, but all other species avoided mined leaves in at least one experiment. However, the way in which the experiment was conducted clearly influenced our ability to detect significant feeding preferences. Contrary to expectations, the paired field trials were very poor discriminators of larval preference. (Note that this was not due to

Table 1. Summary of results of laboratory and field preference tests using *Betula pubescens* leaves damaged by eriocranid leafminers. Several species of birch feeding herbivores were used; all tests are with fourth or fifth instar larvae unless otherwise stated. The upper tests are onm the leaf area eaten in the trials; the lower tests are on the number of leaves eaten. All significant preferences are for undamaged leaves.

Species		Level of experiment			
		2 leaves	2 leaves	shoot	branch
Coleophora serratella		(*)	ns	**	
		(*)	ns	(*)	
November Moth		ns	ns	ns	
(*Epirrita dilutata*)		ns	ns	ns	
Winter moth		ns	(*)	*	
(*Operophtera brumata*)		ns	*	(*)	
Common Quaker	(3rd)	**	ns	*	***
(*Orthosia stabilis*)		*	ns	ns	*
	(5th)	***			
		ns			

(*)   marginal significance (0.1>P>0.05)
*   0.05>P>0.01
**   0.01>P>0.001
*** P<0.001

a lack of replication and hence statistical power; the behaviour of the caterpillars was more variable in the paired field trials. Why this should be the case is unclear.)

Failure of third and fourth instar November moths to show any preference directly contradicts previous results with this species (Hartley and Lawton, 1987). These latter experiments tested fifth instars with eriocranid damage from *B. pendula* trees. It is unclear whether different instars (unlikely), different tree species (possibly), or something else is responsible for the different results.

These difficulties aside, this series of experiments suggest that in general, a range of birch-feeding caterpillars prefer not to feed on eriocranid leaf-mined leaves, but that the results are rather variable, and depend both upon the way in which the experiment is conducted, and on the species used.

Conclusion

Despite all these experiments wlth several different approaches, we are still not able to decide on the importance of RIRs to birch-feeding insects; we cannot even clearly link chemical changes to insect behaviour, let alone population dynamics. [Note we have also carried out some longer-term population dynamic studies with caterpillars of A*pocheima pilosaria*. Again these experiments gave contradictory results; in one case *Apocheima* survival was reduced by damaged leaves (Fowler and MacGarvin, 1986) and in another case it was not (Bergelson and Lawton 1988)].

One of the reasons is that the rapid induction of phenolics in damaged birch foliage has proven to be complex. Chemical changes vary depending on the type of damage inflicted, and different methods of analysis produce different results. This, together with the small magnitude of the induced changes compared to the large between- and within-tree variation present before damage, makes the results very difficult to interpret.

One intriguing result which has emerged from a generally negative picture is that birch appears able to 'recognize' different types of damage, and respond differently to each. Furthermore, the response to insects is larger than that to artificial damage, and is associated with the induction of phenolics in undamaged leaves. This is due to *in situ* synthesis via an increase in PAL activity, rather than to passive transport of phenolics from the site of damage.

Despite these features of damage-induced responses being consistent with their being a defence against insects, we find rather variable and inconsistent effects on leaf choice by birch herbivores. Eriocranid leaf-mined leaves are the most consistently avoided but, even here, the responses are variable. Against this confusing background, it may be appropriate to consider alternative hypotheses. First, some of the variability may be attributable to our inability to control adequately for tree genotype, 'stress', nutrient status and so on. Secondly, the results may be variable and equivocal because RIRs are not, primarily, an evolved response to insect attack. Effects on insects are incidental.

Work on the signals by which birch distinguishes insect attack from artificial damage is only just beginning (Hartley and Lawton, 1990), but it seems likely that the mechanisms will be similar to those well known to plant pathologists. For example, elicitors capable of inducing PAL in plant cells are found in some fungal–plant interactions (Hahlbrock *et al.*, 1981; DeLorenzo *et al.*, 1987; Habereder, Schroder and Ebel, 1989) and phytoalexin production is known to be stimulated by plant and fungal cell wall fragments (Hargreaves and Bailey, 1978; Darvill and Albersheim, 1984). Furthermore, the comparison of insect and artificial damage described here has striking parallels with the differences between infection and wounding noted by plant pathologists. For example, fungal attack induces PAL at sites remote from damaged tissue or fungal hyphae, and this does not occur with wounding alone (Thorpe and Hall, 1984). Also the increase in PAL is larger following innoculation than after wounding alone (Bhattachryya and Ward, 1986). Speculating, we believe that fungi and other microorganisms associated with insects may be the reason why birch's reaction to herbivores and to artificial damage is so different. Indeed the primary role of damage-induced changes in birch phenolics may be antifungal or antimicrobial, rather than directed principally at insect herbivores.

References

AMRHEIN, N. (1979). Biosynthesis of cyanidin in buckwheat hypocotyls. *Phytochemistry* 18, 585–589.
AMRHEIN, N., GOEDEKE, K. H. and GERHARDT, J. (1976). The estimation of

phenylalanine ammonia lyase (PAL) activity in intact cells of higher plant tissues. *Planta* **131**, 33–40.

BAILEY, J. A., ROWEL, P. M. and ARNOLD, E. M. (1980). Temporal relationship between host cell death, phytoalexin production and fungal inhibition during hypersensitive reactions of *Phaseolus vulgaris. Physiological Plant Pathology* **17**, 329–343.

BHATTACHARYYA, M. K. and WARD, E. W. B. (1988). Phenylalanine ammonia lyase acticity in soybean hypocotyls and leaves following infection with *Phytophora megasperma* f.sp. *glycinea. Canadian Journal of Botany* **66**, 18–23.

BERGELSON, J., FOWLER, S. and HARTLEY, S. (1986). The effects of foliage damage on case-bearing moth larvae, *Coleophora serratella*, feeding on birch. *Ecological Entomology* **11**, 241–250.

BERGELSON, J. M. and LAWTON, J. H. (1988). Does foliage damage influence predation on the insect herbivores of birch? *Ecology* **69**, 434–445.

DARVILL, A. P. and ALBERSHEIM, P. (1984). Phytolexins and their elicitors: a defense against microbial infection in plants. *Annual Review of Plant Physiology* **35**, 243–275.

DELORENZO, G., RANUCCI, A., BELLINCAMPI, D., SALVI, G. and CERRONE, F. (1987). Elicitation of PAL in *Daucus carota* by oligogalacturonides released from sodium polypectate by homologous polygalacturonase. *Plant Science* **51**, 147–150.

EDWARDS, P. J. and WRATTEN, S. D. (1987). Ecological significance of wound-induced changes in plant chemistry. In: *Insects–Plants*, pp. 213–215 (eds V. Labeyrie, G. Fabres and D. Lachaise). Dordrecht: W. Junk.

EDWARDS, P. J., WRATTEN, S. D. and GIBBERD, R. (1990). The impact of inducible phytochemicals on food selection by insect herbivores and its consequences for the distribution of grazing damage. In: *Phytochemical Induction by herbivores*, (eds D. W. Tallamy and M. J. Raupp). New York: John Wiley in press.

EDWARDS, P. J., WRATTEN, S. D. and GREENWOOD, S. (1986). Palatability of British trees to insects: constitutive and induced defenses. *Oecologia* **69**, 316–319.

ESNAULT, R., CHIBBAR, N., LEE, D., VAN HUYSTEE, B. and WARD, E. W. B. (1987). Early differences in the production of mRNAs for PAL and chalcone synthase in resistant and susceptible cultivars of soybean innoculated with *Phytophthora megasperma. Physiological and Molecular Plant Pathology* **30**, 293–297.

FOWLER, S. V. and LAWTON, J. H. (1985). Rapidly induced defenses and talking trees: the Devil's advocate position. *American Naturalist* **126**, 181–195.

FOWLER, S. V. and MacGARVIN, M. (1986). The effects of leaf damage on the performance of insect herbivores on birch, *Betula pubescens. Journal of Animal Ecology* **55**, 565–573 .

HABEREDER, H., SCHRODER, G. and EBEL, J. (1989). Rapid induction of phenylalanine ammonia lyase and chalcone synthase mRNAs during fungus infection of soybean (*Glycine max* L.) roots or elicitor treatment of soybean cell cultures at the onset of phytoalexin synthesis. *Planta* **177**, 58–65.

HAHLBROCK, K., LAMB, C. J., PURWIN, C., EBEL, J., FAUTZ, E. and SCHAFER, E. (1981). Rapid response of suspension-cultured parsley cells to the elicitor

154   S. E. HARTLEY AND J. H. LAWTON

from *Phytophora megasperma*: induction of the enzymes of general pheny-lpropanoid metabolism. *Plant Physiology* 67, 768–773.

HARGREAVES, J. A. and BAILEY, J. A. (1978). Phytoalexin production by hypocotyls of *Phaseolus vulgaris* in response to constitutive metabolites released by damaged bean cells. *Physiological Plant Pathology* 13, 89–100.

HARTLEY, S. E. (1987). Rapidly induced chemical changes in birch foliage: their biochemical nature and impact on insect herbivores. PhD thesis. University of York.

HARTLEY, S. E. (1988). The inhibition of phenolic biosynthesis in damaged and undamaged foliage and its effect on insect herbivores. *Oecologia* 76, 65–70 .

HARTLEY, S. E. and FIRN, R. D. (1989). Phenolic biosynthesis, leaf damage and insect herbivory in birch (*Betula pendula*). *Journal of Chemical Ecology* 15, 275–283 .

HARTLEY, S. E. and LAWTON, J. H. (1987). The effects of different types of damage on the chemistry of birch foliage and the responses of birch feeding insects. *Oecologia* 74, 432–437.

HARTLEY, S. E. and LAWTON, J. H. (1990). Biochemical aspects and significance of the rapidly induced accumulation of phenolics in birch foliage. In: *Phytochemical Induction by Herbivores* (eds D. W. Tallamy and M. J. Raupp). New York: John Wiley, in press.

HAUKIOJA, E. and NEUVONEN, S. (1987). Insect population dynamics and induction of plant resistance: the testing of hypotheses. In: *Insect Outbreaks*, pp. 411–432 (eds P. Barbosa and J. C. Shultz). New York: Academic Press.

HAUKIOJA, E., NEUVONEN, S., HAHNIMAKI, S. and NIEMELA, P. (1988). The autumnal moth in Fennoscandia. In: *Dynamics of Forest Insect Populations: Patterns, Causes and Implications*, pp. 165–177 (ed. A. A. Berryman). New York: Plenum Press.

LAWTON, J. H. (1986). Food-shortage in the midst of apparent plenty: the case for birch feeding insects. In: *Proceedings of the Third European Congress of Entomology*, pp. 219–228 (ed. H. W. Velthius). Amsterdam: Nederlandse Entomolgische Verening.

MARTIN, J. S. and MARTIN, M. M. (1982). Tannin assays in ecological studies: lack of correlation between phenolic, proanthocyanidin and protein-precipitating constituents in mature oak foliage of six oak species. *Oecologia* 54, 205–211.

MARTIN, J. S. and MARTIN, M. M (1983). Tannin assays in ecological studies. Precipitation of ribulose-1,5-bis phosphate carboxylase/oxygenase by tannic acid, quebracho, and oak foliage extracts. *Journal of Chemical Ecology* 9, 285–294.

MOLE, S. and WATERMAN, P. (1987). A critical analysis of techniques for measuring tannins in ecological studies. II Techniques for biochemically defining tannins. *Oecologia* 72, 148–156.

NIEMELA, P., ARO, E.M. and HAUKIOJA, E. (1979). Birch leaves as a resource for herbivores. Damage-induced increase in leaf phenols with trypsin-inhibiting effects. *Report of the Kevo Subarctic Research Station* 15, 37–40.

RISCH, S. J. (1985). Effects of induced chemical changes on feeding preference tests. *Entomologica Experimentalis et Applicata* 39, 81–84.

SILKSTONE, B. E. (1987). The consequences of leaf-damage for subsequent insect grazing on birch (*Betula* spp.). A field experiment. *Oecologia* **74**, 149–152.

THORPE, J. R. and HALL, J. L. (1984). Chronology and elicitation of changes in peroxidase and phenylalanine ammonia lyase activities in wounded wheat leaves in reponse to innoculation by *Botrytis cinerea*. *Physiological Plant Pathology* **25**, 363–379.

WRATTEN, S. D., EDWARDS, P. J. and DUNN, I. (1984). Wound-induced changes in the palatability of *Betula pubescens* and *Betula pendula*. *Oecologia* **61**, 372–375.

15

The Consequences of Natural, Stress-induced and Damage-induced Differences in Tree Foliage on the Population Dynamics of the Pine Beauty Moth

ALLAN D. WATT

Institute of Terrestrial Ecology, Bush Estate, Penicuik, Midlothian, EH26 0QB, Scotland, UK.

Introduction

The major general contribution to insect ecology of the last 15 years' research on insect–plant relationships has been to view the plant no longer as a uniform entity but as a food resource which varies in quality in both space and time, and to appreciate the effect that this variability can have on insect abundance. This paper reviews recent research on the effect of host-plant variability on one insect herbivore, the pine beauty moth (*Panolis flammea* (D. & S.)), a pine-feeding noctuid, in Scotland.

There are at least three sources of variation in habitat quality which may affect the growth and survival of *P. flammea* larvae: within-tree variability, between-tree variability and variability in habitat quality caused by the feeding damage of pine beauty moth larvae themselves.

Previous studies on a range of insects have shown that each of these categories of plant variability can have substantial effects on insect growth, survival, development and fecundity, effects which may have an impact on insect abundance.

Population Dynamics of Forest Insects
© Intercept Ltd, PO Box 716, Andover, Hampshire, SP10 1YG, UK

Neodiprion sertifer, for example, is unable to survive on the current year's foliage (Ikeda, Matsumara and Benjamin, 1977), but other insects require young foliage for their survival (Bevan, 1987). Detailed studies of some insects which feed on the current year's foliage of conifers and other trees have revealed that larval growth and survival rise and fall as buds or shoots develop and grow (e.g. Day, 1984) and this is thought to strongly affect year to year changes in abundance of a number of insects (e.g. Feeny, 1976). The larvae of the eastern spruce budworm (*Choristoneura fumiferana*) survive better on flowering than on non-flowering balsam fir shoots, prompting the view that trees become more susceptible when they reach maturity (Blais, 1952). Concerning larger scale effects, there are many studies which strongly suggest that site conditions, through their influence on foliage chemistry, influence larval survival and insect abundance (White, 1984). Moreover, research on a wide range of tree foliage feeders has suggested that several patterns of population change, but especially population cycles, are due to damage-induced changes in tree foliage (Haukioja, 1980).

Thus previous research on tree-dwelling insects has shown that the insect–plant interaction has the potential to drive their population dynamics and that several different aspects of the insect–plant interaction should be considered. Accordingly, a broad front of research on the pine beauty moth–pine relationship was started in 1983. Some of the results of this research are presented below, followed by a consideration of how this research can be brought together by looking at the effect of plant foliage chemistry on insect growth and survival. Finally, some of the results are discussed in relation to the population dynamics of the pine beauty moth in Scotland.

The effects of within-tree variability in pine foliage

At any single time in the season, the pine tree provides foliage of variable quality (as measured by larval growth and survival). Larvae reared at the top of the pine crown survive and grow better than those reared at the base (Watt, unpublished). The 'quality' of pine foliage also depends on the age of the larvae; young larvae are unable to feed on the previous years' foliage but older larvae feed on both old and current years' foliage (Watt, 1987a).

If foliage quality over a whole season is considered, then the degree of phenological coincidence between egg hatch and bud burst has a marked effect on insect performance: the growth and survival of young *P. flammea* larvae rise sharply in April and May, and decline in June and July (Watt, 1987b; *Figure 1*). However, the growth and survival of *P. flammea* larvae are unaffected by whether they feed on flowering or non-flowering shoots (Watt, unpublished).

The fact that young *P. flammea* larvae are restricted to the current year's foliage is unlikely to have any significant effect on the abundance of *P. flammea*. This restriction means that the limit set by intraspecific competition is much lower than had the previous years' foliage been able to support young larvae. However, the feeding requirements of young larvae are so small compared with mature larvae that the ceiling set by the feeding activity of older larvae is much lower than that theoretically set by the younger instars. In contrast, the degree of phenological

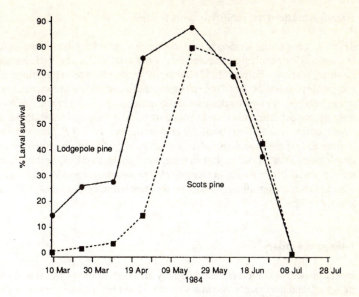

Figure 1. The effect of pine shoot stage on the survival of *P. flammea* larvae (● lodgepole pine; ■ Scots pine) (after Watt 1987).

coincidence between egg hatch and bud burst may play a significant role in year to year variability in *P. flammea* numbers (Watt 1987b)

The effects of between-tree variability in pine foliage

One aspect of between-tree variability has been considered in particular detail: the possible effect of site conditions through foliage quality on *P. flammea* abundance. This research was prompted by the hypothesis that lodgepole pine is stressed by being grown in deep unflushed peat, a soil principally associated with pine beauty moth outbreaks (Watt and Leather, 1988). In summary, this hypothesis has not been supported by research: the growth and survival of *P. flammea* larvae was not found to be higher on experimentally stressed trees, or on trees growing naturally in deep peat compared to trees growing in other soils where tree growth is better (Watt 1986, 1988, 1989b). Moreover, analysis of the chemical composition of the foliage of lodgepole pine growing in deep peat indicates that it is nutritionally poorer for *P. flammea* larvae than the foliage of trees growing elsewhere (Watt, 1989a). This aspect is considered further in the 'Population dynamics' section.

Damage-induced variability in pine foliage

There is conflicting evidence on the presence of an antiherbivore defence or response in lodgepole pine induced by previous defoliation by *P. flammea* larvae. Leather, Watt and Forrest (1987) found that *P. flammea* reared on three-year-old lodgepole pine (on 2 out of 4 provenances) defoliated during the previous year, had lower survival or growth rates than those reared on previously undefoliated trees. Subsequent experiments, also on young trees, have failed to repeat these findings (Watt, unpublished). Moreover, the growth and survival of *P. flammea* larvae on mature lodgepole pine does not appear to be significantly affected by previous defoliation (Watt, Leather and Forrest, unpublished). Thus, herbivore-induced defences may play a role in preventing *P. flammea* on young trees, but they do not appear to have a significant impact on the population dynamics of *P. flammea* on mature trees.

Foliage chemistry

Insect growth and survival are primarily affected by the physical and chemical nature of their host plants, and many attempts have been made to identify the major plant chemicals which influence insect performance (e.g. Mattson, 1980). Only a small range of chemicals (nitrogen, phosphorus, water and tannin) have been studied in relation to *P. flammea* on pine. These were chosen for study because previous work had demonstrated their importance in other insect–plant relationships (e.g. White, 1984; Mattson and Scriber, 1986). Chemical analyses of plant foliage have been carried out in conjunction with several of the laboratory and field experiments referred to above. For example, the seasonal rise and fall in the survival of *P. flammea* on the developing shoots of lodgepole and Scots pine may be attributed to seasonal changes in foliar water content (Watt 1987b; *Figure 2*) ($r = 0.78$, $P<0.05$). Similarly, between-site and between-year variability in the growth and survival of *P. flammea* are positivity correlated with foliar nitrogen and negatively correlated with foliar tannin content (*Figure 3*) (survival/nitrogen: $r = 0.68$, $P<0.05$; weight/nitrogen: $r = 0.66$, $P<0.05$; survival/tannin: $r = 0.70$, $P<0.05$; weight/tannin: $r = 0.91$, $P<0.001$). These correlations were derived from chemical and insect performance data obtained from *P. flammea* studies on both lodgepole pine and Scots pine growing in a range of soil types and including plots of lodgepole pine thinned and treated with fertilizer (Watt, 1989a, 1989b). The extent to which these correlations imply that nitrogen, water and tannin affect larval growth and survival is unknown: other chemicals should be considered and direct studies on the beneficial or detrimental effects of a range of chemicals is needed. For example, recent work suggests that pine terpene chemistry is important, at least in relation to the adult behaviour of *P. flammea* (Leather, Watt and Forrest, 1987). However, the research summarized here cannot be taken in isolation: it adds to the growing body of evidence that nitrogen in particular is the major influence on the growth and survival of phytophagous insects (Mattson, 1980; Scriber and Slansky, 1981).

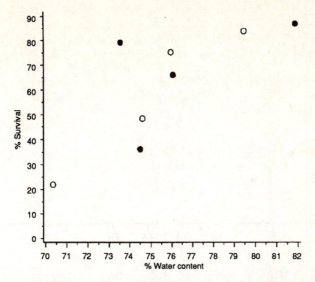

Figure 2. The relationship between the survival of *P. flammea* larvae on Scots and lodgepole pine over a series of shoot stages, and the water content of the foliage (● lodgepole pine; ○ Scots pine) (data from Watt, 1987b).

Although it can be shown that food sources rich in nitrogen lead to better larval growth and survival (e.g. *Figure 3*) it should not be assumed that differences in plant nutritive condition and insect performance will inevitably result in differences in insect abundance. In the concluding section, therefore, the impact of differential rates of insect survival are examined, first theoretically and second in relation to other factors affecting the abundance of the pine beauty moth.

Population dynamics

Although many factors have been shown to affect insect growth and survival, rarely have the full consequences of these factors on insect abundance been explored (Fowler and Lawton, 1985). One method by which the impact of variability in insect survival can be studied is simulation modelling (e.g. Barlow and Dixon, 1980). A simple simulation model has already been built for *P. flammea* on lodgepole pine (Watt, 1987a; Watt and Leather, 1988) and this can be used to examine the effect on *P. flammea* population dynamics of the range of effects on larval survival which have been outlined above. This model was built from information obtained from life table studies (Watt and Leather, 1988). At its simplest the model simulates population behaviour from average levels of fecundity, larval and pupal mortality. This model suggests that a 20% increase or decrease in larval survival does not dramatically affect the outbreak frequency of *P. flammea* (*Figure 4a*). This simple model does not, however, adequately describe *P. flammea*

(a)

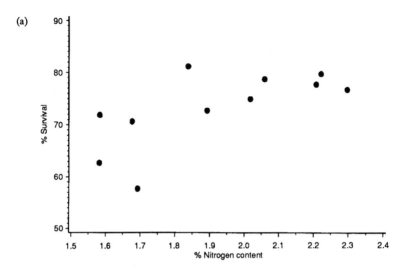

(b)

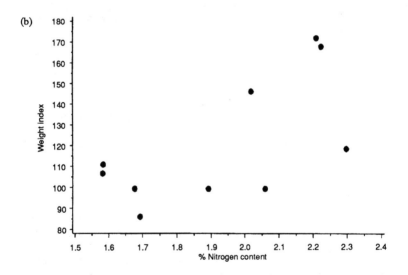

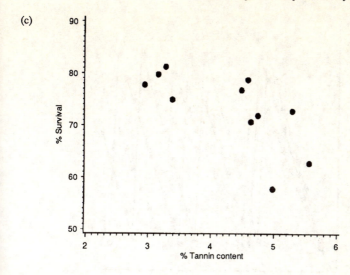

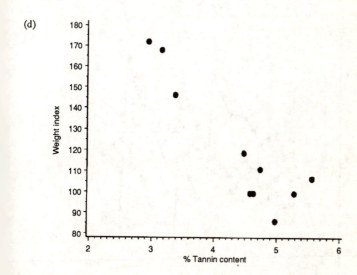

Figure 3. The relationship between the growth (b, d) and survival (a, c) of *P. flammea* larvae on Scots and lodgepole pine at different sites in a single forest 1984–6, and the nitrogen (a, b) and tannin (c, d) contents of the foliage (data from Watt, 1989a, 1989b).

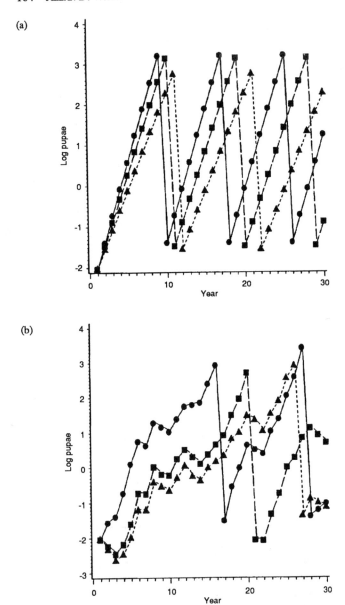

Figure 4.   Simulated fluctuations in the abundance of *P. flammea* pupae at an outbreak lodgepole pine site in Scotland (outbreaks controlled by insecticide application), to demonstrate the effects of three different levels of early larval survival (● 30%, ■ 40%, ▲ 50%). (a) Simple model with fixed late instar mortality and fixed fecundity; (b) inversely density-dependent predation of late instar larvae and randomly varying fecundity (to simulate the effect of spring weather variability).

population dynamics in Scotland where year to year variability in population growth rates is marked (Watt, Leather and Stoakley, 1989). Some of this variability may be attributed to variability in fecundity which is correlated with spring temperatures (Leather, Watt and Barbour, 1985). A further impediment to accurate modelling is year to year variability in larval mortality some of which appears to be due to inversely density-dependent predation of mature larvae (Watt and Leather, 1988). When these two factors are incorporated in the model it demonstrates that a difference of 20% in larval survival has a potentially very significant effect on the abundance of *P. flammea* (*Figure 4b*). Thus, it appears that host-plant variability can have an important role in the population dynamics of *P. flammea*. This conclusion should, however, be taken with some caution: a more realistic simulation model of *P. flammea* is needed because, although the current model appears to describe the population dynamics of *P. flammea* at outbreak sites fairly well, it does not encapsulate the difference between the dynamics of *P. flammea* at outbreak and non-outbreak sites. Future models should include the effects of weather on all stages of the life cycle of *P. flammea*, and should contain a more complete representation of density-dependent mortality.

An alternative way of assessing the role of the host plant in the population dynamics of *P. flammea* is to examine the available data on the effects of the host

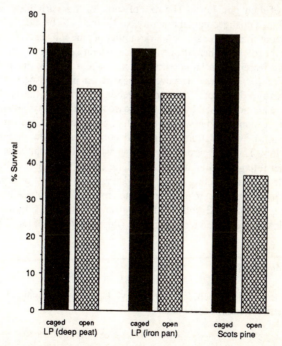

Figure 5. The survival of *P. flammea* larvae on Scots pine and on lodgepole pine growing in different soils inside predator exclusion cages (caged) and exposed to predation (open) (data from Watt, 1987, 1989a). Note that the difference between each pair of survival figures gives an estimate of mortality due to predators.

plant and other mortality factors in relation to data on the abundance and outbreak history of *P. flammea*. Data on the effects of the host plant and predation have been collected to see why *P. flammea* outbreaks are associated with lodgepole pine rather than Scots pine (in Scotland), particularly lodgepole pine growing in deep unflushed peat (see above). The available data on host-plant-associated mortality contrast sharply with the data on overall mortality. Larval growth and survival (in the absence of natural enemies) tends to be greater on Scots pine than on lodgepole pine and *P. flammea* larvae grow and survive least well on lodgepole pine growing in deep peat (Watt, 1988). However, the mortality among (manipulated and natural) cohorts exposed to natural enemies shows the opposite pattern (Watt, 1989, *Figure 5*), i.e. mortality due to natural enemies is greater on Scots pine than on lodgepole pine. It therefore appears that, for the pine beauty moth in Scotland, differences in larval survival caused by differences in host plant nutritive quality are not responsible for the observed patterns in pine beauty moth abundance.

Conclusions

P. flammea larvae are affected by spatial and temporal variability in the quality of the foliage of their pine hosts. The role of some of this variability, particularly damage-induced changes in plant quality, still remains to be clearly understood. Research on the chemical basis for the variability of pine foliage on *P. flammea* has highlighted the roles of foliar nitrogen, water and tannin. Simulation modelling has shown that differences in larval survival can have a significant impact on the abundance of *P. flammea*. However, research on plant-associated mortality, abundance and predation of *P. flammea* in Scotland has shown that spatial patterns of abundance are not caused by host-plant-associated mortality: situations where *P. flammea* grow and survive well as a result of good host plant nutrition may also be situations where predation is high and consequently the abundance of *P. flammea* is low. Thus, it is wrong to view one aspect of the population ecology of this and other foliage-feeding insects in isolation when making conclusions about the principle factors affecting insect abundance. This is most clearly shown by studies on spatial variation in plant-associated mortality and predation (*Figure 5*; this volume, Chapters 23 and 27; Larsson and Tenow, 1984;) but the same principle also applies at the temporal scale (this volume, Chapter 27).

Acknowledgements

The author thanks I. Barnett, C. Beetham, M. Docherty, R. Evans, I. Firkins and A. McFarlane for technical support.

References

BARLOW, N. D. and DIXON, A. F. G. (1980). *Simulation of Lime Aphid Population Dynamics*. Wageningen: Pudoc.

BEVAN, D. (1987). *Forest Insects*. London: HMSO .

BLAIS, J. R. (1952). The relationship of the spruce budworm (*Choristoneura fumiferana*, Chem.) to the flowering condition of balsam fir (*Abies balsamea* (L.) Mill.) *Canadian Journal of Zoology* **30**, 1–28.

DAY, K. (1984). Phenology, polymorphism and insect–plant relationships of the larch bud-moth, *Zeiraphera diniana* (Guenee) (Lepidoptera: Tortricidae), on alternative conifer hosts in Britain. *Bulletin of Entomological Research* **74**, 47–64.

FEENY, P. (1976). Plant apparency and chemical defense. In: *Biochemical Interactions between Plants and Insects, Recent Advances in Phytochemistry 10*, 1–40 (eds J. W. Wallace and R. L. Mansell). New York: Plenum.

FOWLER, S. V. and LAWTON, J. H. (1985). Rapidly induced defences and talking trees: the devil's advocate position. *American Naturalist* **126**, 181–195.

HAUKIOJA, E. (1980). On the role of plant defences in the fluctuations of herbivore populations. *Oikos* **35**, 202–213.

IKEDA, T., MATSUMARA, F. and BENJAMIN, D. M. (1977). Chemical basis for feeding adaptation of pine sawflies *Neodiprion rugifrons* and *Neodiprion swainei*. *Science* **197**, 487–499.

LARSSON, S. and TENOW, O. (1984). Areal distribution of a *Neodiprion sertifer* (Hym., Diprionidae) outbreak on Scots pine related to stand condition. *Holartic Ecology* **7**, 81–90.

LEATHER, S. R., WATT, A. D. and BARBOUR, D. A. (1985). The effect of host-plant and delayed mating on the fecundity and life-span of the pine beauty moth, *Panolis flammea* (Denis & Schiffermuller) (Lepidoptera: Noctuidae): their influence on population dynamics and relevance to pest management. *Bulletin of Entomological Research* **75**, 641–651.

LEATHER, S. R., WATT, A. D. and FORREST, G. I. (1987). Insect-induced chemical changes in young lodgepole pine (*Pinus contorta*): the effect of previous defoliation on oviposition, growth and survival of the pine beauty moth, *Panolis flammea*. *Ecological Entomology* **12**, 275–281.

MATTSON, W. J. (1980). Herbivory in relation to plant nitrogen content. *Annual Review of Ecology and Systematics* **11**, 119–161.

MATTSON, W. J. and SCRIBER, J. M. (1986). Nutrition ecology of insect folivores of woody plants: nitrogen, water, fiber and mineral considerations. In: *The Nutritional Ecology of Insects, Mites and Spiders*, pp. 105–146 (eds F. Slansky Jr. and J. G. Rodriguez). New York: John Wiley.

SCRIBER, J. M. and SLANSKY, F. (1981). The nutritional ecology of immature insects. *Annual Review of Entomology* **26**, 183–211.

WATT, A. D. (1986). The performance of the pine beauty moth on water-stressed lodgepole pine plants: a laboratory experiment. *Oecologia* **70**, 578–579.

WATT, A. D. (1987a). Pine beauty moth outbreaks: the influence of host species, plant phenology, soil type, plant stress and natural enemies. In: *Population Biology and Control of Pine Beauty Moth*, pp. 21–25 (eds S. R. Leather, J. T. Stoakley and H. F. Evans). London: HMSO.

WATT, A. D. (1987b). The effect of shoot growth stage of *Pinus contorta* and *Pinus sylvestris* on the growth and survival of *Panolis flammea* larvae. *Oecologia* **72**, 429–433.

WATT, A. D. (1988). Effects of stress-induced changes in plant quality and host-plant species on the population dynamics of the pine beauty moth in Scotland: partial life tables of natural and manipulated populations. *Journal of Applied Ecology* **25**, 209–221.

WATT, A. D. (1989a). The chemical composition of pine foliage in relation to the population dynamics of the pine beauty moth, *Panolis flammea,* in Scotland. *Oecologia* **78**, 251–258.

WATT, A. D. (1989b). The growth and survival of *Panolis flammea* larvae in the absence of predators on Scots pine and lodgepole pine. *Ecological Entomology* **14**, 225–234.

WATT, A. D. and LEATHER, S. R. (1988). The pine beauty moth in Scottish lodgepole pine plantations. In: *Dynamics of Forest Insect Populations*, pp. 243–266 (ed. A. A. Berryman). New York: Plenum.

WATT, A. D., LEATHER, S. R. and STOAKLEY, J. T. (1989). Site susceptibility, population development and dispersal of the pine beauty moth in a lodgepole pine forest in northern Scotland. *Journal of Applied Ecology* **26**, 147–157.

WHITE, T. C. R. (1984). The abundance of invertebrate herbivores in relation to the availability of nitrogen in stressed food plants. *Oecologia* **63**, 90–105.

16

Douglas-fir (*Pseudotsuga menziesii*) Population Variation in Terpene Chemistry and its Role in Budworm (*Choristoneura occidentalis* Freeman) Dynamics

REX G. CATES AND JIPING ZOU

Chemical Ecology Laboratory, Department of Botany and Range Science, 495 WIDB, Brigham Young University, Provo, Utah 84602 USA

Introduction

Only recently has the role of variation in the production of defensive secondary metabolites on herbivores and pathogens been addressed. Qualitative and quantitative variation in primary nutrition and secondary metabolites are proposed to (1) decrease predictability or apparency of plants to their natural enemies; (2) increase heterogeneity and variability of potential resources to herbivores and pathogens; (3) significantly influence interference interactions for suitable resources among natural enemies of plants; and, (4) increase herbivore and pathogen apparency to their natural enemies (Denno and McClure, 1983; Schultz, 1983). In addition, variation may either result in mixing and/or dilution of the gene pool of herbivores,

Population Dynamics of Forest Insects
© Intercept Ltd, PO Box 716, Andover, Hampshire, SP10 1YG, UK

and thereby, influence the evolution of resistance to defensive compounds, and potentially be an important tool in agriculture and silvicultural management of pests (Cates and Redak, 1988).

Hypotheses dealing with the role of chemical variation in reducing plant and tissue apparency to herbivores, increasing resource heterogeneity to herbivores, and the potential to be used in silviculture management, have been explored in our research using the Douglas-fir (*Pseudotsuga menziesii*)/western spruce budworm (*Choristoneura occidentalis* Freeman) system as a model (Cates and Redak, 1988; Wulf and Cates, 1987; Cates, Redak and Henderson, 1983). In a continuing effort to study these aspects, the objectives here were to determine: (1) if budworm, when placed on populations of Douglas-fir that they normally are not exposed to, would be adversely affected when compared to growth on their 'native' population; (2) if a significant degree of among-population variation in terpene chemistry existed in these Douglas-fir populations; and (3) if the budworm populations would respond differently to Douglas-fir volatiles in laboratory experiments. Finally, using another population of Douglas-fir we attempt to determine the actual pattern in volatile production between supposedly resistant and susceptible trees.

Methods

SITE LOCATIONS AND BUDWORM REARING EXPERIMENTS IN THE FIELD

Thirty-six Douglas-fir trees of similar age, height and crown diameter were selected at sites near Missoula, Montana (MT), Island Park, Idaho (ID), and the Oquirrh Mountains, Utah (UT). Trees among sites varied between 50 and 70 years of age but within sites trees were less than 5 years different in age. Third instar budworm larvae were collected from Montana and Idaho sites and transported back to their 'natural' site and to each of the 'foreign' sites. Budworm larvae 'transported' to a foreign site and then returned within two days and reared on their native trees were not significantly different (ANOVA, $P \leq 0.05$) from non-transported larvae which were reared on their native trees. The budworm population at the Oquirrh Mountain site was extremely sparse during this study such that insufficient numbers of larvae could be collected to carry out reciprocal transplant experiments of this population to the Montana and Idaho sites. Larvae were transported in ice-cooled containers among sites in capped 6 oz cups, and were reared to pupation in nylon screen bags on the current year's growth on branches that were about 45 cm long (Cates and Redak, 1988). Approximately 250–500 larvae from each budworm population were placed on trees at each site. Pupae were collected from trees, taken to the laboratory at BYU, and adults were allowed to emerge in an environmentally controlled growth chamber with a photoperiod of 16 hours of light, temperature maintained at 20°C, and relative humidity at 45%.

In order to minimize site, tree and foliage differences, budworm were placed on foliage of the same developmental stage, in the mid-crown of each tree, and on trees that had identical bud break times. Larvae were placed on the trees when current year's foliage had expanded 2.5 cm. Foliage collected for chemical analysis was collected at the same developmental stage for each population. An additional check of the problem concerning potential developmental differences in foliage among populations was to rear in the laboratory budworm from each population on agar diets containing controlled amounts of terpenes (see below).

TISSUE COLLECTION AND TERPENE CHEMISTRY

No significant differences were detected in the volatile chemistry of foliage collected inside versus outside the bags (ANOVA, $P \leq 0.05$). Therefore, all chemical analyses were done on tissue collected on an adjacent branchlet of the same branch holding the bag. Current year's growth was analysed for volatiles on a model 5890 Hewlett Packard capillary gas chromatograph equipped with a 100% methyl polysiloxane capillary column and autosampler-injector. Injector and detector temperature programs and detailed methods are given in Cates and Redak (1988). Identification of compounds was accomplished by mass spectrometry (Hewlett Packard GCMS, model number 5995), and by co-chromatography of standards using different columns. The internal standard gamma-terpinolene was used to quantify the unidentified volatiles. Data are expressed as mg/g fresh weight and were statistically analysed by ANOVA.

POPULATION RESPONSE TO TERPENES INCORPORATED INTO AGAR DIETS

Budworm were reared on agar diets containing terpenes and acetates following the method of Robertson (1979) as modified by Redak and Cates (1984). Adjusting the nitrogen content of the diet to more closely resemble that of Douglas-fir foliage was the major modification. The concentration used for each compound in the agar diets was determined from concentrations found in the current year's foliage of Douglas-fir. The 10 highest concentrations of samples from Montana and Idaho trees were averaged and served as the guideline for the concentrations used in the agar diets. Verification of the actual concentration of each compound in the agar diet was done by gas chromatography. Based on these data, agar diets were changed every other day in order to maintain the desired concentration. Exceptionally high mortality occurred in females reared on terpinolene (Idaho females only) and bornyl acetate, and for males of both populations on bornyl acetate. Consequently, the sample size for statistical analysis was too small (*Table 4*).

CHEMICAL PATTERNS OF RESISTANT AND SUSCEPTIBLE TREES

One of the objectives of this paper was to determine how patterns in the production of volatiles between resistant and susceptible trees differed. This was approached using trees from a New Mexico population for which terpene chemistry and adult budworm dry-weight production on the same trees had been monitored for three years. For each of the approximately 50 trees that were studied from the same population in 1981, 1982 and 1983, a Simpson's index was calculated. The formula used to calculate Simpson's index for each tree was: $D = -\log \ p_i^2$; where $D =$ Simpson's index, and $p_i =$ the proportion of the individual chemical i in the individual tree's chemistry community (Pielou, 1976). Trees were then ranked from the lowest to the highest Simpson's index. For each year the six trees with the lowest index and the six with the highest index were selected. ANOVA was used to test whether the indices were significantly different between trees judged resistant (small Simpson's index) or susceptible (large Simpson's index). ANOVA also was used to determine if female dry-mass production within a given year was different between the presumably resistant and susceptible groups.

Results

REARING BUDWORM ON DIFFERENT DOUGLAS-FIR POPULATIONS

When male and female budworm native to the Montana site were reared on trees either from the Idaho site or the Utah site they produced significantly less biomass (*Table 1*). The reduction in budworm biomass production, when compared to their native trees in Montana, was greatest when Montana budworm were reared on the Utah trees. This reduction in dry-weight production amounted to 4.6 times for males and 4.1 times for females.

Table 1. Dry-weight production (mg) of Montana adult western spruce budworm reared on their native trees in Montana and on the foreign tree population at the Idaho site and at the Utah site.

Reciprocal transplant	Male* $\bar{x}$ sd	P	Female* $\bar{x}$ sd	P
Montana BW on Montana Trees	16.1 ± 3.7		32.3 ± 6.0	
Montana BW on Idaho Trees	11.6 ± 2.4	0.0001	23.9 ± 7.9	0.0001
Montana BW on Utah Trees	3.5 ± 0.06	0.0001	7.8 ± 2.4	0.0001

* Within each sex, vertical means between Montana budworm on Montana trees versus those on Idaho trees, and versus those on Utah trees, respectively, were compared using ANOVA.

Male and female budworm native to the Idaho site, when reared on trees from the Montana site, did not show any significant decrease in biomass production (*Table 2*). However, when reared on the Utah trees biomass production was significantly reduced for both sexes. Males produced 2.4 times less biomass while females produced 2.6 times less biomass at the Utah site.

AMONG-POPULATION COMPARISON OF TERPENE AND VOLATILE CHEMISTRY

Since the trees at each of the 'foreign' sites significantly reduced the dry-weight production of the Montana budworm, and the trees at the Utah site had a significant adverse affect on Idaho budworm, the volatile chemistry of the trees from each site

Table 2. Dry-weight production (mg) of Idaho adult western spruce budworm reared on their native trees in Idaho versus those reared on foreign tree populations at the Montana site and at the Utah site.

Reciprocal transplant	Male* $\bar{x}$ sd	P	Female* $\bar{x}$ sd	P
Idaho BW on Idaho Trees	12.0 ± 3.5		26.6 ± 8.0	
Idaho BW on Montana Trees	12.1 ± 2.3	NS	27.6 ± 4.3	NS
Idaho BW on Utah Trees	5.0 ± 0.08	0.0001	10.1 ± 4.1	0.003

* Within each sex, vertical means between Idaho budworm on Idaho trees versus those on Montana trees, and versus those on Utah trees, respectively were compared using ANOVA.

was subjected to statistical analysis. The chemistry of the Montana trees was significantly different ($P \leq 0.05$) in all but 5 of the 16 compounds analysed when compared to the chemistry of trees from the Idaho site (*Table 3*). Comparing the chemistry of Montana trees to that of Utah trees revealed that only one of the 16 compounds was not significantly different. An important group of compounds that distinguished the Montana–Utah comparison, but less so the Idaho–Utah comparison, was alpha-humulene along with the six unknown compounds (*Table 3*). The Idaho tree chemistry differed from that of the Utah trees in all but three of the 16 compounds.

Table 3. Comparison of terpene chemistry* of Douglas-fir populations used in reciprocal transplant studies of budworm.

	Douglas-fir Population Montana		Idaho		Utah		P* MT vs ID	MT vs UT	ID vs UT
Compound	$\bar{x}$	sd	$\bar{x}$	sd	$\bar{x}$	sd			
Tricyclene	0.25	0.08	0.52	0.07	0.35	0.10	0.0001	0.0001	0.0001
Alpha-pinene	1.03	0.34	1.84	0.26	1.53	0.42	0.0001	0.0001	0.0027
Camphene	1.68	0.56	3.36	0.78	2.38	0.73	0.0001	0.0001	0.0001
Beta-pinene	0.51	0.24	0.76	0.41	0.32	0.21	0.0003	0.0017	0.0001
Myrcene	0.71	0.03	0.15	0.05	0.18	0.05	0.0001	0.0001	0.0001
Limonene	0.37	0.12	0.79	0.18	0.85	0.22	0.0001	0.0001	0.1375
Terpinolene	0.03	0.02	0.1	0.06	0.07	0.02	0.0001	0.0001	0.0014
Bornyl acetate	1.99	0.73	3.9	1.33	2.6	0.77	0.0001	0.0014	0.0001
Junipene	0.03	0.03	0.08	0.1	0.03	0.02	0.0001	0.8869	0.0002
Alpha-humulene	0.12	0.04	0.14	0.20	0.04	0.03	0.2619	0.0002	0.0001
Unknown 3	0.10	0.06	0.13	0.22	0.24	0.1	0.2984	0.0001	0.0006
Unknown 5	0.08	0.07	0.09	0.16	0.16	0.05	0.5014	0.0001	0.0092
Unknown 6	0.15	0.06	0.15	0.19	0.28	0.18	0.9014	0.0001	0.0004
Unknown 8	0.05	0.03	0.07	0.13	0.15	0.07	0.1074	0.0001	0.0001
Unknown 9	0.13	0.08	0.28	0.42	0.37	0.18	0.003	0.0001	0.1154
Unknown 10	0.12	0.1	0.87	1.6	0.98	0.65	0.0001	0.0001	0.5660

* Statistical analysis of mg/g fresh weight by ANOVA.

BUDWORM POPULATION RESPONSE TO VOLATILES IN AGAR DIETS

Adults from Montana and Idaho budworm populations differed significantly in their biomass production when larvae were reared in the laboratory on agar diets containing terpenes and acetates (*Table 4*). Montana adult female budworm showed significant ($P \leq 0.10$) reduced dry weight production when larvae were reared on seven of the 10 volatiles. Females from the Idaho population, however, showed reduced dry-weight production only when larvae were reared on geranyl acetate and bornyl acetate. For males from the Montana population, all the terpenes reduced biomass production significantly ($P \leq 0.10$). Males from Idaho were adversely affected by only bornyl acetate.

Table 4. Dry-weight production (mg) of Montana and Idaho budworm when reared on terpenes and acetates incorporated into an agar diet.

	Population					
	Montana		$P*$	Idaho		$P*$
Sex/compound	$\bar{x}$	sd		$\bar{x}$	sd	
Female						
Tricyclene	0.024	0.005	0.0001	0.033	0.005	0.70
Alpha-pinene	0.033	0.007	0.28	0.034	0.008	0.44
Beta-pinene	0.029	0.004	0.007	0.035	0.006	0.23
Camphene	0.02	0.006	0.0001	0.028	0.004	0.36
Myrcene	0.031	0.009	0.06	0.028	0.005	0.28
Limonene	0.031	0.005	0.06	0.028	0.003	0.34
Terpinolene	0.03	0.006	0.089	0.025	—	—**
Bornyl acetate	0.033	—	—**	0.034	—	—**
Geranyl acetate	0.033	0.004	0.56	0.025	0.005	0.05
Citronellyl acetate	0.035	0.006	0.99	0.032	0.005	0.74
Male						
Tricyclene	0.011	0.004	0.0001	0.013	0.003	0.63
Alpha-pinene	0.014	0.002	0.021	0.014	0.003	0.41
Beta-pinene	0.015	0.002	0.073	0.015	0.003	0.25
Camphene	0.011	0.002	0.0001	0.012	0.0005	0.66
Myrcene	0.013	0.002	0.0005	0.11	0.0008	0.20
Limonene	0.014	0.002	0.07	0.012	0.003	0.36
Terpinolene	0.013	0.003	0.013	0.012	0.002	0.38
Bornyl acetate	0.014	—	—**	0.011	—	—**
Geranyl acetate	0.015	0.002	0.10	0.013	0.002	0.94
Citronellyl acetate	0.15	0.004	0.09	0.014	0.003	0.54

* Comparison of budworm dry weight of control vs. agar diet with terpene by ANOVA.
** Only 1–3 adults of the initial 35 larvae survived resulting in a sample size too small for statistical analysis. In addition, by chance only 3 Idaho females larvae were initially put on the terpinolene diet. Consequently, no statistical comparison could be made.

TERPENE CHEMISTRY OF ADJACENT DOUGLAS-FIR POPULATIONS

Significant qualitative and quantitative differences in terpene and phenolic chemistry across a variety of habitat types have been noted for many populations of Douglas-fir across western United States (Cates and Redak, 1988; Horner, Cates and Gosz, 1987; Cates, Redak and Henderson, 1983; von Rudloff and Rehfeldt, 1980). It appears also that significant differences in several of the terpenes that are known to adversely influence budworm larval growth, mortality, and adult biomass production exists among local Douglas-fir populations. For example, within a stand of Douglas-fir in New Mexico, two populations located within about a kilometer of one another differed in at least some of the terpenes produced (*Table 5*). Significant quantitative differences occurred in myrcene, ocimene, terpinolene, bornyl acetate, and total acetates between the two populations. For the western spruce budworm, all of these with the exception of ocimene have been shown to adversely affect budworm success (Cates and Redak, 1988; Cates, Henderson and Redak, 1987; Cates, Redak and Henderson, 1983; Zou and Cates, in preparation).

Table 5. Comparison of the terpene chemistry of current years needles between two adjacent populations of Douglas-fir (after Cates, Horner, and Gosz in preparation)

| Compound | Douglas-fir population | | | | |
	$\bar{x}$	se	$\bar{x}$	se	P
Alpha-pinene	1.07	0.02	1.05	0.03	NS
Beta-pinene	0.07	0.05	0.06	0.03	NS
Camphene	1.9	0.03	1.9	0.05	NS
Myrcene	0.13	0.03	0.1	0.03	0.05
Ocimene	0.1	0.02	0.03	0.01	0.05
Terpinolene	0.2	0.02	0.06	0.02	0.001
Tricyclene	0.28	0.02	0.3	0.02	NS
Bornyl acetate	2.2	0.08	1.6	0.06	0.05
Citronellyl acetate	0.9	0.05	0.4	0.02	0.05
Cadinene	0.09	0.01	0.09	0.01	NS
Total acetates	2.2	0.01	1.6	0.04	0.05

* Statistical analysis of mg/g fresh weight by ANOVA. NS = $P>0.05$; $n = 50$ trees/site.

TERPENE PRODUCTION AS RELATED TO TREE RESISTANCE AND SUSCEPTIBILITY

Different levels of resistance and susceptibility exist among individuals within a population in addition to the among- population resistance described above and elsewhere (Cates and Redak, 1988). For example, in every case studied where trees were tested for levels of resistance using a field budworm bioassay, those trees with a high Simpson's index produced significantly larger male and female budworm compared to trees with smaller Simpson's indices (Cates and Redak, 1988). Here we describe one of the common patterns in production of terpenes, as quantified by Simpson's index of diversity, that exists between resistant and susceptible trees in a population in New Mexico.

For 1981, trees that were resistant to budworm, as bioassayed by female dry-mass production, had a mean Simpson's index of 0.69 whereas trees that produced larger female budworm averaged 0.81 ($P = 0.001$). Female budworm on trees with low Simpson's indices averaged 18.7 mg dry-mass biomass whereas those on trees with higher indices averaged 28.4 mg ($P = 0.001$). For 1981, the most significant difference between resistant and susceptible trees, in volatile chemistry, was the increased production of camphene, linalool and especially bornyl acetate in resistant trees. In addition, smaller differences in increased levels of alpha-pinene, beta-pinene and limonene existed in the resistant-tree terpene pattern (*Figure 1*).

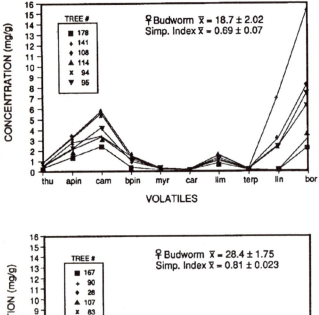

Figure 1. Patterns in terpene production between resistant (upper panel) and susceptible (lower panel) trees (1981). For each set of trees, female budworm dry weights and Simpson's index values are given (mean and sd). thu = thujene, apin = alpha-pinene, cam = camphene, bpin = beta-pinene, myr = myrcene, car = carene, lim = limonene, terp = terpinolene, lin = linalool, bor = bornyl acetate.

For nearly the same set of 50 trees analysed in 1982, a similar pattern emerged in that resistant trees were quantitatively higher in alpha-pinene, beta-pinene, camphene, and bornyl acetate as compared to susceptible trees (*Figure 2*). Also some resistant trees were higher than some susceptible trees in limonene and sesquiterpene 4. Additionally, both sets of trees produced three compounds in 1982 which were not in sufficient quantity in 1981 to be detected by gas chromatography. These were cadinene and two other unidentified compounds all of which were

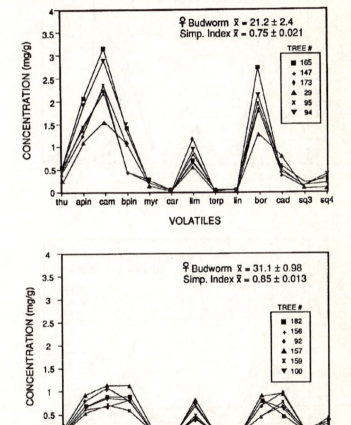

Figure 2. Patterns in terpene production between resistant (upper panel) and susceptible (lower panel) trees (1982). For each set of trees, female budworm dry weights and Simpson's index values are given (mean and sd). thu = thujene, apin = alpha-pinene, cam = camphene, bpin = beta-pinene, myr = myrcene, car = carene, lim = limonene, terp = terpinolene, lin = linalool, bor = bornyl acetate, cad = cadinene, and sq 3 and sq 4 = unidentified sesquiterpenes.

sesquiterpenes. Particularly evident is the quantitatively much higher production of camphene and bornyl acetate in resistant trees as compared to the susceptible group. Equally evident is the overall lower production of terpenes in the current year's foliage during 1982 as compared to 1981 (*Figures 1 and 2*). Simpson's indices and female budworm biomass production were significantly different

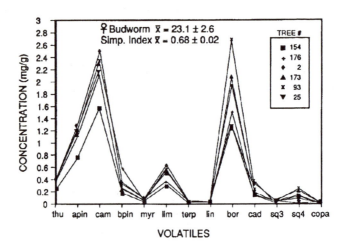

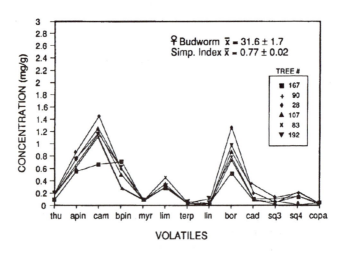

Figure 3. Patterns in terpene production between resistant (upper panel) and susceptible lower panel) trees (1983). For each set of trees, female budworm dry weights and Simpson's index values are given (mean and sd). thu = thujene, apin = alpha-pinene, cam = camphene, bpin = betapinene, myr = myrcene, car = carene, lim = limonene, terp = terpinolene, lin = linalool, bor = bornyl acetate, cad = cadinene, sq 3 and sq 4 = unidentified sesquiterpenes, and copa = copanene.

between the two groups at about the same degree of difference noted in 1981 (*P* = 0.001).

For 1983, resistant trees showed the same pattern in terpene production as was observed for resistant trees in 1981 and 1982 (*Figure 3*). Compared with susceptible trees, resistant trees were higher in alpha-pinene, camphene and bornyl acetate. In addition, copanene, a compound that was not detected in 1981 or 1982 foliage, was quantified for both sets of trees by gas chromatography. Since the tissue concentrations were small, and similar for both resistant and susceptible trees, it is doubtful that copanene was an effective deterrent or feeding stimulant to the budworm. Although not significant, some resistant trees produced more limonene but less beta-pinene than some susceptible trees. Average female budworm dry-weight production and Simpson's index of diversity were significantly lower for the resistant trees compared to susceptible trees, a pattern that was observed for 1981 and 1982 data as well.

The patterns describing female dry-weight production and Simpson's indices for resistant and susceptible trees were consistent among all three years (*Figures 1, 2 and 3*). Interestingly, the trees that were observed to be resistant or susceptible for 1981 were not the same trees in these same two categories for 1982 but several reappeared in the 1983 data set. No trees switched from one category to the other in any of the years. Results reported elsewhere showed that relatively resistant trees in 1981 remained resistant in 1982 and 1983 (Cates and Redak, 1988).

Discussion

HOST-RACE EVOLUTION IN BUDWORM AND DOUGLAS-FIR

Data presented here suggest that various levels of host-race evolution have occurred between Douglas-fir and the western spruce budworm (*Tables 1 and 2*). When budworm from Montana were moved to the Idaho or Utah population of trees, male and female dry-weight production was significantly reduced when compared to their performance on native trees. Budworm from the Idaho population of Douglas-fir showed significant decreases in dry-weight production when reared on the Oquirrh Mountains, Utah trees. However, Idaho budworm, when reared on trees at the Montana site, did not show reduced biomass production when compared to their performance on native trees.

It appears that the Montana budworm were not as well adapted to the Idaho trees as were the Idaho budworm to the Montana trees. Nine of the 16 volatiles reported in *Table 3* were in higher concentration in Idaho tree foliage as compared to the foliage of Montana trees. Five of the remaining seven were not significantly different between the Montana and Idaho populations. A simultaneous comparison of the volatile chemistry of all three populations using discriminant analysis showed that all populations were significantly different, but that Montana and Idaho populations were more similar to one another and the Utah population was very different from the other two populations (Cates and Zou, unpublished). These data further suggest that this Idaho budworm population may be well adapted to the Douglas-fir chemistry of the Montana population. However, neither budworm population was adapted to the Utah trees and/or the Utah site.

The widespread populations of Douglas-fir on which budworm were reared (*Table 3*), as well as closely adjacent populations in New Mexico (*Table 5*), differed significantly in their volatile chemistry. Furthermore, seven major field studies, including this report, involving over 660 trees and 15 populations testify to the significant level of within- and among-population variation in terpene, acetate and nitrogen content of Douglas-fir current year's growth (summarized in part in Cates and Redak, 1986, 1988). All three populations reported here, and the 12 other populations studied elsewhere, differ in some aspect of their terpene, acetate and nitrogen chemistry, a pattern hypothesized to exist in natural systems by Cates and Rhoades (1977), Feeny (1976) and Rhoades and Cates (1976).

BUDWORM POPULATION RESPONSE TO TERPENES INCORPORATED INTO AGAR DIET

Precautions were taken to minimize site, tree and foliage differences among the sites used in this study. For example, budworm were placed on foliage of the same developmental stage, in the mid-crown of each tree, and on trees that had identical budbreak times. These and several other alternatives may have contributed to the differences observed between budworm populations when they were transferred to 'foreign' Douglas-fir populations.

To address these concerns, at least in part, budworm from Montana and Idaho were reared on agar diets containing terpenes and acetates typical of the content found in the current year's foliage of Douglas-fir. These laboratory studies indicate that the dry-weight production of female and male budworm from the Montana population were more affected by the presence of terpenes and acetates than were budworm from the Idaho population (*Table 4*). Other synthetic diet studies using these same two populations have demonstrated that larval mortality, growth rate, pupal weight and/or adult dry weight production were significantly and adversely affected by naturally occurring concentrations of camphene, myrcene, terpinolene, bornyl acetate and citronellyl acetate (Zou and Cates, unpublished). In addition, these same studies suggest that depending on the budworm population some terpenes, such as alpha- and beta-pinene, may serve as feeding cues or stimulants (Cates, Henderson, and Redak, 1987). Taken together, these and other data suggest strong host-race formation in the Douglas-fir/western spruce budworm system (Whilhite and Stock, 1983).

TERPENE PATTERNS AS RELATED TO RESISTANCE AND SUSCEPTIBILITY

Little is known about how individual compounds, or an interaction among compounds, contribute to natural tree resistance. One pattern commonly observed in Douglas-fir/budworm interactions is that of resistant trees possessing 1, 2 or possibly 3 compounds in high concentration (*Figures 1, 2, and 3*). Susceptible trees invariably possess a qualitative and quantitative pattern that is near the population average in compound concentration (*Figures 1, 2, and 3*; Cates and Redak, 1988).

This study supports genetic control over the production of terpenes (Gambliel and Croteau, 1984; Hanover, 1975), but also suggests that shifts in carbon allocation occurs due to loss of tissue from herbivores. This loss may result in lowered production of secondary metabolites as observed in foliage of New Mexico trees sampled in 1982 and 1983 compared to foliage concentrations observed in 1981. However, the basic pattern of relatively high concentrations of camphene and bornyl acetate in resistant versus susceptible trees was maintained among years. Based on other field studies and laboratory feeding trials (Cates, Henderson, and Redak, 1987) it appears that camphene and bornyl acetate contributed significantly to resistant trees over a three-year period.

Natural enemies, foliage quality, site conditions for tree growth, the vagaries of weather, the interaction among these factors and other selection pressures, have been suggested as important factors in the regulation of natural populations of herbivores. In the case of the western spruce budworm, interdisciplinary studies that incorporate appropriate hypotheses and experimental designs are needed to determine the role of each of these factors in budworm dynamics. The role of foliage quality in the sudden demise of budworm populations merits investigation. For example, in Montana, budworm decreased from 45 larvae per square foot of trap surface in 1987 to about 4.5 larvae per square foot in 1989 (Carlson, personal communication). Foliage quality changes in Douglas-fir are suspected as a major contributor to the observed population decrease.

Acknowledgements

We thank Drs Clinton Carlson, James Gosz, Rick Redak and John Horner, Mr J. Harper, Mr S. Spencer, Mr Todd Cates, and several field and laboratory technicians, along with dedicated undergraduates in biology, for their help in gathering data used in this report. This work was partially supported by an NSF Ecosystem Studies Grant (BSR-8305991), grants from the Canada/United States Spruce Budworms Program (CANUSA), and professional development funds from Brigham Young University, all to RGC and colleagues.

References

CATES, R. G. and REDAK, R. (1986). Between-year population variation in resistance of Douglas-fir to the western spruce budworm. In: *Natural Resistance of Plants to Insects*, pp. 106–115 (eds M. Green and P. Hedin). Washington, DC: American Chemical Society.

CATES, R. G. and REDAK, R. (1988). Variation in the terpene chemistry of Douglas-fir and its relationship to western spruce budworm success. In: *Chemical Mediation of Coevolution*, pp. 317–344 (ed. K. Spencer). New York: Pergamon Press.

CATES, R. G. and RHOADES, D. (1977). Patterns in the production of antiherbivore chemical defenses in plant communities. *Biochemical Systematics and Ecology* **5**, 185–193.

CATES, R. G., REDAK, R. and HENDERSON, C. (1983). Patterns in defensive natural product chemistry: Douglas-fir and western spruce budworm interactions. In: *Mechanisms of Plant Resistance to Insects*, pp. 3–19 (ed. P. Hedin).Washington DC: American Chemical Society.

CATES, R. G., HENDERSON, C. and REDAK, R. (1987). Responses of the western spruce budworm to varying levels of nitrogen and terpenes. *Oecologia* **73**, 312–316.

DENNO, R. and McCLURE, M. (1983). *Variable Plants and Herbivores in Natural and Managed Systems*, pp. 717. New York: Academic Press.

FEENY, P. (1976). Plant apparency and chemical defense. *Recent Advances in Phytochemistry* **10**, 1–40.

GAMBLIEL, H. and CROTEAU, R. (1984). Pinene cyclases I and II. *Journal of Biological Chemistry* **259**, 740–748.

HANOVER, J. (1975). Physiology of tree resistance to insects. *Annual Review of Entomology* **20**, 75–95.

HORNER, J., GOSZ, J. and CATES, R. G. (1987). Tannin, nitrogen, and cell wall composition of green vs senescent Douglas-fir foliage: Within- and between-stand differences in stands of unequal density. *Oecologia* **72**, 515–519.

REDAK, R., and CATES, R. G. (1984). Douglas-fir (*Pseudotsuga menziesii*) - spruce budworm (*Choristoneura occidentalis*) interactions: The effect of nutrition, chemical defense, tissue phenology, and tree physical parameters on budworm success. *Oecologia* **62**, 61–67.

ROBERTSON, J. (1979). Rearing the western spruce budworm. Washington, DC: USDA Forest Service. GPO- 1979-699-944.

RHOADES, D. and CATES, R. G. (1976). Toward a general theory of plant antiherbivore chemistry. *Recent Advances in Phytochemistry* **10**, 168–213.

SCHULTZ, J. (1983). Impact of variable plant defensive chemistry on susceptibility of insects to natural enemies. In: *Plant Resistance to Insects*, pp. 37–55 (ed. P. Hedin). Washington DC: American Chemical Society.

VON RUDLOFF, E. and REHFELDT, G. (1980).Chemosystemic studies in the genus *Pseudotsuga*. IV. Inheritance and geographical variation in the leaf oil terpenes of Douglas-fir from the Pacific Northwest. *Canadian Journal of Botany* **58**, 546–556.

WILHITE, E. and STOCK, M. (1983). Genetic variation among western spruce budworm (*Choristoneura occidentalis*) outbreaks in Idaho and Montana. *Canadian Entomologist* **115**, 41–54.

WULF, W. and CATES, R. G. (1987). Site and Stand Characteristics. In: *Western Spruce Budworm*, pp. 89–115 (eds M. Brookes, R. Campbell, J. Colbert, R. Mitchell and R. Stark). Washington, DC: USDA.

17

Interactions between Host-plant Chemistry and the Population Dynamics of Conifer Aphids

N. A. C. KIDD, S. D. J. SMITH, G. B. LEWIS AND C. I. CARTER*

School of Pure and Applied Biology, University of Wales, Cardiff, CF1 3TL, UK
*Forestry Commission Research Station, Alice Holt Lodge, Farnham, Surrey, UK

Introduction
Conifer aphids: ecology and life cycles
The influence of host-plant quality on aphids
The nature of host-plant quality to aphids
Do aphids affect host-plant quality?
      Beneficial effects
      Detrimental effects
Conclusions
Acknowledgements
References

Introduction

During recent years there has been a growing understanding of the complexities involved in the chemical interactions between insect herbivores and the trees on which they feed. Most of this work has been carried out on either leaf-chewing or wood-boring insects (e.g. Cates, 1983; Cook and Hain, 1988). Sap-sucking insects on trees have, on the other hand, been relatively neglected in this respect, despite numerous relevant studies on species infesting non-woody plants (e.g.van Emden, 1972). This becomes all the more surprising when one considers the intimate relationships such parasitic insects have with their hosts, influencing plant physiology through feeding and salivary secretions and, in turn, responding to subtle changes in plant chemistry.

    In this paper we illustrate the variety and complexity of chemical interactions which can occur between one group of sap-sucking insects, the aphids, and their host trees, using examples from our own and other work on species that infest conifers. Much of our work is as yet unpublished and is presented here in summary only.

Population Dynamics of Forest Insects
© Intercept Ltd, PO Box 716, Andover, Hampshire, SP10 1YG, UK

Conifer aphids: ecology and life cycles

Conifer aphids mainly infest the needles or the bark of stems and twigs, where they feed by extracting phloem sap. Some species also feed on the roots, for at least part of their life cycle, but this habit has been little studied. Those aphids which specialize in needle-feeding tend to be small compared with those feeding on bark, their sizes differing by as much as two orders of magnitude. This may be partly explained by the need of bark-feeding aphids to have very long and robust mouth parts, in order to locate the relatively deep phloem tissue of twigs and stems.

The needle-feeders, such as *Elatobium abietinum* (Walk.) on Sitka spruce and *Eulachnus agilis* (Kalt.) on Scots pine may induce defoliation of their hosts, while other species may reduce or disturb tree growth or increase host susceptibility to other pathogens (Carter and Maslen, 1982; Holopainen and Soikkeli, 1984; Kidd, 1989). There are strong commercial incentives, therefore, to understand the interaction between the aphid and its host plant and how this might affect infestation levels.

We have studied three species in some detail on Scots pine, each with different feeding and behavioural characteristics. *Schizolachnus pineti* (Fabr.), a needle-feeder, and *Cinara pini* (L.), a bark feeder on older twigs, characteristically form dense contact-clustered colonies, while *Cinara pinea* (Mordv.), another bark feeder, lives in looser aggregations confined to the foliated shoots of the current and previous year. The three aphids therefore tend to be spatially separated on a tree, exploiting phloem sap from different locations. In Britain, all three species have relatively simple life cycles. They reproduce parthenogenetically during the summer, show alate/apterous dimorphism and switch to the production of sexual forms in the autumn, the females laying eggs on the needles. They are thus holocyclic, although *S. pineti* adults can frequently survive and reproduce during mild winters.

The influence of host-plant quality on aphids

Growth, development and reproduction of aphids may vary considerably in response to seasonal variations in host-plant quality. On trees, aphid performance is usually highest in spring and early summer, when leaves and shoots are actively growing, and lowest in mid-summer after flushing has ceased. This relationship has been shown to apply to all of the conifer aphids studied so far. In *C. pini*, for example, nymphal growth rates on field trees can decline by as much as 56% between June and September, with a corresponding increase of 62% in development time. This produces adults 17% smaller in late summer, with a 38% reduction in fecundity. These changes are accompanied by a characteristic seasonal pattern of distribution within the host (Larsson, 1985; Smith, unpublished). Early in the season the youngest shoots (0–2 years old) are colonized, but from July onwards, once shoot elongation is complete, the colonies tend to move on to older internodes (2–4 years old) and into the lower part of the canopy. With the appearance of sexual forms in September, the colonies return to the younger internodes for egg

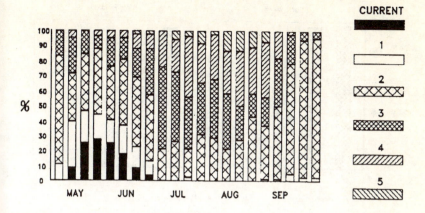

Figure 1.   Seasonal distribution of *C. pini* on the 0–5-year-old internodes of *P. sylvestris*.

laying (*Figure 1*). The only time that the current and one year shoots are colonized is during the period of shoot elongation and then by only 45% of the total population.

In *C. pinea* adults decline in size from around 12 mg in spring to less than 5 mg in August. These smaller adults also produce fewer offspring and have a longer delay before the onset of reproduction. Mortality is also highest amongst summer nymphs and this is associated with an increased restlessness and tendency for nymphs to move frequently between feeding sites, with a greater likelihood of falling out of the canopy. Confirmation that these observations are a direct result of a seasonal decline in tree quality to the aphids was obtained by rearing nymphs concurrently on both actively flushing and previously flushed trees in the laboratory. Adults reared from mature foliage were considerably smaller and suffered a higher mortality than those reared on actively growing shoots (Kidd, 1985).

On young potted trees in the laboratory, the growth rates and fecundity of *S. pineti* also decline, in this case by over 50% between bud burst and shoot maturity, while development times increase by 27% (Lewis, 1987).

More recently, there has been a growing awareness of the considerable variation which may occur in the quality of different trees to the aphids. This is reflected in consistent differences in infestation levels between trees and between years (Kidd, 1985, 1989; Whitham, 1983). In *C. pinea*, for example, the average growth rates of aphids on different trees may vary as much as eight fold in early summer (Kidd and Lewis, unpublished).

During the period of shoot elongation, *C. pini* shows between-tree differences of up to 63% in nymphal growth rates, 40% in adult weight, 38% in fecundity and 25% in development time. In the laboratory, *S. pineti* shows similar between-tree variation with nymphal growth rates and adult weights varying up to two fold during bud burst (Lewis, 1987). This variation becomes even more marked when

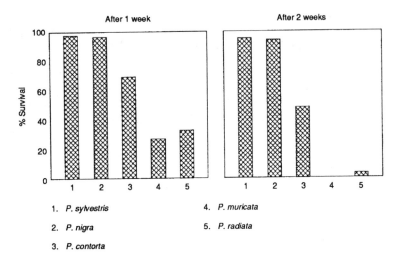

Figure 2.   Survival of *S. pineti* nymphs on different *Pinus* species.

differences in *S. pineti* performance, as measured by reproduction, survival, growth and development, on five pine species are compared. The pine species varied from being 'favourable' to 'unacceptable', as demonstrated by nymphal survival (*Figure 2*).

In *C. pini* the level of infestation (peak number) varied as much as 14-fold between trees. Only 25% of trees supported maximum aphid performance, as measured by nymphal growth rates, and this correlated strongly with infestation level. However, differences in infestation levels between trees were only apparent during the period of shoot elongation, populations reverting to low background numbers later in the season.

Variations in *S. pineti* performance between trees and between host species is also reflected at the population level in the field. While populations on Scots pine show a consistent seasonal pattern, total infestation levels vary by up to three fold. Populations showed a similar pattern on *P. nigra*, but failed to become established on *P. contorta*, *P. muricata* and *P. radiata* (Lewis, 1987).

For *C. pinea*, a computer simulation model has been used to assess the relative importance of different factors in driving population changes. In this species, seasonal and between-tree differences in plant quality are critical factors in determining annual peak infestation levels, with predation also playing an important role (see Chapter 28) (Kidd, 1985, 1989).

The nature of host-plant 'quality' to aphids

Aphid performance has most frequently been interpreted in terms of the nutritional quality of phloem sap. Work with artificial diets and the detailed analysis of plant amino acids has shown that the overall concentrations of amino-nitrogen may be less important to aphids than the number and relative amounts of individual amino acids (van Emden, 1972; van Emden and Bashford, 1971). Moreover, non-nutrient substances may also influence a plant's suitability for aphid feeding.

However, the task of assessing which chemicals present in plants affect aphid performance is not an easy one. Even more difficult is that of determining the relevant concentrations involved. So far, studies have tended to focus on the analysis of crude concentrations of whole-leaf extracts with attempts to correlate those with either plant acceptability or aphid growth and survival. As far as we know, this has not yet been done with phloem sap itself, mainly because of the technical problems involved in sap extraction. Thus, whether these crude measures of the chemical concentrations present in plants resemble those directly experienced by aphids is open to question. Moreover, statistical correlations between chemical concentrations in extracts and aphid performance do not necessarily reflect any underlying causation. We are firmly of the opinion that such experiments require further corroboration by rearing aphids on artificial diets containing the relevant substances in varying concentrations.

These problems are particularly acute with conifer aphids, where there is the added complexity of an extremely diverse allelochemistry. Moreover, with such specialist aphids, chemical cues for host recognition undoubtedly have a role to play, thus complicating the task of finding suitable artificial diets.

These problems notwithstanding, we are beginning to understand something of the chemical environments required by the different aphid species on Scots pine. In our own work, direct chemical analysis of shoot and stem phloem was made possible by the tendency of this tissue to remain attached to the underside of the cortex, when the bark layers are peeled away from the wood. This allowed the

Table 1. The association (- negative or + positive) between phloem chemistry (amino acids and phenolics) and growth rates of *C. pinea* in the field (using stepwise multiple regression).

	May	June	July	August
Correlated amino acids and phenolics	-tryptophan	+methionine	+phenolics	+phenolics
	+glycine	-phenylalanine	+methionine	-glycine
	-tyrosine	(+cysteine)*	+threonine	+serine
	-histidine		-arginine	-cysteine
	+threonine		+glutamine	-valine
				+alanine
Percentage variation accounted for	57	87 (98)*	70	79

(* With inclusion of cysteine.)

phloem tissue to be removed and analysed separately. Needles, however, could not be treated this way, so analyses were carried out on whole-needle extracts. In *C. pinea* multiple regression analysis has been used to determine which phloem constituents best explain the variation in nymphal growth rates between trees throughout the season (*Table 1*). During May 57% of variation in growth rates was accounted for by five amino acids, two of which were positively related and three negatively. In June, 87% of the variation could be accounted for by two amino acids, methionine (+) and phenylalanine (-). This increased to 98% with the inclusion of cysteine (+). In July, methionine was still important, but phloem phenolics also became a significant component, being positively related to aphid growth rates. In August, phloem phenolics remained important, accounting together with five amino acids, for 79% of the variation. These results suggest that one or more phloem phenolics may act as phagostimulants and their declining concentration in mid-summer could account for the relatively poor performance of *C. pinea* at this time of the year. Investigations using artificial diets are currently under way to determine which phenolic substances may be important in this respect.

The importance of allelochemicals is further illustrated by *S. pineti* where variation in total or individual amino acids could not account for differences in performance between and within host species (Lewis, 1987). The greatest abundance and diversity of phenolic and volatile compounds is found in Scots pine, the preferred host species. *P. nigra*, also found to sustain *S. pineti* populations, has a secondary plant chemistry which closely resembles *P. sylvestris*. Pines unsuitable for colonization, on the other hand, have a much reduced or altered secondary plant

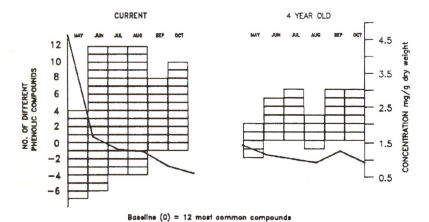

Figure 3.   Seasonal changes in the number of phenolic compounds (boxes) and total amino acid concentration (———) in the current and 4-year-old bark, cortex and phloem of Scots pine. Results for phenolics are expressed as the number of compounds in excess of (positive values) and missing from (negative values) the 12 most common (baseline = 0).

chemistry. This suggests that *S. pineti* may be using one or more allelochemic substances as host recognition cues.

Seasonal changes in nutritional and allelochemical profiles may also partially account for the pattern of distribution seen in *C. pini* (*Figure 1*). In May, when amino acid concentrations in the outer tissues (bark, phloem and cortex) of the new growth are relatively high, 50% of characteristic phenolic compounds are absent (*Figure 3*) and the terpene composition shows the greatest deviation from that of more mature growth. Later, when shoots have expanded, the total amino acid concentration dramatically decreases (to resemble levels in older growth), while the number of phenolic compounds increases two fold. This is largely accounted for by a three fold increase in new non-characteristic phenolic compounds, the majority of which are unique to current and one-year growth. It is only later in the season that a relatively stable composition of phenolic and terpenoid compounds is achieved. It is possible therefore that the current and one-year growth is rendered unsuitable for aphid colonization as the season progresses, by a changing balance between nutrition and allelochemistry. Any advantage afforded by the higher amino acid concentration is outweighed by unfavourable phenolic and terpene compositions.

Do aphids affect host-plant quality?

Aphids may affect the quality of their host-plant resources, either to the benefit or detriment of future generations of aphids.

BENEFICIAL EFFECTS

Evidence has accumulated from a number of studies that aphid aggregations can act as 'physiological sinks', by redirecting phloem-transported nutrients to the colony from other parts of the host plant (Murdie, 1969; Way and Cammell, 1970; Dixon and Wratten, 1971). Aphids reared in aggregations may benefit, therefore, through faster growth rates and a greater size and fecundity, although competition between aphids may mask the advantage at higher densities. Thus, the contact-clustered aggregations of *S. pineti* and *C. pini* might be expected to improve the nutritive quality of needle and bark phloem to aphids, at least temporarily. In fact, we have detected local increases of up to 47% in amino acid concentrations at feeding sites of *C. pini* colonies, resulting in increased nymphal growth rates of 31%. In *S. pineti*, nymphal growth rates were found to increase with colony size, but only up to a maximum of three aphids/colony, after which growth rates declined. At higher densities it would appear that the advantage of colonial life is provided by other factors such as protection from predation (Kidd, 1982). Feeding by *S. pineti* also results in an increase in needle length in the year of infestation, thus enlarging the available space for colonization.

The above effects are carried to an extreme where senescence and defoliation are induced by aphid feeding, for example, in *E. abietinum* on Sitka spruce and *E. agilis* on Scots pine. Senescing needles have a high nutritive value to aphids and are actively searched for by *E. agilis*, when its numbers are low (Kidd, Lewis and

Howell, 1985). At higher densities, the aphids can create these conditions themselves (Kearby and Bliss, 1969), but this can lead to defoliation at very high aphid densities, representing a loss of resource to the aphids. It is clear, therefore, that aphid feeding tends to have beneficial effects only when colony sizes are small, higher densities affecting the aphids adversely.

DETRIMENTAL EFFECTS

In this section, we attempt to answer two questions. Do conifer aphids affect trees in such a way as to make them less suitable for future colonization? In particular, is there any evidence that the aphids can induce defensive reactions in trees similar to those induced by leaf-chewing insects (e.g. Schultz and Baldwin, 1982)?

On Scots pine we have found that high densities of *S. pineti* can reduce the quality of foliage to later generations of the aphid (Lewis, 1987). Such previous infestation can reduce nymphal growth rates by as much as 43% with development prolonged by 17%. This results in a reduction in adult weight of over 36%, with a consequent curtailment of reproduction. In the absence of aphids, these negative changes in foliage quality disappear after about three months, the needles again becoming capable of supporting optimum aphid performance. The chemical changes which underlie this aphid-induced decline in foliage quality involve increases in both the number of phenolic compounds and the total concentrations of amino acids. Thus, despite an apparent improvement in food quality (increased amino acids), the aphids still fare badly on these altered needles. However, there are also marked changes in the *balance* of amino acids, some increasing in concentration (methionine sulphoxide, asparagine, glutamic acid, histidine, tryptophan and arginine) and some decreasing (aspartic acid, tyrosine, phenylalanine, GABA and glycine). These findings call into question the importance of total amino acids as a measure of host quality to conifer aphids.

Further evidence for an aphid-induced response by the host is provided by the distribution patterns of *C. pini*, where, regardless of any seasonal changes in plant quality, previously highly infested sites are not recognized in the same or the following year. If aphids are artificially confined on these areas, growth rates remain low and survival is severely impaired. Analysis of physical and biochemical parameters has shown significant changes in bark and phloem as a result of aphid feeding which could in turn affect their performance and behaviour. Histological sections through the outer tissues of infested and uninfested stems show localized physical damage at the point of stylet insertion, with lignification of the cortex parenchyma cells. At very high levels of infestation produced on laboratory saplings this lignification was found to be extensive and may have constituted a barrier to further stylet penetration. As with *S. pineti*, the nutritional quality of these fed sites appeared to be enhanced, as measured in terms of total amino acid levels (see above), despite their unacceptability for further aphid feeding. Here again, the balance of amino acids could have become unfavourable to the aphids as glutamic acid, glycine, valine, tyrosine, asparagine and ornithine all significantly increased in concentration in the bark and phloem. However, these raised levels did not persist into the following year, although the sites remained uncolonized. This strongly suggests that non-nutrient components, such as increased lignification may have been primarily responsible for rendering the sites unsuitable to the aphids. It is perhaps also significant that aphid feeding was shown to influence the terpene and phenolic composition of phloem and bark material. On the basis of six volatile monoterpenes, discriminant analysis was able to separate between uninfested sites and those infested in the same or previous year. No difference in the total phenolic concentration was found between previously fed and unfed sites, but one new phenolic compound was detected in previously infested tissue. All of these chemical changes were restricted to the local area within the site of infestation.

In choice tests between previously infested and uninfested feeding sites, it was clear that aphids on previously fed sites rejected them with or without short probes, suggesting that site discrimination was influenced by allelochemics such as volatile terpenes at the plant surface. Thus, from the evidence accumulated, the physical and allochemical changes appear to be of greater importance than the changes in amino acids in the movement and relocation of *C. pini* colonies. Extended studies are needed, however, to determine the duration of this induced effect before the implications for the population dynamics of *C. pini* can be fully understood.

Conclusions

Early attempts to interpret aphid performance used crude measures of leaf amino nitrogen concentrations (Dixon, 1970). What has become increasingly apparent is that the intimate interaction between aphid and host is a complex one involving numerous chemical components. As our studies on *S. pineti* and *C. pini* have shown, the prevailing allelochemical profiles of host tissue may often outweigh the importance of amino acid concentrations in determining aphid performance, a point first emphasized over a decade ago by van Emden (1978) in relation to aphids on brassicas.

Besides responding passively to variations in host-plant chemistry, conifer aphids are also capable of inducing chemical changes in host tissue which may be either to their benefit or detriment. From the evidence presented here, we can conclude that some detrimental effects appear to involve the type of chemical changes often considered 'defensive' in relation to other herbivores (Bryant and Kuropat, 1980; Schultz and Baldwin, 1982), but we have to be cautious in assuming such a function for many of these changes. For example, the increased number of phenolic compounds induced by both *S. pineti* and *C. pini* feeding, could simply reflect the mobilization of these substances in the pathway leading to lignin synthesis as a barrier to aphid feeding. Thus, the defensive reaction by the tree would in this case be increased lignification of cell walls, rather than the production of phenolic toxins or antifeedants. Similarly, the increased concentrations of certain amino acids as a result of aphid 'damage' may simply reflect their use in biochemical pathways involved in defence or tissue repair.

Nevertheless, there is some evidence that allelochemicals are involved in host recognition and defence. The speed with which *C. pini* rejects potential feeding sites that have been previously infested, suggests discrimination on the basis of secondary plant substances, whilst *S. pineti* appears to discriminate between species in a similar fashion. *C. pinea*, as far as we know, does not induce changes in secondary plant chemistry as a result of feeding, at least at the densities prevailing in the field (see Chapter 28). But its performance is positively infuenced by the concentration of phloem phenolics. Evidence is now emerging that one or more of these may act as a phagostimulant (Kidd, unpublished), improving aphid performance in spring and early summer when phloem phenolics are relatively abundant, but reducing growth rates and survival later in the season, when their concentrations decline. This could be interpreted as a defensive adaptation by the plant—by withdrawing the phagostimulant at a critical phase in the aphid's population increase, heavy infestation is prevented. Thus, defence against herbivory could commonly involve the timely removal of host recognition cues and phagostimulants, as well as the production of toxins and antifeedants.

Acknowledgements

We thank Mark Jervis for his critical comments on the manuscipt. The work was financed by NERC grants to NACK and NERC CASE Studentships to SDJS and GBL.

References

BRYANT, J. P. and KUROPAT, P. J. (1980). Selection of winter forage by subarctic browsing vertebrates: the role of plant chemistry. *Annual Review of Ecology and Systematics* **11**, 261–285.

CARTER, C. I. and MASLEN, N. (1982). Conifer Lachnids. *Forestry Commission Bulletin*, **58**. pp. London: HMSO.

CATES, R. G. (1983). Patterns in defensive natural product chemistry: Douglas fir and western spruce budworm interactions. In: *Plant Resistance to Insects*, pp. 3–21 (ed. P.A. Hedin). Washington DC: American Chemical Society.

COOK, S. P. and HAIN, F. P. (1988). Resistance mechanisms of loblolly and shortleaf pines to southern pine beetle attack. In: *Mechanisms of Woody Plant Defences against Insects*, pp. 295–305 (eds W. J. Mattson, J. Levieux and C. Bernard-Dagan). New York, Berlin: Springer-Verlag.

DIXON, A. F. G. (1970). Quality and availability of food for a sycamore aphid population. In: *Animal Populations in Relation to their Food Resources*, pp. 271–287 (ed. A. Watson). Oxford: Blackwell.

DIXON, A. F. G. and WRATTEN, S. D. (1971). Laboratory studies on aggregation, size and fecundity in the black bean aphid, *Aphis fabae* Scop. *Bulletin of Entomological Research* **61**, 97–111.

HOLOPAINEN, J. and SOLKKELI, S. (1984) Occurrence of *Cinara pini* (Homoptera, Lachnidae) in Scots pine (*Pinus sylvestris*) seedlings with disturbed growth. *Annales Entomologici Fennici* **50**, 108–110.

KEARBY, W. H. and BLISS, M. (1969). Field evaluation of three granular systemic insecticides for control of the aphids, *Eulachnus agilis* and *Cinara pinea* on Scotch pine. *Journal of Economic Entomology* **62**, 60–62.

KIDD, N. A. C. (1982). Predator avoidance as a result of aggregation in the grey pine aphid, *Schizolachnus pineti*. *Journal of Animal Ecology* **51**, 397–412.

KIDD, N. A. C. (1985). The role of the host plant in the population dynamics of the large pine aphid, *Cinara pinea*. *Oikos* **44**, 114–122.

KIDD, N. A. C. (1989). The large pine aphid on Scots pine in Britain. In: *Dynamics of Forest Insect Populations*, pp. 111–128 (ed. A. A. Berryman). New York: Plenum.

KIDD, N. A. C., LEWIS, G. B. and HOWELL, C. A. (1985). An association between two species of pine aphid, *Schizolachnus pineti* and *Eulachnus agilis*. *Ecological Entomology* **10**, 427–432.

LARSSON, S. (1985). Seasonal changes in the within-crown distribution of the aphid *C. pini* on Scots pine. *Oikos* **45**, 217–222.

LEWIS, G. B. (1987). Regulating interactions between pine aphid colonies (*Schizolachnus pineti*) and host plant growth. PhD thesis, University of Wales.

MURDIE, G. (1969). The biological consequences of decreased size caused by crowding or rearing temperatures in apterae of the pea aphid, *Acyrthosiphon pisum* Harris. *Transactions of the Royal Entomological Society of London* **121**, 443–455.

SCHULTZ, J. C. and BALDWIN, I. T. (1982). Oak leaf quality declines in response to defoliation by gypsy moth larvae. *Science* **217**, 149–151.

van EMDEN, H. F. (1972). Aphids as phytochemists. In: *Phytochemical Ecology*, pp 25–43 (ed. J.B. Harborne) London, New York: Academic Press.

van EMDEN, H. F. (1978). Insects and secondary plant substances—an alternative viewpoint with special reference to aphids. In: *Biochemical Aspects of Plant and Animal Coevolution*, pp. 309–323 (ed. J.B. Harborne). London: Academic Press.

van EMDEN, H. F. and BASHFORD, M. A. (1971). The performance of *Brevicoryne brassicae* and *Myzus Persicae* in relation to plant age and leaf amino acids. *Entomologia Experimentalis & Applicata* **14**, 349–360.

WAY, M. J. and CAMMELL, M. (1970). Aggregation behaviour in relation to food utilisation by aphids. In: *Animal Populations in Relation to their Food Resources*, pp. 229–247 (ed. A. Watson). Oxford: Blackwell.

WHITHAM, T. G. (1983). Host manipulation of parasites: within-plant variation as a defense against rapidly evolving pests. In: *Variable Plants and Herbivores in Natural and Managed Systems* pp. 15–41. (eds R. F. Denno and M. S. McClure). New York: Academic Press.

18
Air Pollution and Tree-Dwelling Aphids

S. McNEILL* AND J. B. WHITTAKER**

Department of Pure and Applied Biology, Imperial College, Silwood Park, Ascot, Berkshire SLB 7PY, UK
**Division of Biological Sciences, Institute of Environmental and Biological Sciences, University of Lancaster, Bailrigg, Lancaster LA1 4YQ, UK*

Introduction
Field observations
Experimental studies in chambers
      Aphids on broadleaved trees and shrubs
      Conifer aphids
Open-air fumigation
Response of the tree
Discussion
Acknowledgements
References

Introduction

There is a large amount of circumstantial evidence that herbivorous insects are affected by atmospheric pollutants. Much of this is reviewed by Alstad, Edmunds and Weinstein (1982), Hain (1987) and Riemer and Whittaker (1989). The consensus view is that populations of many insect herbivores are increased in the presence of moderate levels of pollutants (though they may be decreased at high levels) and so may be contributing to the general symptoms of forest decline (Krause, 1989).

Indeed, one of the fundamental questions is, 'Does air pollution make trees more susceptible to insect attack?' If it does, then it may do so by changing the physiology of the food plants and/or the insects themselves, affecting host finding by the insect, their ability to detect pheromones, fecundity, growth rates and mortality. However, experimental evidence for the mechanisms involved has only recently been gathered (reviewed by Whittaker and Warrington, 1990) and most of the evidence relates to herbivores of non-woody plants. Hain, writing as recently as 1987, was able to say that 'there is at present no evidence directly linking air pollutants and tree decline'.

Here we review the situation with regard to aphids on trees. In doing so we restrict our comments to gaseous atmospheric pollutants derived from combustion

Population Dynamics of Forest Insects
© Intercept Ltd, PO Box 716, Andover, Hampshire, SP10 1YG, UK

of fossil fuels, because most of the experimental evidence relates to these. The questions that we wish to address are these:

1. Is there experimental evidence linking pollution to population growth in aphids and do we understand the mechanisms involved?
2. Does insect damage make trees more susceptible to atmospheric pollution?
3. Are changes in aphid numbers likely to be sustained in the field, and if so, how may they affect tree growth?

Field observations

The first reports of numerical changes in insect populations which were ascribed to air pollution were in the 1830s (Cramer, 1951). Since then there have been many attempts to relate population changes to pollution gradients. The majority of these refer to trees (usually conifers) and shrubs as food plants. Amongst the effects on Homoptera on trees there are more records of adelgids and coccids (particularly bark-feeding species) than aphids (Riemer and Whittaker, 1989). Usually the claim is made that these sap-feeding insects become more numerous downwind of strongly polluting industrial sources (Hillman, 1972; Charles and Villemant, 1977; Katayev, Golutrin and Selikhovkin, 1983; Heliovaara and Väisänen, 1989) or alongside major roads (Przybylski, 1979; Bolsinger and Flückiger, 1984; Braun and Flückiger, 1985).

However, there are exceptions to this. Berge (1973), for example, found that woolly aphids were less numerous on *Abies concolor* which showed signs of SO_2 and fluoride damage than on healthy trees and Galecka (1984) attributed reduced numbers of *Aphis frangulae* on elder buckthorn in a polluted area to disruption of the synchrony between hatching and bud burst. Villemant (1981) found that some aphids (*Cinara pini* and *Protolachnus agilis*) increased and others (*Schizolachnus tomentosus* and *Pineus pini*) decreased in areas of high pollution.

Experimental studies in chambers

Since the pioneer work of Dohmen, McNeill and Bell (1984), a number of experimental studies have demonstrated that gaseous atmospheric pollutants, apparently acting through changes in plant physiology and biochemistry can cause increases in mean relative growth rate (MRGR) and hence in reproduction and population growth of aphids. We shall restrict our discussion to experimental studies on woody species.

APHIDS ON BROADLEAVED TREES AND SHRUBS

Braun and Flückiger (1984a, 1985) and Bolsinger and Flückiger (1984) have investigated the effect of motorway pollution on aphid population development using potted shrubs adjacent to, and at 300 m distances from a motorway. Those on the motorway verge were enclosed in cabinets which allowed ingress of ambient air (that is, polluted with SO_2, NO_x and ozone amongst others) or air filtered by activated carbon and Purafil filters which removed most of the pollutants. Tree and shrub species used were hawthorn (*Crataegus*) (Braun and Flückiger, 1984a, 1985) and Guelder Rose (*Viburnum opulus*) (Bolsinger and Flückiger, 1984). Populations of introduced aphids were censused every two weeks over a two–three month period.

In addition, Braun and Flückiger (1984a) factored out the effect of natural enemies close to the motorway and at 300 m, by having some cages with open access and others closed. Whilst there was a reduced efficiency of natural enemies close to the motorway, the large increase in *Aphis pomi* on *Crataegus* could not be explained by this, and was found in unfiltered chambers, but not in filtered ones which were both sealed against natural enemies. Although these studies are complicated by the demonstration that drought and de-icing salt both had an effect on aphid populations close to the road (Braun and Flückiger, 1984b), this did not affect their conclusion that enhanced aphid numbers on the trees were associated with gaseous pollutants and that natural enemies or other mortality factors, such as intraspecific competition, did not succeed in cancelling out this effect. Similar observations were reported for *Aphis fabae* on *Viburnum opulus* on the motorway verge (Bolsinger and Flückiger, 1984) and for *Phyllaphis fagi* on beech (Flückiger and Braun, 1986).

These experiments are especially valuable because the increased aphid numbers were linked to increased free amino acids (Flückiger, Oertli and Baltensweiler, 1978; Bolsinger and Flückiger, 1989) and in particular to a significant increase in glutamine relative to sugar content (Braun and Flückiger, 1985). Bolsinger and Flückiger (1989) followed up these observations by rearing aphids on artificial diets designed to match the changes taking place in phloem exudates in the experimental plants. Aphids reared on a diet like that in the plants grown in unfiltered air close to the motorway had increased growth rates and shorter development times compared with those on 'filtered air' diets, thus supporting their conclusion and providing a mechanism for it.

This technique of filtering the air entering one set of chambers and comparing growth of aphids in filtered compared with ambient air was also used by Dohmen (1988) with chambers containing *Macrosiphum rosae* on rose bushes on a roof in central Munich (estimated mean concentrations in ambient air; SO_2: (9 nl l⁻¹ (range 2–70 nl l⁻¹); NO_2:20 (7–137) nl l⁻¹; O_3:12 (5–66) nl l⁻¹). Growth of *Macrosiphum rosae* was significantly higher (by an average of 20%) on rose bushes grown in ambient compared with filtered air.

Although ozone was undoubtedly a component of the atmospheric cocktail in all these experiments, the separate effects of the gases could not be detected. Coleman and Jones (1988), however, have exposed Eastern Cottonwood trees (*Populus deltoides*) to acute ozone stress (187 nl l⁻¹) and measured the effect on the aphid *Chaitophorus populicola*. They found that aphid performance was not significantly different on fumigated and control plants.

A series of short-term (five-day) experiments by Warrington, Moore and Whittaker (unpublished) (*Table 1*), on a range of broad-leaved trees, failed to detect any effect of ozone (70 nl l⁻¹) on *Eucallipterus tiliae* on lime. SO_2 (40 nl l⁻¹) did not significantly affect MRGR of *Drepanosiphum platanoidis* on sycamore at

Table 1.   Mean relative growth rates (MRGR) over 5 days, of several aphid species exposed to simultaneous fumigation of their host trees in closed chambers

Aphid species	Tree species	Pollutant	Mean gas conc. $nl\,l^{-1}$	Date of start of experiment	MRGR ± S.E.	% change in MRGR from control	P
Euceraphis punctipennis Zett.	Betula pubescens	NO_2	9	12 Aug	0.242 ± 0.006	—	
			57		0.279 ± 0.005	+15.3	NS
			105		0.315 ± 0.006	+30.2	<0.05
			157		0.263 ± 0.005	+8.7	NS
Eucallipterus tiliae L.	Tilia x europaea	O_3^*	5	10 Aug	0.253 ± 0.001	—	
			72		0.253 ± 0.001	0	NS
Drepanosiphum platanoidis Schrank	Acer pseudoplatanus	SO_2	3	29 May	0.076 ± 0.004	—	
			40		0.086 ± 0.004	+13.2	NS
			3	12 June	0.159 ± 0.010	—	
			40		0.169 ± 0.018	+6.3	NS
			3	1 Sept	0.137 ± 0.013	—	
			40		0.137 ± 0.011	0	NS
			3	7 Sept	0.228 ± 0.019	—	
			40		0.207 ± 0.014	-9.2	NS
Phyllaphis fagi L.	Fagus sylvatica	SO_2	0	9 June	0.095 ± 0.008	—	
			40		0.121 ± 0.008	+26.3	<0.05
			0	19 June	0.081 ± 0.006	—	
			100		0.099 ± 0.006	+22.2	<0.05
			0	26 Aug	0.101 ± 0.107	—	
			40		0.107 ± 0.009	+5.9	NS
			0	5 Sept	0.106 ± 0.005	—	
			40		0.109 ± 0.010	+2.8	NS

* Ozone generated by electric discharge but N_2O_5 was low (appox. 0.02 moles per mole of O_3) (Neighbour, Pearson and Mehlhorn, 1990).

any time between June and September but MRGR of *Phyllaphis fagi* was increased in the early part of June by SO_2 but not later in the season and that of *Euceraphis punctipennis* on *Betula pubescens* was significantly increased by NO_2 in August.

Garsed, Farrar and Rutter (1979) argued that broadleaved trees (*Betula pendula*, *B. pubescens* and *Acer pseudoplatanus*) seemed less susceptible to atmospheric pollutants than conifers.

CONIFER APHIDS

Experiments on a range of aphid/conifer/air pollutant systems are largely being conducted in closed chambers but there is at least one large open field fumigation system in which aphids are being studied. Experiments have involved the following aphid/host plant systems: (a) *Elatobium abietinum* on Sitka or Norway spruce; (b) *Schizolachnus pineti* on Scots pine; (c) *Cinara pilicornis* on Sitka spruce; and (d) *Cinara pini* on Scots pine. Experimental studies have involved one or more of the following gases with each of these systems: sulphur dioxide, nitrogen dioxide and ozone. The experimental regimes are summarized in *Table 3*.

The results of the studies so far available suggest several relatively clear-cut patterns in terms of increased and decreased MRGR in the responses of the aphids to plants pre-fumigated with these gaseous pollutants at moderately high concentrations (100 nl l^{-1}) for short periods (up to 7 days) in charcoal filtered air in closed chambers, the control treatment being charcoal filtered air alone.

Sulphur dioxide and nitrogen dioxide

As has been seen in the case of aphids feeding on herbaceous hosts, pre-fumigation with these gases at a range of concentrations and for various periods (Dohmen, McNeill and Bell, 1984; Warrington, 1987; McNeill, *et al.*, 1987; Houlden, *et al.*, 1990) generally, but not always, leads to an increase in aphid performance compared to filtered air controls. The speed of the response, however, is often much slower, at least in the case of NO_2, than that seen in herbaceous plants where the maximum response occurs at around seven hours exposure.

In experiments, however, in which *E. abietinum* nymphs were fed on the needles of Sitka spruce exposed to 100 nl l^{-1} of SO_2 for periods between 0.5 and 48 h the peak response in terms of increased MRGR over filtered air controls occurred after exposures of only 3 h (McNeill, *et al.*, 1987). The peak response showed an increase of 19.4% above controls falling back to a steady 12.4% increase after 16 h exposure, this level continuing throughout the rest of the experiment (*Figure 1*). These experiments were conducted in April to early June when the trees supported rapidly growing populations of aphids. Experiments conducted at other times, when the aphid populations are growing much more slowly, show a much reduced response in absolute terms, but not in relative terms, which rapidly becomes non-significant as the new shoots expand.

Experiments in progress using all four of the aphid species mentioned above, fumigating with NO_2 at 100 nl l^{-1}, have shown increases in four-day MRGR when fumigation time exceeds 24 h, the increases are significant in all cases except for

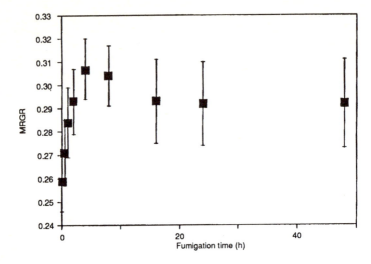

Figure 1.   Three-day MRGR of the *Elatobium abietinum* nymphs on Sitka spruce pre-fumigated with 100 nl l^{-1} SO$_2$. Vertical bars indicate 95% confidence intervals.

Cinara pini feeding on Scots pine. The time response curve is similar in all cases with the aphids showing increasing percentage stimulation of MRGR over controls with increasing pre-fumigation exposure to NO$_2$ up to 168 hours. The rate of increase in relative performance slows steadily at longer exposure times (*Figure 2*) when increases in MRGR of between 18% and 58% have been measured. Similar increases in aphid MRGR post-fumigation have been observed also at lower concentrations of NO$_2$ but the response only begins to become significant at longer fumigation times (e.g. after 7 days at 50 nl l^{-1}).

Ozone

Although experimental measurements of aphid responses to trees exposed to ozone are at a very early stage, the results so far indicate that the responses are neither so marked nor so straightforward as those to sulphur dioxide or nitrogen dioxide. This situation was further complicated until very recently when it was realized that ozone generated from air by electric discharge generators is often contaminated by N$_2$O$_5$ (McLeod *et al.*, 1989). This problem, however, does not arise in the studies reported in this section as the ozone generator used was run on oxygen and not air. Furthermore, in those few experiments where ozone was generated from air by the action of UV light, water filters showed no evidence of contamination by nitrogen oxides.

Ozone as a pollutant is naturally episodic in its occurrence and is usually at its greatest during periods of drought/temperature stress as it is largely a secondary product of photochemical decay of nitrogen oxides and hydrocarbons. Studies of its effect on the aphid/conifer systems have therefore concentrated on two types of fumigation: (a) continuous exposure to relatively high levels (100 nl l^{-1}) over short

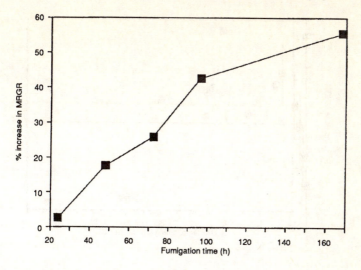

Figure 2. Percentage increase in 3-day MRGR of *Elatobium abietinum* nymphs on Sitka spruce pre-fumigated with 100 nl l⁻¹ NO₂.

periods, as in the SO_2 and NO_2 experiments, and (b) exposure to similar levels presented as a series of short episodes (8 h in 24 h) over longer periods. Work is being carried out with all four insect/host combinations as before and so far there have been no significant responses in aphid MRGR with any of the insects using plant material exposed to episodic fumigation over periods of up to 7 days, except in the case of *C. pilicornis* feeding on new shoots of Sitka spruce where a significant decrease in MRGR was recorded at fumigation temperatures above 23°C. Continuous fumigation, on the other hand, has produced a variety of responses.

In the case of *Elatobium* and *C. pilicornis* feeding on last year's shoots of Sitka spruce and *C. pini* on Scots pine there is no response by the insects to pre-fumigation at periods up to 96 h at 100 nl l⁻¹ ozone. In the case of *S. pineti* on new shoots of Scots pine, however, there was a significant decrease in performance after 48 h pre-fumigation (control 0.196 ± 0.005; fumigated 0.153 ± 0.011; 10 replicates; $P<0.01$).

C. pilicornis, although showing no response to ozone on old shoots, has shown significant positive and negative responses on new shoots. This response shows a strong interaction with the maximum temperatures experienced by the tree during fumigation, the measurements of aphid MRGR being carried out post-fumigation in 20°C controlled temperature rooms. After exposure to ozone at 100 nl l⁻¹ in closed chambers in a greenhouse where the maximum temperatures were below 20°C there is a significant increase in the MRGR of the aphids after exposures of more than 8 h, rising after 24 h exposure to increases of up to 63% over controls (*Table 2*). However, if the temperature at which the trees are fumigated rises to a daily maximum beyond 23°C then the result is a significant decrease in the MRGR

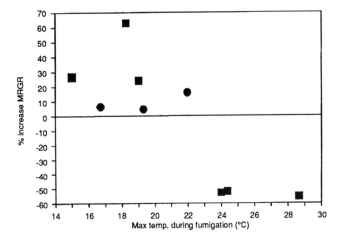

Figure 3.   Percentage increase in 4-day MRGR of *Cinara pilicornis* nymphs on new shoots of Sitka spruce pre-fumigated continuously with 100 nl l⁻¹ O₃. ■ = significant increase or decrease in MRGR versus controls (*P*<0.05). ● = non-significant change in MRGR versus controls.

Table 2.   MRGR of *Cinara pilicornis* on Sitka spruce fumigated with 100 nl l⁻¹ ozone (max. daily temp. 20°C).

Fumigation (hrs)	MRGR controls ±95% CI	n	MRGR 100 nl l⁻¹ O₃ ±95% CI	n	% increase in MRGR	
4	0.118 ± 0.021	22	0.134 ± 0.023	23	13.6	NS
8	0.171 ± 0.023	10	0.216 ± 0.021	13	25.3	**
24	0.099 ± 0.026	13	0.161 ± 0.027	12	62.6	**

with both continuous and episodic fumigations, intermediate conditions giving non-significant results (*Figure 3*). The logarithmic survivorship (*K*) of the aphids decreases significantly on both fumigated and non-fumigated trees with exposure of the trees to increasing daily maximum temperatures.

Open-air fumigation

All of the above work has been carried out in relatively small chambers on small trees in pots. It is notoriously difficult to extrapolate such results to trees in the natural environment where any tendency for populations of aphids to increase due to reactions to pollutant stress may be countered by density-dependent responses in natural enemy populations.

A large-scale open-air fumigation experiment, the Liphook Open-Air Forest Fumigation Project, set up by the Technology and Environment Centre of National Power Company should provide data to answer these problems. In this experiment the effects of raising the background levels of ozone and sulphur dioxide in a carefully controlled manner, singly and in combination in a factorial design, over several seasons are being measured on a range of conifer species. A detailed description of the design and operation of this experiment is given in McLeod *et al.* (1989). The responses of a range of insect populations on Sitka spruce, Norway spruce and Scots pine are being investigated in this system. These studies include measurements of individual and population performance of several aphid species as well as studies of interactions with their natural enemies.

Response of the Tree

Much of the experimental evidence reviewed is of aphid response over a few days, and the emphasis is on changes in aphid MRGR. These are, however, significantly positively correlated with increases in nymph production and population growth. Warrington and Whittaker (1990) allowed populations of *Elatobium abietinum* to build up for two months on Sitka spruce trees which were enclosed in solar dome glasshouses and subjected to SO_2 fumigation (30 nl l^{-1}) and/or drought. Aphid populations and tree growth parameters in each of these treatments in a factorial design were compared with controls. The concentration of SO_2 used had no significant effect on the trees. SO_2 in the presence of aphids, however, led to reduced leader length and root biomass which was greater in both cases than reductions caused by adding together the separate effects of aphids and SO_2. At the end of the experiment aphids were up to three times as abundant on the SO_2 fumigated trees as on the controls. Drought alone had no significant effect on aphid numbers but drought plus SO_2 made the trees less acceptable to the aphids than those treated with SO_2 and watering but still more acceptable than trees grown in filtered air whether watered or droughted. The effect of low concentration of SO_2 on the aphids was thus more important than the effects of drought, even though the latter was severe and reduced plant growth significantly.

Discussion

Short-term, closed-chamber fumigations on potted trees show that aphid performance is consistently enhanced by exposure of the host to SO_2 or NO_2 over relatively short periods (*Table 3*). However, aphid response is usually slower on trees than on the herbaceous plants which have been investigated (Houlden *et al.*, 1990). For a given concentration of pollutant the effect on aphid MRGR generally increases as pre-fumigation exposure of the plant increases up to a maximum of approximately one week's exposure. There is also a trade-off between concentration of the fumigant and period of exposure; low concentration and long fumigation being equivalent to higher concentration but shorter fumigation.

The responses to ozone fumigation indicate a more complex set of interactions (*Table 3*) with positive, negative and no effect being shown depending on the feeding site (old or new shoots) or environmental conditions (high or low maximum temperature). These interactions suggest that it is likely that ozone exposure would be a much less important pollutant in relation to aphid performance in the field as compared to NO_2 and SO_2 exposure. The results showing that only aphids feeding on new shoots respond significantly means that there will be only a relatively short period in the spring/early summer when ozone could be important; and even here, the interaction with temperature and probably drought stress would tend to constrain any positive effects as ozone episodes are associated with periods of dry, sunny, still weather.

As yet, we have no experimental evidence to show how important mixtures of these gases may be, though it is known that mixtures can have a much more serious effect on plants (including tree physiology and growth) than similar concentrations of individual gases (Freer-Smith, 1984).

Table 3. Summary of the experimental regimes and results of closed chamber experiments measuring the effects of the pre-fumigation of host trees on the MRGR of conifer aphids.

Gas regime	Aphid	Host tree	Effects on MRGR
SO_2 100 nl l^{-1} continuously for 0.5–48 h:			
	Elatobium abietinum	*Picea sitchensis*	+ve
NO_2 100 nl l^{-1} continuously for 24–168 h:			
	E. abietinum	*P. sitchensis*	+ve
	Cinara pilicornis	*P. sitchensis*	+ve
	Schizolachnus pineti	*Pinus sylvestris*	+ve
	C. pini	*P. sylvestris*	NS
			(but strong +ve trend)
O_3 100 nl l^{-1} continuosuly for 4–96 h:			
	E. abietinum	*P. sitchensis*	NS
	C. pilicornis	*P. sitchensis* (old shoots)	NS
	C. pilicornis	*P. sitchensis* (new shoots)	+ve ⎫ temperature NS ⎬ dependent -ve ⎭ (after 48 h)
	S. pineti	*P. sylvestris*	-ve
	C. pini	*P. sylvestris*	NS
O_3 100 nl l^{-1} episodically (8h in 24 h) for 24–96 h:			
	E. abietinum	*P. sitchensis*	NS
	C. pilicornis	*P. sitchensis* (new shoots)	–ve at high temp. (low temp. experiments in progress)
	S. pineti	*P. sylvestris*	NS
	C. pini	*P. sylvestris*	NS

Other contributors to this symposium have shown how increased aphid numbers and rates of feeding may affect tree growth. Dixon (1971), for example has shown how *Drepanosiphum platanoidis* reduces timber production of sycamore and Carter (1977) observed a conspicuous reduction in annual rings following defoliation caused by high densities of *Elatobium abietinum*. In our study, increased numbers of *E. abietinum* in the presence of SO_2 caused a reduction in leader length as well as in root biomass. A recent review (Welter, 1989) summarizes what is known about the effects of phloem feeders and plant–gas exchange. Unfortunately only one of these studies is of a woody plant, pecan (Wood, Tedders and Thompson, 1985; Wood and Tedders, 1986), but it shows no change or a decrease in photosynthetic rates per unit area, resulting from clogged phloem vessels and interveinal damage by salivary toxins. There was also a 25% reduction in leaf respiratory rates. It is also possible that there may be a direct consequence of insect feeding damage which may result in increased uptake of gaseous pollutants comparable with that found for typhlocybine damage on sycamore by Warrington, Cottam and Whittaker (1989).

In conclusion, the available evidence from laboratory fumigations does support the field observations that populations of sap-feeding insects are likely to increase on trees exposed to gaseous atmospheric pollutants. One explanation of this is that these changes are plant-mediated and that changes in amino acid concentrations and balance are an important factor (McNeill *et al.*, 1987; Bolsinger and Flückiger, 1989).

Acknowledgements

We wish to thank Vic Brown at Imperial College for allowing us to quote results from experiments in progress and/or in press. The work is supported by an NERC (Special Topic) grant to the authors. In addition, Dr McNeill is in receipt of a contract from NP Tec. for insect studies associated with the Liphook Open-Air Forest Fumigation Project. We are very grateful to these organizations for their support.

References

ALSTAD, D. N., EDMUNDS, G. F. and WEINSTEIN, L. H. (1982). Effects of air pollutants on insect populations. *Annual Review of Entomology* 27, 369–384.

BERGE, H. (1973). Beziehungen zwischen Baumschadlingen und Immissionen. *Anzliger für Schaedlingskdunde Pflanze Umweltschutz* 46, 155–56.

BOLSINGER, M. and FLÜCKIGER, W. (1984). Effects of air pollution at a motorway on the infestation of *Viburnum opulus* by *Aphis fabae*. *European Journal of Forest Pathology* 14, 256–260.

BOLSINGER, M. and FLÜCKIGER, W. (1989). Ambient air pollution induced changes in amino acid pattern of phloem sap in host plants-relevance to aphid infestation. *Environmental Pollution* 56, 209–216.

BRAUN, S. and FLÜCKIGER, W. (1984a). Increased population of the aphid *Aphis pomi* at a motorway: Part 1, Field evaluation. *Environmental Pollution* **33**, 107–120.

BRAUN, S. and FLÜCKIGER, W. (1984b). Increased population of the aphid *Aphis pomi* at a motorway. Part 2, The effect of drought and deicing salt. *Environmental Pollution* **36**, 261–270.

BRAUN, S. and FLÜCKIGER, W. (1985). Increased population of the aphid *Aphis pomi* at a motorway: Part 3, The effect of exhaust gases. *Environmental Pollution* **39**, 183–192.

CARTER, C. I. (1977). Impact of green spruce aphid on growth: can a tree forget its past? *Forestry Commission Research and Development Paper* **116**, 1–8.

CHARLES, P. J. and VILLEMANT, C. (1977). Modifications des niveaux de populations d'insectes dans les jeunes plantations des pins sylvestres de la forêt de Roumare (Seine Maritime) soumises à la pollution atmosphérique. *Comptes Rendus des Séances de l'Académie d' Agriculture de France* **63**, 502–510.

COLEMAN, J. S. and JONES, C. G. (1988). Acute ozone stress on Eastern Cottonwood *Populus deltoides* Bartr. and the pest potential of the aphid *Chaitophorus populicola* Thomas (Homoptera: Aphidae). *Environmental Entomology* **17**, 207–212.

CRAMER, H. H. (1951). De geographischen Gründlagen des Massen wechsels von *Epiblema tedella* Cl. *Forstwissenschaftliches Centralblatt* **70**, 42–53.

DIXON, A. F. G. (1971). The role of aphids in wood formation. I. The effect of the sycamore aphid, *Drepanosiphum platanoidis* (Schr.) (Aphididae), on the growth of sycamore, *Acer pseudoplatanus* L. *Journal of Applied Ecology* **8**, 165–179.

DOHMEN, G. P. (1988). Indirect effects of air pollutants: changes in plant/parasite interactions. *Environmental Pollution* **53**, 197–207.

DOHMEN, G. P., McNEILL, S. and BELL, J. N. B. (1984). Air pollution increases *Aphis fabae* pest potential. *Nature* **307**, 52–53.

FLÜCKIGER, W. and BRAUN, S. (1986). Effects of air pollutants on insects and hostplant/insect relationships. In: *Effects of Air Pollution on Terrestrial and Aquatic Ecosystems. Working Party III: How are the effects of Air Pollutants on Agricultural Crops Influenced by the Interaction with Other Limiting Factors?* Proceedings of a workshop jointly organized within the framework of the Concerted Action. Commission of the European Communities and the National Agency of Environmental Protection, Riso National Laboratory, Denmark. 23–25 March 1986.

FLÜCKIGER, W., OERTLI, J. J. and BALTENSWEILER, W. (1978). Observations of an aphid infestation on hawthorn in the vicinity of a motorway. *Naturwissenschaften* **65**, 654–655.

FREER-SMITH, P. H. (1984). The responses of six broadleaved trees during long term exposure to SO_2 and NO_2. *New Phytologist* **97**, 49–62.

GALECKA, B. (1984). Phenological development of *Frangula alnus* Mill. in an industrial region and the number of *Aphis frangulae*, Kalt. *Polish Ecological Studies* **10**, 141–155.

GARSED, S. G., FARRAR, J. F. and RUTTER, A. J. (1979). The effects of low concentrations of sulphur dioxide on the growth of four broadleaved tree species. *Journal of Applied Ecology* **16**, 217–226.

HAIN, F. P. (1987). Interactions of insects, trees and air pollutants. *Tree Physiology* 3, 93–102.

HELIOVAARA, K. and VÄISÄNEN, R. (1989). Invertebrates of young Scots pine stands near the industrialised town of Harjavalta, Finland. *Silva Fennica* 23, 13–19.

HILLMANN, R. C. (1972). Biological effects of air pollution on insects, emphasizing the reactions of the honeybee (*Apis mellifera* L.) to sulphur dioxide. Unpublished thesis, Pennsylvania State University.

HOULDEN, G., McNEILL, S., AMINU-KANO, M. and BELL, J. N. B. (1990). Air pollution and agricultural aphid pests. I Fumigation experiments with SO_2 and NO_2 (In press).

KATAYEV, O. A., GOLUTRIN, G. I. and SELIKHOVKIN, A. V. (1983). Changes in arthropod communities of forest biocoenoses with atmospheric pollution. *Entomological Reviews* 62, 20–29.

KRAUSE, G. H. M. (1989). Forest decline. In: *Toward a More Exact Ecology* (eds P. J. Grubb and J. B. Whittaker). Symposium of the British Ecological Society 30. Oxford: Blackwell Scientific.

McLEOD, A. R., BROWN, K. A., SKEFFINGTON, R. A. and ROBERTS, T. M. (1989). Studies of ozone effects on conifers and the formation of oxides of nitrogen during ozone generation. In: *Air Pollution and Forest Decline* (eds J. B. Bucher and I. Bucher-Wallin). Proceedings of 14th International Meeting for Specialists in Air Pollution Effects on Forest Ecosystems. Birmensdorf: IUFRO.

McNEILL, S., AMINU-KANO, M., HOULDEN, G., BULLOCK, J. M. CITRONE, S. and BELL, J. N. B. (1987). The interaction between air pollution and sucking insects. In: *Acid Rain-Scientific and Technical Advances*, (eds R. Perry, R. M. Harrison, J. N. B. Bell and J. W. Lester). London: Selper.

NEIGHBOUR, E. A., PEARSON, M. and MEHLHORN, H. (1990). Purafil-filtration prevents the development of ozone-induced frost injury: A potential role for nitric oxide. *Atmospheric Environment* 24A, 711–715.

PRZYBYLSKI, Z. (1979). The effect of automobile exhaust gases on the arthropods of cultivated plants, meadows and orchards. *Environmental Pollution* 19, 157–161.

RIEMER, J. and WHITTAKER, J. B. (1989). Air pollution and insect herbivores: observed interactions and possible mechanisms. *Insect–Plant Interactions* 1, 73–105.

VILLEMANT, C. (1981). Influence de la pollution atmosphèrique sur les populations d'aphides du pin sylvestre en Forêt de Roumaere (Seine Maritime). *Environmental Pollution* 24, 245–262

WARRINGTON, S. (1987). Relationship between SO_2 dose and growth of the pea aphid, *Acyrthosiphon pisum*, on peas. *Environmental Pollution* 43, 155–162.

WARRINGTON, S. and WHITTAKER, J. B. (1990). Interactions between Sitka spruce, the green spruce aphid, sulphur dioxide pollution and drought. Evironmental Pollution (in press).

WARRINGTON, S., COTTAM, D. A. and WHITTAKER, J. B. (1989). Effects of insect damage on photosynthesis, transpiration and SO_2 uptake by sycamore. *Oecologia* 80, 136–139.

WHITTAKER, J. B. and WARRINGTON, S. (1990). Effects of atmospheric pollutants on interactions between insects and their food plants. In *Pests, Pathogens and Plant Communities* (eds J. J. Burdon and S. R. Leather). Oxford: Blackwell Scientific (in press).

WELTER, S. C. (1989). Arthropod impact on plant gas exchange. *Insect–Plant Interactions* **1**, 135–150.

WOOD, B. W., TEDDERS, W. L. and THOMPSON, J. M. (1985). Feeding influence of 3 pecan aphid species on carbon exchange and phloem integrity of seedling pecan foliage. *Journal of the American Society of Horticultural Science* **110**, 393–397.

WOOD, B. W. and TEDDERS, W. L. (1986). Reduced net photosynthesis of leaves from mature pecan trees by three species of pecan aphid. *Journal of Entomological Science* **21**, 355–360.

19

Changes in Population Dynamics of Pine Insects Induced by Air Pollution

KARI HELIÖVAARA* AND RAUNO VÄISÄNEN**

*Finnish Forest Research Institute, Vantaa, SF01301, Finland
**Water and Environment Research Institute, Helsinki, Finland

Introduction
Patterns
A case study on a pollutant gradient
    Field studies
    Experimental studies
    Chemical analyses
    Patterns observed
    Experimental results
Discussion
Acknowledgements
References

Introduction

Recent laboratory studies, summarized in this paper, have shown that air pollution may influence major population parameters of forest insects. Changes in these parameters, if similar in the field, may play an important role in triggering insect outbreaks or their decline. Because air pollution affects both insects and their host trees, and may have both direct and indirect effects, early speculations about these effects are not well supported by relevant quantitative studies. Field observations have also shown variable and contradictory patterns depending on the species, pollutants and their concentrations, climatic and edaphic conditions, and study design (Alstad, Edmunds and Weinstein, 1982; Templin, 1982; Katayev, Golutvin and Selikhovkin, 1983; Führer, 1985).

Due to the mainly applied interest of forest entomologists attention has been paid to traditional aspects such as stand density, stand age, management and control of local pest outbreaks. Most often interest has been concentrated on local forest damage caused by insects, which has sometimes been observed in heavily polluted areas, although the elevation of general air pollution level affects nature everywhere. The latter more diffuse issue can be studied along pollutant gradients which provide an opportunity for the calibration of spatial and temporal occurrence of forest pests in relation to pollution.

Population Dynamics of Forest Insects
© Intercept Ltd, PO Box 716, Andover, Hampshire, SP10 1YG, UK

Patterns

Pollution can affect both the spatial distribution of outbreaks and the outbreak frequency. This can take place on wide biogeographical as well as on local scales. Slight increases in the abundance of pests on the biogeographical scale are much more difficult to observe than local outbreaks, though the latter may have less economic importance.

In theory, air pollutants can indirectly increase the abundance of forest insects. Direct benefit from higher pollutant concentrations is less likely. However, direct toxic effects may partly mask the indirect advantageous effects, or vice versa. Pollutant stress can induce changes in tree biochemistry. For instance, altered concentrations of soluble nitrogen, sugars and defensive secondary compounds may improve the survival, reproduction or growth of insects.

It has been suggested that some parasitoids are more sensitive to pollution than their hosts (Alstad, Edmunds and Weinstein, 1982). If this is applicable to forest insects, decrease in parasitization may be an important factor in increasing pest survival. Similar mechanisms may work through predators and diseases. Changes in between-species competition may take place as well. For instance, bark beetle species feeding on ambrosia fungi intolerant to pollutants may decrease in favour of species without fungal symbiosis.

Changes in stand and tree structure, undervegetation, and litter may lead to unexpected alterations in insect population dynamics. For example, needle loss has been shown to decrease populations of large spiders due to increased predation (Gunnarsson, 1988), and this may decrease spider predation on pest insects. The loss of undergrowth improves overwintering conditions and survival of the pine bark bug (Heliövaara and Väisänen, 1986a). Similarly, when the shading vegetation is lacking, the development of cocoons of pine sawflies is faster and it may decrease mortality (Olofsson, 1987).

Forest pests do not form any ecological or taxonomic entity. Thus, some pests may decrease in polluted areas, while others may not. Toxic pollutants can decrease insects by increasing mortality, and decreasing growth and reproduction. When the concentrations of toxic compounds exceed a threshold level, their significance in population dynamics of insects becomes crucial. Indirect detrimental effects are diverse and difficult to predict.

A case study on a pollutant gradient

FIELD STUDIES

The occurrence of several pine insects along a local pollutant gradient was studied by utilizing 9-km-long transects, each consisting of nine sample plots around a distinctive factory complex. The field studies were carried out in Harjavalta, South-west Finland (61° 20′N, 22° 10′E) on a wide ridge with Scots pines (*Pinus sylvestris* L.). The factory complex situated in the area produces copper, nickel, fertilizers and sulphuric acid. The pollutants emitted include sulphur oxides, and heavy metals. The amounts of copper, nickel and cadmium emitted during 1986 were 305 000, 73 000 and 2000 kg, respectively.

EXPERIMENTAL STUDIES

The effect on defoliators of different pollutant concentrations in pine needles was studied using the sawflies *Gilpinia virens* (Klug), *G. frutetorum* (Fabricius), *G. socia* (Klug), *Microdiprion pallipes* (Fallén), *Neodiprion sertifer* (Geoff.) and *Diprion pini* (L.), and the moths *Panolis flammea* (Denis & Schiffermüller) and *Bupalus piniaria* (L.). They were reared in the laboratory from first larvae to pupae/adults. Each larva or larval colony always received its nutrition from the same numbered pine. After the larval development the pupae or cocoons were measured and weighed. The material yielded 2990 larvae. Rearing procedures are described in Heliövaara and Väisänen (1989a).

Effects of food quality on mortality and developmental rate of *D. pini* were investigated. The hatching times of adults were recorded after hibernation when the cocoons had been brought back into the laboratory. Relationships between female size and fecundity, egg viability and embryonal development were studied in detail in *N. sertifer*; 15 568 eggs laid by unfertilized females were divided into eight categories (for details, see Heliövaara, Väisänen and Varama, 1990).

CHEMICAL ANALYSES

Quantitative variation in the elemental composition of selected pines was studied along the gradient. The needles were sampled in June to obtain a realistic picture of their mineral nutrient and metal composition from a herbivore's point of view. After determining the dry weight and ash weight, the concentrations of total P, K, Mg, Mn, Fe, Al, Cu, Zn, Na and Ca were determined on an ARL ICP 3580 Spectrometer. C, H and N were analysed on a LECO CHN analysor. Levels of copper, iron, nickel and cadmium were also analysed from the pupae/cocoons of insects reared on the needles by flame and flameless atomic absorption spectrometry.

PATTERNS OBSERVED

The number of aphids (mainly *Cinara pini* (L.)), pine bark bugs (*Aradus cinnamomeus* (Panzer)), diprionids (*Diprion pini* (L.)) and tortricid moths (*Blastesthia turionella* (L.), *B. posticana* (Zetterstedt), *Rhyacionia pinicolana* (Doubleday) and *Retinia resinella* (L.)), was highest in the moderately polluted pine stands (Heliövaara and Väisänen, 1986a, 1989b). An outbreak of *D. pini*, extending approximately 1000 ha, was observed in this area in summer 1989. In general, herbivores, excluding aphids and bark beetles, were scarce in the immediate vicinity of the industrial plants. The pest species were scarce at an extended distance from the pollutant source. A few positive but no negative correlations were found between herbivores, suggesting that insects derive only slight if any

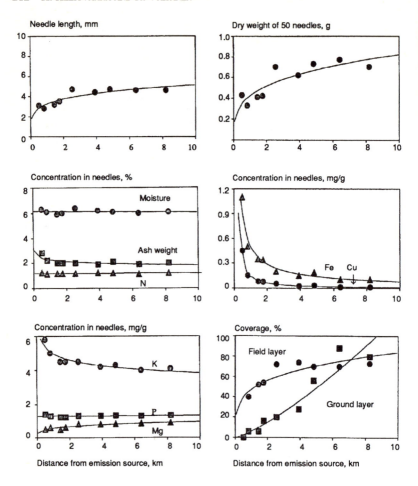

Figure 1.   Characteristics of pines and vegetation in the pollution gradient.

advantage from the previous occurrence of other species in the same host tree (Heliövaara and Väisänen, 1989b).

Heavy metal (Cu, Fe) concentrations in needles decreased exponentially with increasing distance from the emission source (*Figure 1*). Concentration of potassium and the ash weight of needles also decreased, but concentrations of nitrogen and phosphorus showed no clear patterns. Magnesium, manganese and calcium concentrations showed an increasing tendency with increasing distance (Heliövaara and Väisänen, 1989c). Some other variables possibly affecting insect population dynamics are given in *Figure 1*.

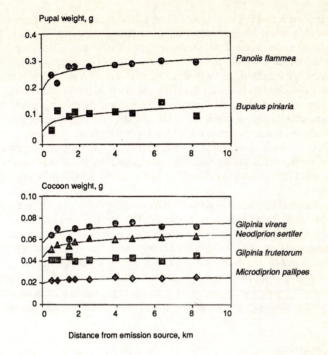

Figure 2.   Pupal/cocoon weight of the reared insect species in relation to the distance from the emission source.

EXPERIMENTAL RESULTS

The size of laboratory reared pupae/cocoons was negatively correlated (in most species to statistically significant degree) with the distance from the industrial plants at which their food was collected. The smallest pupae were those reared on needles collected from heavily polluted stands (Heliövaara and Väisänen, 1989a; Heliövaara, Väisänen and Kemppi, 1989). The highest reduction (49%) in the weight was observed in *B. piniaria* females. The same trend was evident for all cases of Lepidoptera and Hymenoptera studied (*Figure 2*).

Regression models suggested that the size differences were largely due to the quality of the food. In most cases the observed variation in size measurements were associated with heavy metal concentrations. For instance, the elemental composition of the needles explained 24–54% of the variation in pupal weight of the two moths. The metal concentrations in the pupae/cocoons, irrespective of the species, were highest in those fed as larvae on needles collected from near the emission source, and decreased exponentially with increasing distance. For instance, in *B. piniarius* the highest copper concentration was 137 and the lowest less than 20 µg g^{-1}. Concentrations of nickel were greater than those of cadmium, and were both exceeded by those of copper and iron. Although all of the species showed variation according to pollution level and distance from the factories, the differences between species were maintained (Heliövaara and Väisänen, 1990a, b). Cadmium accumulated in *N. sertifer, G. frutetorum* and, especially, in *M. pallipes*, but not at all in *B. piniaria*. The relative cadmium increases were larger as the distance from the factory complex increased. A similar increasing tendency was also obvious for other metals, particularly in copper. In contrast to those of cadmium or copper, concentrations of iron and nickel in pupae/cocoons never exceeded those found in needles (Heliövaara and Väisänen, in 1990a, b).

D. pini reared as larvae on polluted needles tended to develop at a slower rate and hatched on an average four days later than those reared on unpolluted needles. Winter mortality was higher in the group fed with polluted needles. Cocoon weight was not related to the hatching time within the groups (Heliövaara and Väisänen, 1990c).

Egg viability tests suggested that *N. sertifer* female size affects the survival and embryonic development of the eggs. The smaller females reared on polluted needles oviposited an average of 69 eggs (SD 28), and the larger ones reared on unpolluted needles 79 eggs (SD 32) per female, the difference being statistically significant ($F = 6.74, P = 0.01$, d.f. = 1206). However, the higher egg viability in the former group masked the effect of reduced female size. This compensation was primarily related to the size of the females, larger individuals producing relatively fewer viable eggs in both groups (Heliövaara, Väisänen and Varama, 1990).

Discussion

Air pollutants can directly damage plant tissues and disrupt normal patterns of resource acquisition and allocation. These disruptions will, in turn, potentially influence the plant's ability to defend itself against pests and pathogens (Lechowicz, 1987). Pollutants can affect needle-feeding insects through changes in the surface wax cuticle and other structures, water balance, amino acid contents, sugars and other organic compounds etc. (Huttunen, 1985). The occurrence of insect pests with specific air pollutants has been reviewed by Smith (1981, 1984a, b), Alstad, Edmunds and Weinstein (1982) and more recently by Hain (1987).

The assumption that tree decline is a consequence of stress is common to all hypotheses concerning forest decline (Hain, 1987). It has been postulated that stressed plants represent a more suitable food source for insect herbivores (White, 1984). However, insects with different feeding characteristics do not respond in the same way to pollutant-induced stress (Larsson, 1990). Several studies have shown that sucking insects seem to benefit from the effects of some pollutants on their host plants (Thalenhorst, 1974; Braun and Flückiger, 1984; Dohmen, 1988; Neuvonen and Lindgren, 1987). This may be associated with elevated amino acid and sugar levels in plant sap. The colonization success of bark beetles has been found to increase when the trees are subjected to stress (Cobb *et al.*, 1968a, b; Stark *et al.*, 1968; Miller, Cobb and Zavarin, 1968). Weakened trees show little resistance to attack due to decreased oleoresin pressure, but at the same time they may provide a poor breeding substrate (Baltensweiler, 1985). Increased concentrations of soluble nitrogen in stressed trees may positively affect defoliators

(White, 1984). However, levels of secondary compounds often increase in stressed trees (Mattson and Haack, 1987), thus having a negative effect as an antagonistic process.

Our observations support the previous findings that aphids and herbivorous bugs show elevated population levels in polluted areas (Heliövaara and Väisänen, 1986a, 1989b). No clear patterns were observed in bark beetles or defoliators. Tortricids living within buds or shoots were negatively affected by heavy pollution simply through reduced pine growth. Feeding characteristics cause differences in heavy metal concentrations of pine herbivores (Heliövaara *et al.*, 1987) which, in turn, may affect insect performance.

The levels of heavy metals in pine needles are related to pollutant load and may be detrimental to insects (Heliövaara and Väisänen, 1989c). The observed changes in insect performance were best explained by the concentrations of heavy metals in their food. Because there are conspicuous differences in metal bioaccumulation in insects on the same trophic level and even in closely related species, the between-species ecological relationships may change and thus influence outbreak patterns.

If the pollutant tolerance of parasitoids is lower than that of their hosts, it may lead to pest outbreaks (Alstad, Edmunds and Weinstein, 1982; Führer, 1985). Studies on *R. resinella* along the present pollution gradient did not show any clear tendency as regards insect parasitoids. Prolonged development observed in *D. pini* due to toxic pollutants can make insects more susceptible to predator attacks and diseases (e.g. Heliövaara and Väisänen, 1990c).

It has been calculated that the recurrence of simultaneous outbreaks of various defoliating pest species has decreased from approximately 14 years in 1780 to about 6 years in 1900. The frequency, extent and severity of the outbreaks have increased over the course of the past 200 years. Furthermore, rare insects have become pests in the second half of the twentieth century or known insects now occur in high densities in areas where they did not earlier (Baltensweiler, 1985). Several insects which were barely known appeared recently as pests on a number of conifers (Sierpinski, 1984; Baltensweiler, 1985). However, there are several plausible explanations for these changes in addition to air pollution. The effects of annual changes in climate and intensive silvicultural practices are major factors affecting pest abundance. Irrespective of the pollutants' significance as pest regulators, these two other factors may mask or obscure the effects of pollutants. For instance, the 1989 outbreak of *D. pini* may have been caused by favourable weather conditions or poor soil quality, as well as by pollution. *A. cinnamomeus* clearly benefits both from pollution and modern forestry practices (Heliövaara and Väisänen, 1986a).

The large-scale effects of air pollution are diffuse and difficult to study, especially when the insect species in question shows considerable population fluctuations. Geographical gradient studies can help to reveal some of these large-scale effects without the need for long-term monitoring. In the present case study, different responses to pollution were observed in insect species, which depended to some degree on their feeding habits. Air pollution may cause unexpected and uncontrolled changes in population dynamics of present and potential pests. This extra variable in the complex system represents an increasing risk of outbreaks.

216   K. Heliövaara and R. Väisänen

Acknowledgements

Financial support from the Finnish Acidification Project (HAPRO) is acknowledged.

References

ALSTAD, D. N., EDMUNDS, G. F., JR. and WEINSTEIN, L. H. (1982). Effects of air pollutants on insect populations. *Annual Review of Entomology* **27**, 369–384.

BALTENSWEILER, W. (1985). 'Waldsterben': forest pests and air pollution. *Journal of Applied Entomology* **99**, 77–85.

BRAUN, S. and FLÜCKIGER, W. (1984). Increased population of the aphid *Aphis pomi* at a motorway: Part 1. Field evaluation. *Environmental Pollution A* **33**, 107–120.

COBB, F. W., JR., WOOD, D. L., STARK, R. W. and MILLER, P. R. (1968a). Effect of injury upon physical properties of oleoresin, moisture content, and phloem thickness. *Hilgardia* **39**, 127–134.

COBB, F. W., JR., WOOD, D. L., STARK, R. W. and PARMETER, J. R., JR. (1968b). Theory on the relationships between oxidant injury and bark beetle infestation. *Hilgardia* **39**, 141–152.

DOHMEN, G. P. (1988). Indirect effects of air pollutants: changes in plant/parasite interactions. *Environmental Pollution* **53**, 197–207.

FÜHRER, E. (1985). Air pollution and the incidence of forest insect problems. *Journal of Applied Entomology* **99**, 371–377.

GUNNARSSON, B. (1988). Spruce-living spiders and forest decline; the importance of needle-loss. *Biological Conservation* **43**, 309–319.

HAIN, F. P. (1987). Interactions of insects, trees and air pollutants. *Tree Physiology* **3**, 93–102.

HELIÖVAARA, K. and VÄISÄNEN, R. (1986a). Industrial air pollution and the pine bark bug, *Aradus cinnamomeus* Panz. (Het., Aradidae). *Journal of Applied Entomology* **101**, 469–478.

HELIÖVAARA, K. and VÄISÄNEN, R. (1986b). Parasitization in *Petrova resinella* (Lepidoptera, Tortricidae) galls in relation to air pollutants. *Silva Fennica* **20**, 233–236.

HELIÖVAARA, K. and VÄISÄNEN, R. (1989a). Reduced cocoon size of diprionids (Hymenoptera) reared on pollutant affected pines. *Journal of Applied Entomology* **107**, 32–40.

HELIÖVAARA, K. and VÄISÄNEN, R. (1989b). Interactions among herbivores in three polluted pine stands. *Silva Fennica* **22**, 283–292.

HELIÖVAARA, K. and VÄISÄNEN, R. (1989c). Quantitative variation in the elemental composition of pine needles along a pollutant gradient. *Silva Fennica* **23**, 1–11.

HELIÖVAARA, K. and VÄISÄNEN, R. (1990a). Between-species differences in heavy metal levels in four diprionids (Hymenoptera) along an air pollutant gradient. *Environmental Pollution* **62**, 253–261.

HELIÖVAARA, K. and VÄISÄNEN, R. (1990b). Heavy metal contents in pupae of the pine moths *Bupalus piniarius* and *Panolis flammea* (Lepidoptera: Geometridae and Noctuidae) near an idustrial source. *Environmental Entomology*, in press.

HELIÖVAARA, K. and VÄISÄNEN, R. (1990c). Prolonged development in *Diprion pini* (Hymenoptera, Diprionidae) reared on pollutant affected pines. *Scandinavian Journal of Forest Research* **5**, 127–131.

HELIÖVAARA, K., VÄISÄNEN, R., BRAUNSCHWEILER, H. and LODENIUS, M. (1987). Heavy metal levels in two biennial pine insects with sap-sucking and gall-forming life-styles. *Environmental Pollution* **48**, 13–23.

HELIÖVAARA, K., VÄISÄNEN, R. and KEMPPI, E. (1989). Change of pupal size of *Panolis flammea* (Lepidoptera; Noctuidae) and *Bupalus piniarius* (Geometridae) in response to concentration of industrial air pollutants in their food plant. *Oecologia* **79**, 179–183.

HELIÖVAARA, K., VÄISÄNEN, R. and VARAMA, M. (1990). Fecundity and egg viability in relation to female body size in *Neodiprion sertifer* (Hymenoptera: Diprionidae). *Holarctic Ecology*, in press.

HUTTUNEN, S. (1985). Interactions of disease and other stress factors with atmospheric pollution. In: *Air Pollution and Plant Life*, pp. 321–356 (ed. M. Treshow). Chichester: John Wiley.

KATAYEV, O. A., GOLUTVIN, G. I. and SELIKHOVKIN, A. V. (1983). Changes in arthropod communities of forest biocoenoses with atmospheric pollution. *Entomological Review* **62**, 20–35.

LARSSON, S. (1990). Stressful times for the plant stress-insect performance hypothesis. *Oikos* **56**, 277–283.

LECHOWICZ, M. J. (1987). Resource allocation by plants under air pollution stress: Implications for plant–pest–pathogen interactions. *Botanical Review* **53**, 281–300.

MATTSON, W. J. and HAACK, R. A. (1987). The role of drought stress in provoking outbreaks of phytophagous insects. In: *Insect Outbreaks*, pp. 365–407 (eds. P. Barbosa and J. C. Schultz). San Diego: Academic Press.

MILLER, P. R., COBB, F. W., JR. and ZAVARIN, E. (1968). Effect of injury upon oleoresin composition, phloem carbohydrates, and phloem pH. *Hilgardia* **39**, 135–140.

NEUVONEN, S. and LINDGREN, M. (1987). The effect of simulated acid rain on performance of the aphid *Euceraphis betulae* (Koch) on silver birch. *Oecologia* **74**, 77–80.

OLOFSSON, E. (1987). Mortality factors in a population of *Neodiprion sertifer* (Hymenoptera: Diprionidae). *Oikos* **48**, 297–303.

SIERPINSKI, Z. (1984). Über den Einfluss von Luftverunreinigungen auf Schadinsekten in polnischen Nadelbaumbeständen. *Forstwissenschaftliches Centralblatt* **103**, 83–91.

SMITH, W. H. (1981). *Air Pollution and Forest Interactions between Air Contaminants and Forest Ecosystems*, 379 pp. New York: Springer-Verlag.

SMITH, W. H. (1984a). Ecosystem pathology: a new perspective for phytopathology. *Forest Ecology and Management* **9**, 193–219.

SMITH, W. H. (1984b). Effects of regional air pollutants on forests in the USA. *Forstwissenschaftliches Centralblatt* **103**, 48–61.

STARK, R. W., MILLER, P. R., COBB, F. W., JR., WOOD, D. L. and PARMETER,

J. R., JR. (1968). Incidence of bark beetle infestation in injured trees. *Hilgardia* **39**, 121–126.

TEMPLIN, E. (1982). On the population dynamics of several pine pests in smoke-damaged forest stands. *Wissenschaftliche Zeitschrift der Technischen Universität Dresden* **11**, 631–637.

THALENHORST, W. (1974). Investigations on the influence of fluor-containing air pollutants upon the susceptibility of spruce plants to the attack of gall aphids *Sacchiphantes abietis* (L.). *Zeitschrift für Pflanzenkrankheiten und Pflanzenschutz* **81**, 717–727.

WHITE, T. C. R. (1984). The abundance of invertebrate herbivores in relation to the availability of nitrogen in stressed food plants. *Oecologia* **63**, 90–105.

Insect–Natural Enemy Interactions

20

Evaluating the Role of Natural Enemies in Latent and Eruptive Species: New Approaches in Life Table Construction

PETER W. PRICE

Department of Biological Sciences, Northern Arizona University, Flagstaff, Arizona 86011-5640 USA

Introduction: historical perspective on insect herbivore population dynamics

Almost 100 years ago, Howard (1897, p. 48) presented a very clear impression of the role of natural enemies in forest insect population dynamics: 'Wherever a plant-feeding species from some cause or from some combination of causes transcends its normal abundance to any great extent, there is always a great multiplication of its natural enemies, and this multiplication is usually so great as to reduce the species to a point even below its normal.' And a little later: 'With all very injurious larvae, however, we constantly see a great fluctuation in numbers, their parasites rapidly increasing immediately after the increase of the host species, overtaking it numerically and reducing it to the bottom of another ascending period of development.' Howard clearly argued that natural enemies were responsible for regulation of the herbivore population.

Indeed, Lotka (1924) was impressed by Howard's studies and quoted the second passage in support of his formalized views of host and parasitoid interactions. These have become known as the Lotka–Volterra predator–prey equations, stimulating enormous and prolonged research energy, and bolstering the view that natural enemies play a major role in regulating insect herbivore populations. This view persists today, and I would like to call this the 'Howard–Lotka Hypothesis' on insect herbivore population regulation.

Population Dynamics of Forest Insects
© Intercept Ltd, PO Box 716, Andover, Hampshire, SP10 1YG, UK

Another important step in understanding insect population dynamics was developed by Morris and Miller (1954) who formalized the approach to gathering field data in order to construct life tables on insect herbivores and formulated analytical methods for recognizing regulating factors. Because the life table started with the cohort of eggs in the tree, there was a built-in bias to concentrate on mortality factors impinging on this cohort (Price *et al.*, 1990). The 'killing power' of a factor, used by Varley and Gradwell (1960, 1968) in their K-factor analysis, captured concisely this emphasis. Natural enemies were commonly observed to be important mortality factors, for example appearing in predictive equations for spruce budworm, *Choristoneura fumiferana* (Clemens) (Morris, 1963, 1969), and this reinforced the view that enemies performed a significant regulatory function in insect herbivore population dynamics. The concept that life table construction and analysis as practiced commonly since 1954 is an adequate approach for understanding forest insect population dynamics, I will call the 'Morris–Miller Hypothesis'.

Studies in my research group have forced upon us views on some herbivore species that counter the Howard–Lotka Hypothesis and the Morris–Miller Hypothesis. Working with a latent, non-eruptive shoot-galling sawfly, *Euura lasiolepis* Smith, an alternative hypothesis has emerged (Price *et al.*, 1990):

1. Plant resources are the most important limiting factor for sawflies and define a carrying capacity that is typically low and stable;
2. Female sawflies will not oviposit in low-quality resources, they retain eggs in the absence of adequate resources, and presumably emigrate in search of adequate resources. Female behaviour in response to plant quality is the most important factor in defining population size and variation;
3. Rapid negative feedback through competition among females for scarce oviposition sites, acting at low population densities, results in stable populations close to the carrying capacity;
4. Variable plant resistance is the second most important factor in population dynamics;
5. Each species of natural enemy has idiosyncratic traits that dictate a passive role in terms of herbivore population regulation. Together they play little or no part in defining herbivore densities or variation in density;
6. This behavioural and ecological scenario is dictated by very old evolved characters in the phylogeny of the sawfly which are ultimately responsible for the proximate ecological factors seen in population dynamics. I will call this the 'Price, Craig and Preszler Hypothesis', although several others have contributed to the development of these arguments (for review see Price *et al.*, 1990). Brief support for this hypothesis is provided in the next section.

Population dynamics of the shoot-galling sawfly

THE SAWFLY AND ITS RESOURCES

Although thousands of shoots are produced each spring in a stand of willows, *Salix lasiolepis* Bentham, only the most rapidly growing shoots, which eventually become the longest, are successfully attacked by the sawfly. These are rare in a population of shoots, but are frequently all attacked (Craig , Price and Itami, 1986; Craig, Itami and Price, 1989a; Price and Clancy, 1986a,). On long shoots larvae survive better, a factor selecting for strong female behavioural preference for long shoots during oviposition (Craig, Itami, and Price, 1989a). Even at low densities of females they compete for oviposition sites on the best shoots and avoid oviposition sites left by previous females (Craig, Itami and Price, 1988, 1989b). This competition between females provides rapid negative feedback that holds densities close to the low carrying capacity. In evolutionary time it seems that this sawfly has met an insurmountable barrier to exploiting shorter shoots, although the nature of this barrier is unknown. Now females persist in avoiding shorter shoot classes, do not lay eggs in poor quality shoots, and presumably have evolved the option of emigration with some probability that new and adequate resources will be colonized. A similar option seems to play a role in the cassava mealybug, *Phenacoccus manihoti* Matile-Ferrero (van Alphen *et al.*, 1989).

Young vigorous plants, or older plants with regrowth after damage, provide a larger carrying capacity for sawflies than older undamaged plants (Craig, Price and Itami 1986; Craig, Itami and Price, 1989a). In wet sites or after winters with heavy precipitation willows grow better than in dry sites, providing more resources for

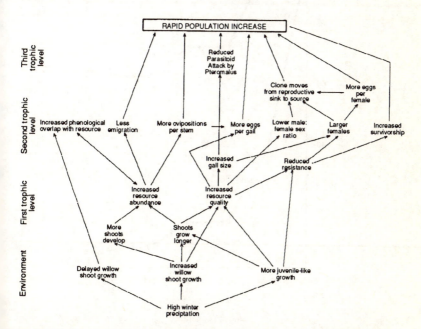

Figure 1. The cascading effects of plant quality on population dynamics on the higher trophic levels involving the shoot-galling sawfly, *Euura lasiolepis* and its natural enemies. Based on research cited in the text and unpublished results. The scenario illustrates beneficial effects of either higher than normal winter precipitation or willows growing near permanent water sources such as springs.

sawflies (Price and Clancy, 1986a; Preszler and Price, 1988). Hence, willows provide a heterogeneous array of resources in space and time based on age of willow clones, damage to clones, patchiness of wet sites, and variation in precipitation. Population size and dynamics are driven by this heterogeneity (*Figure 1*), where plant quality has cascading effects up the trophic system.

Experimental studies using potted willow plants under high and low water treatments have simulated field conditions effectively (Price and Clancy, 1986; Preszler and Price, 1988). Life tables were constructed to capture the female behavioural response to plant quality by starting the table with the cohort of eggs in the searching females. It then became immediately apparent that the maternal response to plant quality was the most important factor in dictating differences in population size between plants in high and low water treatments (*Figure 2*). After

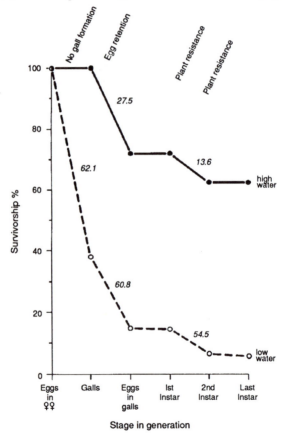

Figure 2. Survivorship curves for sawflies on willows with high-water treatment and low-water treatment, with the causes of loss from the cohort included. The cohort of eggs starting the survivorship curve represents eggs in searching females in order to record the effects of female decisions to lay or withhold eggs in response to plant quality. Based on data in Preszler and Price (1988).

female behaviour, the next most important factor was host plant resistance. Newly emerged larvae died in water-stressed plants causing 54.5% interval loss of the cohort in addition to the 85.1% loss due to maternal response to plant quality (*Figure 2*). All this loss to the cohort occurred before any natural enemies attacked the generation. Clearly, the population dynamics of the sawfly were driven by site and plant quality.

THE ROLE OF NATURAL ENEMIES

We have now studied in some detail all natural enemies of the sawfly: parasitoids, ants and predatory birds. Neither separately nor in concert do they change the fundamental dynamics dictated by the plant–herbivore interaction.

A three-year caging study by Robert Woodman in my research group, lasting three sawfly generations, showed no impact by all parasitoids combined on sawfly population size (*Figure 3*). This is because small parasitoids like pteromalids and eurytomids cannot reach larvae in large galls produced on vigorous plants (Price and Clancy, 1986b). Hence, when plants are most favourable for sawflies, small

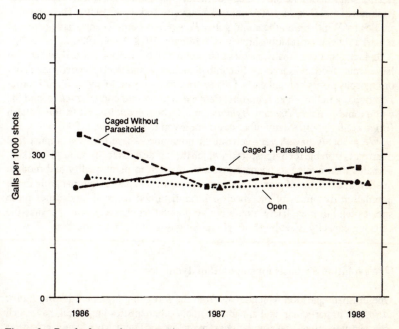

Figure 3. Results from a three-generation caging study on the effects of the parasitoid community on abundance of *Euura lasiolepis* galls, conducted by Robert Woodman. Each of six willow clones was divided into a caged third excluding parasitoids, another third with parasitoids, and another open and uncaged. No significant differences were observed among treatments.

parasitoids are least effective, resulting in a negatively density-dependent response of parasitoids to host density in space (Price, 1988). For larger ichneumonid parasitoids, such as *Lathrostizus euurae* (Ashmead), galls toughen too rapidly for extensive attack, seriously restricting the response of parasitoids (Craig, Itami and Price, 1990).

The major predators, ants and mountain chickadees (*Parus gambeli* Ridgway), commonly attack only in wet sites where sawfly densities are high, with a capacity to compensate for predation because of reduced competition for oviposition sites.

How common is this dynamical pattern?

The general dynamical properties we have described for *Euura lasiolepis* also apply to related sawflies. Although in less detail, we have established this for the bud galler, *Euura mucronata* (Hartig) Man. (Churchill) (Price, Roininen and Tahvanainen, 1987a,b; Roininen, Price and Tahvanainen, 1988), the shoot galler, *Euura exiguae* Smith (Price, 1989), and an undescribed midrib galler (unpublished data). At least 14 species of sawfly gallers probably fit this pattern because they respond positively to shoot length of the host plant (Price *et al.*, 1990). Similar relationships dominate the ecology of the cynipid gallers *Diplolepis fusiformans* (Ashmead) and *D. spinosa* (Ashmead) (Caouette and Price, 1989). The extensive studies by Whitham on the aphid galler *Pemphigus betae* Doane, have revealed remarkably similar relationships (e.g. Whitham, 1978, 1979, 1980).

In fact, females of many species of insect herbivore oviposit directly where the larvae must feed and survive, often into plant tissue, and tend to select young and/or vigorous plants or plant parts. Larvae are commonly endophytic. Shoot-boring Lepidoptera in the genus *Eucosma*, *Petrova*, *Rhyacionia* and *Dioryctria*, and the curculionid genera *Pissodes*, *Hylobius* and *Cylindrocopturus* all fit this pattern (Price *et al.*, 1990). Many other cases are given in Price *et al.* (1990).

We probably have a very general phenomenon, where herbivores must find plant parts suitable for oviposition in which the larvae must feed and survive *in situ*. Females then evolve to become selective relative to plant quality and module quality, and the role of plant resource quality comes to play a dominant role in the population dynamics of the species. In undisturbed forested landscapes these species tend to have latent, non-eruptive population characteristics because the carrying capacity is set by relatively rare high-quality plant resources.

The evolutionary basis for population dynamics

One ultimate question which arises from the foregoing argument is why *Euura lasiolepis* in particular, and apparently many other species in general, have such a restrictive ecology that populations remain low and latent. My research group has developed an answer, summarized in *Figure 4* (see also Price *et al.*, 1990; Price 1990). Briefly, ancient evolved phylogenetic constraints dictate the way in which the herbivore relates to the host plant. This relationship develops through a suite of adaptive responses to these constraints, which in aggregate we call the 'adaptive

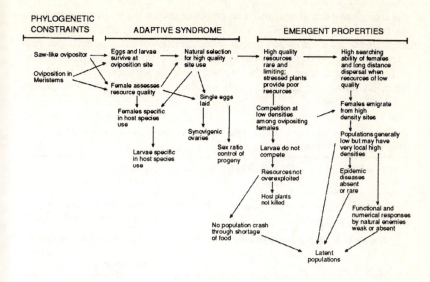

Figure 4. Relationship among phylogenetic constraints, the adaptive syndrome, and emergent properties, resulting in typically latent populations of *Euura lasiolepis*.

syndrome'. These evolved characters of the species then dictate the general ecology of the species and its population dynamics: the 'emergent properties'. Thus, the population dynamics of a species must be nested within the evolved characteristics of the species in order for us to understand pattern and process. I would argue that proponents of the Morris–Miller Hypothesis have concentrated on proximate ecological factors and contemporary time, without addressing the ultimate evolutionary forces that dictate ecology.

Another ultimate question is how eruptive insect herbivore species have escaped the ecological strictures imposed on latent species. The answer again follows an argument based on phylogenetic constraints, adaptive syndromes and emergent properties (Price *et al.* 1990; Price, 1990). We have used the spruce budworm as an example, but the scenario, if generalized, fits eruptive forest Lepidoptera very well. The fundamental evolved differences from latent species is that the female does not evolve to evaluate host plant quality for larvae. Females oviposit where the larvae do not feed, or are relatively indiscriminate during oviposition because they lay large clutches of eggs, even though they may show discrimination regarding host plant species, host plant size, and some other plant characteristics. Some examples are given in *Table 1*. As a result larvae evolve to have a generalized capacity to mature on a wide variation of foliage quality, resulting in a high carrying capacity in mature stands of trees. The evolved basis for population dynamics is permissive in the sense that there is the ecological capacity for very high densities of herbivores to develop periodically (see Price *et al.*, 1990, for details). This scenario does not account for genera containing eruptive species that also include non-eruptive species. Without detailed studies, the best explanation seems to be that species in such genera are potentially eruptive, but for some reason special factors overpower this potential. The most likely factor would be natural enemies as in cases discussed in the next section.

I would argue, then, that the evolved basis for differences in population dynamics among latent and eruptive species has nothing to do with natural

Table 1.   Examples of forest insects in the Lepidoptera and Hymenoptera that exhibit eruptive population dynamics.

Herbivore species	Host plant genus	Feeding location	Oviposition site selection and larval feeding linkage
1. *Operophtera brumata* (L)	*Quercus* etc.	On leaves	Poor. Females wingless, lay eggs in winter when foliage is absent.
2. *Lymantria dispar* (L)	*Quercus* etc.	On leaves	Poor. Oviposition on bark, females do not fly, larvae disperse passively.
3. *Malacosoma disstria* Huber	*Populus*	On leaves	Poor. Oviposition on bark, larva feeds following spring.
4. *Alsophila pometaria* (Harr)	*Quercus* etc.	On leaves	Poor. Female wingless, oviposition in winter, passive dispersal of larvae.
5. *Neodiprion swainei* Middleton	*Pinus*	On leaves	Poor. Female poor flyer, proovigenic with heavy load of eggs, oviposition on young needles, larvae feed on old needles
6. *Orgyia pseudotsugata* (McDunnough)	*Pseudotsuga*	On leaves	Poor. Clusters of 150 eggs laid on female's cocoon, larvae disperse passively.
7. *Phryganidia californica* Packard	*Quercus*	On leaves	Poor. Female weak flyer, 60 eggs per mass not necessarily on leaves larvae will feed on.
8. *Coloradia pandora* Blake	*Pinus*	On leaves	Poor. Eggs in clusters on needles or bark. Most feeding in second year of life cycle.
9. Paleacrita vernata (Peck)	*Ulmus* etc.	On leaves	Poor. Females wingless. Eggs under bark, before foliage is available.
10. Lambdina punctata (Hulst)	*Quercus*	On buds & leaves	Poor. Eggs laid in leaf litter, and under bark scales.

enemies, and female oviposition behaviour in response to plant resources is central. The role of natural enemies is an epiphenomenon or an emergent property, rather distant from the central driving forces resulting in the differences in dynamical types.

The continuum of impact by natural enemies

How, then, do we evaluate the role of natural enemies in forest insect populations? How do we find pattern in where enemies should have weak to strong influence? How do we develop predictions that can be tested on the role of natural enemies? The answers are not simple.

In any geographic region a continuum of impact by natural enemies probably exists. My research group has established that natural enemies play an insignificant role in the population dynamics of *Euura lasiolepis*. This is probably true for many related species, and many other latent species. Many such species respond positively to young, vigorous plants that will occur in early woody-plant succession in small disturbed patches of primaeval forest. Such herbivore populations will be widely scattered and hard to find for natural enemies as well as being endophytic and time-consuming for attack. Many attributes of latent species conspire against an active role for natural enemies in the population dynamics of their hosts. On the other hand, there is good evidence for some herbivore species that natural enemies are of paramount importance in the population dynamics of their hosts. Biological control programmes on exotic insect herbivores, re-associating them with natural enemies from their native location, offer particularly convincing evidence. One such case concerns the Mountain Ash Sawfly, *Pristiphora geniculata* (Hartig), which in Europe has been rare because of heavy parasitoid impact (Eichhorn and Pschorn-Walcher, 1978). Escaping natural enemies in Canada it became epidemic, but introduction of natural enemies from Europe resulted in a rare herbivore in its new location (Quednau, 1984). A similar case concerns the cassava mealybug (van Alphen *et al.*, 1989). Thus, in any one region we can probably find a continuum of effects by natural enemies from innocuous to devastating. Insect herbivores in mature forests live in a stable environment over a prolonged period and are likely to become abundant periodically. Thus, there is a greater probability of natural enemies having a significant impact on herbivore populations. However, in most species we do not yet know their real role in population dynamics (Price 1987). Are they an epiphenomenon on the dynamics, as in *Euura lasiolepis* and bark beetles (Berryman, 1982) or are they central to dynamics? Much more longer-term experimental work is required to resolve this question.

To what extent can the 'ghost of enemies past' be invoked as an alternative hypothesis to account for the basic differences between latent and eruptive species? I do not see how natural enemies could drive species into the extremes of the continuum. Someone else needs to formulate a detailed mechanistic explanation if it is to compete against the one presented here. Herbivores relate to host plants across 100% of the insect population, so plants act as a potent selective force. Even if natural enemies are common they cannot match this power of selection.

With a new emphasis in life table construction that captures the female's behavioural response to plant quality by direct measurement we will certainly place in better perspective the relative importance of ovipositional behaviour, plant quality and natural enemies. In the past, ovipositional responses of females to plant quality have been measured indirectly and crudely, and have commonly been mixed with other potential factors involved with population dynamics, such as adult mortality and emigration; compare life tables on winter moth (Varley, Gradwell and Hassell, 1973) and spruce budworm (Morris and Miller, 1954). With enough new life tables on many species, both latent and eruptive, perhaps patterns will emerge. Perhaps female behaviour and plant quality will be critical for latent endophagous species, and natural enemies will be critical for eruptive exophagous species, although I doubt that a simple dichotomy exists.

Recognizing that the ecology of a species is strongly affected by its evolutionary history, and that the core of this history lies in the plant–herbivore relationship, provides indications of predictive patterns on latent and eruptive species beyond that which can be revealed by life table analysis. More analysis will reveal how general these patterns are, but it will be much harder to generalize about the role of natural enemies as emergent properties of the system. I suspect we will need to recognize many idiosyncratic characteristics of natural-enemy communities which must depend on long-term phylogenetic inertia in co-speciating hosts and parasitoids, many ecological opportunities through time for host shifting, and current ecological factors. To understand pattern and process in the role of natural enemies on forest insects is a great challenge, but I am optimistic that a strong evolutionary approach to population dynamics, coupled with revised life table construction, will yield real knowledge in the long term.

Acknowledgements

I thank the organizers of the conference for the opportunity to present and publish these ideas; my present and former students for providing me with a unique education, and the US National Science Foundation for research support through grants DEB-8021754, BSR-8314594 and BSR-8705302.

References

BERRYMAN, A. A. (1982). Population dynamics of bark beetles. In: *Bark Beetles in North American Conifers* pp. 264–314, (eds J. B. Mitton and K. B. Sturgeon). Austin: University of Texas Press.

CAOUTTE, M. R. and PRICE, P. W. (1989). Growth of Arizona rose and attack and establishment of gall wasps, *Diplolepis fusiformans* (Ashmead) and *D. spinosa* (Ashmead) (Hymenoptera: Cynipidae). *Environmental Entomology* **18**, 822–28.

CRAIG, T. P., PRICE, P. W. and ITAMI, J. K. (1986). Resource regulation by a stem-galling sawfly on the arroyo willow. *Ecology* **67**, 419–25.

CRAIG, T. P., ITAMI, J. K. and PRICE, P. W. (1988). Plant wound compounds from oviposition scars used as oviposition deterrents by a stem-galling sawfly. *Journal of Insect Behavior* **1**, 343–56.

CRAIG, T. P., ITAMI, J. K. and PRICE, P. W. (1989a). A strong relationship between oviposition preference and larval performance in a shoot-galling sawfly. *Ecology*, **70**, 1691–1699.

CRAIG, T. P., ITAMI, J. K. and PRICE, P. W. (1989b). Intraspecific competition and facilitation by a shoot-galling sawfly. *Journal of Animal Ecology* **59**, 147–159.

CRAIG, T. P., ITAMI, J. K. and PRICE, P. W. (1990). The window of vulnerability of a shoot-galling sawfly to attack by a parasitoid. *Ecology*, (in press).

EICHHORN, O. and PSCHORN-WALCHER, H. (1978). Biologie und Parasiten der Ebereschen-Blattwespe, *Pristiphora geniculata* Htg. (Hym: Tenthredinidae). *Zeitschrift für Angewandte Entomologie* 85, 154–167.

HOWARD, L. O. (1897). A study of insect parasitism: a consideration of the parasites of the white-marked tussock moth, with an account of their habits and interrelations and with descriptions of new species. *United States Department of Agriculture Bureau of Entomology Technical Series* 5, 1–57.

LOTKA, A. J. (1924). *Elements of Physical Biology*. Baltimore: Williams and Wilkins.

MORRIS, R. F. (ed.) (1963). The dynamics of epidemic spruce budworm populations. *Memoirs of the Entomological Society of Canada* 31, 1–332.

MORRIS, R. F. (1969). Approaches to the study of population dynamics. In: *Forest Insect Population Dynamics*, pp. 9–28 (ed. W. E. Walters). Washington DC: United States Forest Service Research Paper NE125.

MORRIS, R. F. and MILLER, C. A. (1954). The development of life tables for the spruce budworm. *Canadian Journal of Zoology* 32, 283–301.

PRESZLER, R. W. and PRICE, P. W. (1988). Host quality and sawfly populations: A new approach to life table analysis. *Ecology* 69, 2012–2020.

PRICE, P. W. (1987). The role of natural enemies in insect populations. In *Insect Outbreaks*, pp. 287–312 (eds P. Barbosa, and J. C. Schultz). New York: Academic Press.

PRICE, P. W. (1988). Inversely density-dependent parasitism: The role of plant refuges for hosts. *Journal of Animal Ecology* 57, 89–96.

PRICE, P. W. (1989). Clonal development of coyote willow, *Salix exigua* (Salicaceae), and attack by the shoot-galling sawfly, *Euura exiguae* (Hymenoptera: Tenthredinidae). *Environmental Entomology* 18, 61–68.

PRICE, P. W. (1990). Insect herbivore population dynamics: is a new paradigm available? In: *Insect–Plant Relationships* (eds T. Jermy and A. Szentesi). Budapest: Hungarian Academy of Sciences, (in press).

PRICE, P. W. and CLANCY, K. M. (1986a). Multiple effects of precipitation on *Salix lasiolepis* and populations of the stem-galling sawfly, *Euura lasiolepis*. *Ecological Research* 1, 1–14.

PRICE, P. W. and CLANCY, K. M. (1986b). Interactions among three trophic levels: Gall size and parasitoid attack. *Ecology* 67, 1593–1600.

PRICE, P. W., ROININEN, H. and TAHVANAINEN, J. (1987a). Plant age and attack by the bud galler, *Euura mucronata*. *Oecologia* 73, 334–337.

PRICE, P. W., ROININEN, H. and TAHVANAINEN, J. (1987b). Why does the bud-galling sawfly, *Euura mucronata*, attack long shoots? *Oecologia* 74, 1–6.

PRICE, P. W., COBB, N., CRAIG, T. P., FERNANDES, G. W., ITAMI, J. K., MOPPER, S. and PRESZLER, R.W. (1990). Insect herbivore population dynamics on trees and shrubs: new approaches relevant to latent and eruptive species and life table development. In: *Insect–Plant Interactions*, vol. 2, (ed. E. A. Bernays). Boca Raton; CRC Press (in press).

QUEDNAU, F. W. (1984). *Pristiphora geniculata* (Htg.), mountain-ash sawfly (Hymenoptera: Tenthredinidae). In: *Biological Control Programmes against*

Insects and Weeds in Canada, 1969–1980, pp. 381–385 (eds J. S. Kelleher and M. A. Hulme). Farnham: Commonwealth Agricultural Bureaux.

ROININEN, H., PRICE, P. W. and TAHVANAINEN, J. (1988). Field test of resource regulation by the bud-galling sawfly, *Euura mucronata*, on *Salix cinerea*. *Holarctic Ecology* **11**, 136–139.

VAN ALPHEN, J. J. M., NEUENSCHWANDER, P., VAN DIJKEN, M. J., HAMMOND, W. N. O. and HERREN, H. R. (1989). Insect invasions: the case of the cassava mealybug and its natural enemies evaluated. *Entomologist* **108**, 38–55.

VARLEY, G. C. and GRADWELL, G. R. (1960). Key factors in population studies. *Journal of Animal Ecology* **29**, 399–401.

VARLEY, G. C. and GRADWELL, G. R. (1968). Population models for the winter moth. In: *Insect Abundance*, pp. 132–142 (ed. T. R. E. Southwood). Oxford: Blackwell Scientific.

VARLEY, G. C., GRADWELL, G. R. and HASSELL, M. P. (1973). *Insect Population Ecology: An Analytical Approach*. Oxford: Blackwell Scientific.

WHITHAM, T. G. (1978). Habitat selection by *Pemphigus* aphids in response to resource limitation and competition. *Ecology* **59**, 1164–1176.

WHITHAM, T. G. (1979). Territorial behaviour of *Pemphigus* gall aphids. *Nature* **279**, 324–325.

WHITHAM, T. G. (1980). The theory of habitat selection: Examined and extended using *Pemphigus* aphids. *American Naturalist* **115**, 449–466.

21
Gypsy Moth Predators: An Example of Generalist and Specialist Natural Enemies

RONALD M. WESELOH

Connecticut Agricultural Experiment Station, New Haven, Connecticut, USA

Introduction

The gypsy moth, *Lymantria dispar* L., is a serious defoliating pest of deciduous trees in both old and new worlds. Its impact is generally greatest in North America, where it was inadvertently introduced during the last century. Wherever the moth occurs, one of its most characteristic features is a propensity to vary between very high numbers, with several hundred thousand to millions of insects per ha and very low numbers, with less than 100 larvae/ha. In this it is not alone. There are many forest insects that have such fluctuating populations and many of them are Lepidoptera (Nothnagle and Schultz, 1987). The mechanisms that drive these fluctuations in gypsy moths and others are largely unknown, although climate, food quality, and even sun spots, have been suggested. Natural enemies have also been considered as important and, at least for the gypsy moth, in recent years it has become increasingly clear that predators are significant mortality factors, particularly of low-level populations (Smith and Lautenschlager, 1981). Researchers have begun to clarify the interactions between these predators and the pest, although many details remain unknown. Thus the gypsy moth system is a model with some relevance to forest insect dynamics in general. Using the gypsy moth as an example, I will show the importance that generalist predators can have on low-density forest pest populations, and contrast this with the actions of a specialist predator at high-density populations. Most of the research has been conducted in North America, so that region will be the focus of my comments.

Population Dynamics of Forest Insects
© Intercept Ltd, PO Box 716, Andover, Hampshire, SP10 1YG, UK

Generalist predators

Generalist predators include small woodland mammals such as mice, voles and shrews; birds, spiders and phalangids, and insects such as ground beetles and ants. These predators have wide feeding preferences and, because the preferred gypsy moth larval and pupal stages are only present for about two months of the year, none can survive on gypsy moths alone. Their reproductive success is largely dependent on alternative foods, and thus they have only functional responses to increases in pest populations. Because functional responses are inherently limited in extent, none of these predators has a significant impact on gypsy moth populations when the latter are high. However, because generalist predators are not dependent on gypsy moths for survival, they can cause significant mortality when moth population densities are very low.

At all population levels, gypsy moth instars one to three remain on leaves or twigs continuously unless they are accidently dislodged, as do large larvae during outbreaks. However, in areas having low-density populations, instars four to six migrate up and down trees daily, resting during the day in leaf litter and on tree trunks and feeding at night on leaves (Leonard, 1970). They also often pupate in the litter (Campbell, Hubbard and Sloan, 1975). These daily movements complicate the assessment of predation, but also mean that many kinds of ground-dwelling vertebrate and invertebrate predators have the opportunity to feed on gypsy moths.

PREDATION OF EGGS

Some information is available about predation of gypsy moth egg masses. Mason and Ticehurst (1984) noted that larvae of the dermestid, *Cryptorhopalum ruficorne*, feed on eggs in about 3% of gypsy moth egg masses in southern Pennsylvania, USA. Also in the USA, ants occasionally detach individual eggs from masses (Campbell, 1975, personal observation) but their impact is probably low. In contrast, birds are important predators of eggs in low-density populations in Hokkaido, Japan (Higashiura, 1980, 1989). They may destroy up to 80% of eggs during winters when snow levels are 1 m deep or more, evidently because other food sources are not available then. There is evidence that they function in a density-dependent manner as well (Higashiura, 1989).

PREDATION OF LARVAE AND PUPAE

At low gypsy moth population densities, vertebrate predators are probably the most important mortality agents of large larvae and pupae. This was dramatically shown by Campbell and Sloan (1977) in the USA. In an experiment replicated six times, they removed small mammals from some plots using a combination of live-trapping and snap-traps. Predation by birds was impeded by placing poultry netting around burlap bands on trees, under which large larvae tend to congregate and pupate. Survival of pupae and resulting egg mass numbers were higher in those plots where mammals (240 egg masses/ha) and birds and mammals (320 masses/ha) were removed than where birds only were excluded (20 masses/ha) or in check areas (40 masses/ha). However, survival rates of large larvae were greatest in plots where the activities of both mammals and birds were diminished. These results show conclusively that small mammals have an impact on low-density gypsy moth populations, but the importance of birds is unclear. The authors concluded that birds feed on large larvae, but small mammals destroy most pupae regardless of the number of larvae destroyed by birds. They suggest that the white-footed mouse, *Peromyscus leucopus*, is an important gypsy moth predator in North America because 59% of the mammals trapped were of this species.

Smith and Lautenschlager (1981) conclude that the most important predators of pupae are vertebrates, especially the white-footed mouse and shorttail shrew, *Blarina brevicauda*. Gypsy moths often pupate in dark tunnels of these mammals just below the litter and white-footed mice are agile tree climbers, making it easy for them to prey on the pest. However, the impact these mammals have on gypsy moth populations varies depending on the abundance of alternative foods (Smith and Lautenschlager, 1981). The white-footed mouse prefers smooth-bodied caterpillars or blueberries to gypsy moths, and more gypsy moths are consumed when such alternative foods are scarce. Any attempt to change the availability of alternative foods in order to modify predator impact must be done carefully so predators will have adequate nutrition for the months when gypsy moths are not available. However, supplements should not be so abundant that predators do not feed on large larvae and pupae when the latter are present.

By exposing pupae in the litter, Smith and Lautenschlager (1981) confirmed the importance of predation by mammals, which was two to eight times as great as by small invertebrates (predators of all kinds killed over 70% of pupae). Interestingly, pupae parasitized by the tachnid, *Blepharipa pratensis*, were eaten about three times as often as were healthy pupae. However, this behaviour is probably of little practical importance because at low population densities parasitization by this fly is usually low (personal observation). The most prominent invertebrate predators were carpenter ants (*Camponotus* spp.) and phalangids (*Leiobunum* spp.) (Smith and Lautenschlager, 1981).

Further information on the impact of mammalian predators has been gathered by Joe Elkinton and his colleagues at the University of Massachusetts, USA (personal communication). They noted that rates of predation of deployed pupae in the litter were positively correlated with abundance of the white-footed mouse. They also found that, in 1986, gypsy moth numbers increased dramatically in places in Massachusetts where mouse abundance was low. Such data provide additional evidence that small mammals may regulate the abundance of low density gypsy moth populations.

Obtaining information on predation of larvae has been more difficult than for pupae because larvae are mobile and their remains are not easily found. Birds from time to time have been implicated as important predators of larvae, but there is little conclusive data. As already mentioned, Campbell and Sloan (1977) performed experiments that indicated that birds affected survival rates of large larvae but no effect was noted in subsequent egg mass counts. Furuta (1982) in Japan found that mortality to gypsy moths caused by birds in a natural forest was high and density-dependent, and he stated that bird predation was the most important factor maintaining gypsy moth populations at low levels. While interesting, his data are largely anecdotal and do not prove regulation occurred. Whelan, Holmes and Smith (1989) found that captive birds of numerous species ate both early and late instar gypsy moth larvae, but preferred non-hairy caterpillars if given a choice. These researchers concluded that migrating, insectivorous birds may have substantial impacts on gypsy moth larval populations, but their results have not been validated under natural conditions. In field tests (Weseloh, 1988), I found that neither early or late instar larvae on tree leaves, particularly high in trees, were preyed on extensively (i.e. in locations where birds might be expected to forage). Predation by birds probably varies substantially because of environmental conditions and availability of alternative foods, and their impact will have to be determined in different situations before anything definitive can be said about their importance.

Polistes wasps have been shown to have a large effect on low-density gypsy moth larval populations in a study by Furuta (1983). He placed gypsy moth larvae on small forest trees near Kyoto, Japan and observed them. Wasps attacked mainly the large caterpillars, and tended to aggregate in plots having the most gypsy moths. Numbers of wasps observed were linearly correlated with disappearance rates of caterpillars, strongly suggesting that the wasps caused the mortality. Also, wasp success rates increased as caterpillar density increased, mainly because the predators had more chances to capture larvae at higher densities.

To obtain information about predation of gypsy moth larvae in the USA, I placed caterpillars tethered with thread in various forest microhabitats. A similar technique was used by Semevskiy (1973) in the Soviet Union. To obtain information on kinds of predators involved, I tethered some inside wire cages of various mesh (0.6, 1.6 and 2.5 cm spaces) (Weseloh, 1988). As already mentioned, larvae of any stage on leaves were not attacked by predators to any great extent. However, large caterpillars placed on tree trunks were heavily attacked by predators that could only enter the large mesh cages (presumably small mammals, such as the white-footed mouse, that climb trees). Those in the litter were removed by both large and small predators. These results are consistent with data on predation of pupae, which is highest in the litter (Campbell and Sloan, 1976).

By placing groups of tethered 5th instar larvae in litter and on tree trunks and checking them hourly for 24 h periods, I found that predation rates were similar both day and night. This means that tethered larvae need only be checked once a day, because predation rates can be corrected for migration by multiplying by the proportion of the day larvae are normally in each microhabitat. Ants were by far the most abundant invertebrate predators observed because they took many hours to subdue and eat prey. Their activity was inversely related to humidity levels in the forest, probably because they do not forage when the litter is wet. The ants mainly attacked larvae placed in the litter. Other predators (presumably vertebrates) removed larvae quickly between one observation time and the next, and they were not influenced by forest humidity (Weseloh, 1988). These (presumed) vertebrates were responsible for about 80% of the predation that occurred, compared to 20% for ants.

By using cages that would admit ants yet not let large caterpillars escape, I found that free-moving large caterpillars were destroyed by ants about half as often as tethered ones were (Weseloh, 1990a). By correcting for this bias, I estimated in one site that 12.1% of gypsy moth caterpillars were killed by predators per day. This resulted in an overall reduction during the larval stages of 99.84%. While variations in this rate occurred, second and third instars were no more or less vulnerable to predators than older ones were (Weseloh, 1990a). (While low, predation of early instars in trees, when coupled with predation of those that accidently fell to the forest floor, was enough to account for the rates on young larvae.) This is in contrast to the conclusions of Campbell (1976) and Elkinton (personal communication), who in life table studies found that mortality of large larvae was highest. Perhaps the relative importance of mortality to different larval stages differs at different sites and times, but this is not known.

I compared the estimated predation rate to the disappearance rate of caterpillars determined from larval numbers estimated at the start and near the end of the larval period (by mark-recapture techniques described in Weseloh, 1987). The daily disappearance rate was 12.9%, a value not much higher than the 12.1% predation rate (Weseloh, 1990a). Thus predation is confirmed for larvae as for pupae as the most important mortality factor in low density populations.

The possible involvement of ants as gypsy moth larval predators led me to more detailed studies of them (Weseloh, 1990b). Gypsy moth larvae of instars one to five were placed, one at a time, on litter in a forest and watched for periods of up to 1 h or until captured. For comparative purposes, instars one to three were used even though of the first three instars only neonates would likely be in forest litter in substantial numbers. Ants were the most numerous predators, making almost all contacts with larvae at an average of once every 20 min. Brown forest ants (*Formica neogagates* and *F. subsericea*) made over 80% of the contacts and had higher successful capture rates (17–20%) than any other ants, including carpenter ants (*Camponotus* spp.). Neonates were successfully captured only about 5% of the time, while the success rate on first instars that had fed was 56%. Larger caterpillars were successfully attacked at rates of 24% (second instar) to 0% (fifth instar). The striking difference between fed and unfed first instars is probably because fed first instars have gaps between body hairs due to stretched intersegmental membranes that ants can exploit. The protection afforded neonates is advantageous because they are dispersed by wind. Many fall to the forest floor over which they must crawl to a tree. If they were as vulnerable as feeding first instars that stay in trees, few would survive the journey.

Thus, it is clear from the above that the gypsy moth instars most vulnerable to ants are often not in the microhabitat that ants usually frequent. The latter apparently account for only a small fraction of the mortality that occurs on gypsy moth larvae. Fortunately the reasons for their ineffectiveness are becoming known and might be overcome. Encouraging ants to forage in trees would be one strategy, possibly by spraying with sugar solutions or encouraging non-damaging levels of aphids or other honeydew producers. Other strategies may become evident to those who understand the present limitations of these potentially important predators.

Generalist predators account for almost all the mortality the gypsy moth suffers at low population densities. This does not mean that they regulate moth populations. Conceivably, they may do no more than retard the growth of pest populations that will eventually and inevitably increase beyond the threshold of predator effectiveness. An outbreak would then be the result. However, with overall mortality being so large, small increases in predation rates may perturb prey populations downward, increasing the lengths of time between outbreaks or perhaps suppressing them altogether. The further study of generalist predators of the gypsy moth is likely to be rewarding and is encouraged.

Specialist predators

In contrast to generalist predators, there is only one specialist predator of gypsy moths for which detailed information is available—the carabid beetle, *Calosoma sycophanta* L. This insect, a native to Europe that was successfully imported into North America, is large and active. Both adults and larvae readily climb trees, and adults feed on caterpillars while immatures attack pupae. *C. sycophanta* has been said to be the primary cause for some gypsy moth declines (Bess, 1961; Campbell, 1975). Its life cycle is given by Burgess (1911). The beetle appears to be active only when gypsy moth numbers are high. Adults overwinter in soil, and remain there until large gypsy moth larvae are available. They then emerge and feed, and females lay eggs in soil that hatch when gypsy moths are pupating. While immature beetles are feeding on pupae, adults enter the soil again to overwinter. Beetles live for at least three to four years.

Because adults are inactive about 11 months of the year they are able to survive on gypsy moths alone and, because of a high reproductive capacity (females may lay 200 or more eggs each summer), they can increase numerically as gypsy moth numbers increase. These characteristics are probably responsible for the beetle's impact when gypsy moths are numerous.

Vasic (1972) in Yugoslavia observed that beetles stay underground in years when gypsy moth numbers are low, and emerge to feed when prey populations are high. How they detect prey density is unknown. However, it is also clear that beetles are strong and agile fliers (Doane and Schaeffer, 1971), and conceivably they could disperse widely. Because *C. sycophanta* does not usually become abundant enough to have a substantial impact on gypsy moth populations until the second or third year of an outbreak, determining its behaviour between outbreaks and its responses to increasing prey numbers is an important research goal.

I am monitoring populations of *C. sycophanta* at two sites. Adult beetles captured in live traps (Collins and Holbrook, 1929) are marked by punching holes in their elytra and then released to obtain information about population levels, survival, and dispersal. Monitoring began during the later stages of an outbreak, at which time there were no more than 400 adult beetles per ha (Weseloh 1985a, b). This was low compared to prey density, which was probably over 100 000 large caterpillars per ha (Weseloh, 1985c). However, in at least one of the sites, beetle larvae destroyed up to 70% of the gypsy moth pupae on tree trunks near the ground and about 50% of all pupae (Weseloh 1985b). Movements of beetles during the outbreak appeared to be mainly local and random (Weseloh 1985a, b). In one site the following year, when gypsy moth numbers were low, beetle adults were still as abundant as before. However, they did not reproduce and dispersal was practically non-existent (Weseloh, 1985a).

These data are consistent with Vasic's (1972) studies, in which he found that beetles were inactive in years when pests were scarce. However, if beetles are able to sense when gypsy moth populations increase it is unclear why they usually have little impact during the first year of an outbreak. It is possible that many adult beetles die between outbreaks, which even in North America usually occur no less than seven to eight years apart. Thus when a new outbreak occurs, the few beetles left would require a year or two of reproduction before their impact becomes noticeable. If all beetles had died, others would need to migrate into the area, and this would also take time if it happened at all. Thus the effect of *C. sycophanta* on gypsy moth populations can be expected to be sporadic, and this does appear to characterize their population dynamics (personal observations).

A possibility for improving the beetle's effectiveness is based on the rather low numbers of adults present at any time during outbreaks and their apparent low dispersal tendencies. Small numbers of beetles could be released in forests where gypsy moths are increasing rapidly, with the hope that many progeny would be produced. Such releases would also be a way of assessing the impact beetles have on the pest. I carried out preliminary releases in 1988 with promising results. Survival rates of caterpillars were higher in plots in which none or only 10 beetles were released, and lower where 20 or 40 beetles were released. However, mortality of pupae by immature beetles was not related to beetle numbers, probably because released beetles did not reproduce (Weseloh, submitted).

A larger, replicated experimental release was done in 1989. Before release, beetles were fed gypsy moth larvae in the laboratory for one week in order to provide females with enough protein to produce eggs. Probably as a result of this regime, pupal mortality and numbers of immature beetles were higher in release plots than in check plots. Thus released beetles did reproduce (Weseloh, submitted). Sampling will continue for one more year to assess long term effects of the releases.

Discussion

Predators that attack the gypsy moth are important at both low and high densities of the pest. At low densities, generalist predators are by far the most prevalent mortality agents. Their impact depends on many factors, particularly the availability of alternative foods and their own numbers.

Their interactions with other mortality agents may be complex. As already stated, the preference of white-footed mice for gypsy moth pupae parasitized by the tachinid parasitoid, *Blepharipa pratensis*, may have little importance for gypsy moth population dynamics. However, other parasitoids, in particular the tachinids *Compsilura concinnata* and *Parasitigena sylvestris*, are often prominent when gypsy moth populations are low (personal observations). Any unknown preferences that generalist predators have for such parasitized larvae may complicate attempts to understand gypsy moth population dynamics.

There is still an imperfect understanding of how these predators respond to changes in prey populations, and thus prey regulatory mechanisms, if they exist, are not known. Factors that cause gypsy moth populations to outbreak have not been linked causally to these predators. Finding what these factors are (and there could be many, such as climate, alternative foods, and activities of humans, that vary in time and space) would aid greatly in understanding the dynamics of gypsy moth populations and is probably the most important research priority involving generalist predators.

Much the same situation exists for other forest Lepidoptera. Price (1987) summarized conclusions of six studies on four species for which rather detailed life tables have been constructed. Predators were considered to be important mortality factors in most cases, and in many cases to be density dependent. However, Price also indicated such studies are deficient because correlations do not prove cause and effect, and I certainly agree. More definitive information can be obtained by artificially excuding natural enemies. This is the approach used by Campbell and Sloan (1977) to demonstrate the importance of small mammals as gypsy moth predators. Holmes, Schultz and Nothnagle (1979) showed how effective birds can be against leaf-feeding caterpillars by using cages that excluded birds but permitted free exchange of insects. Similar exclosure techniques have been used for other insects with similar results (some described by Price, 1987). However, these may give biased results unless the prey are able to enter and leave the exclosures easily. If they cannot, populations inside cages may be related more to lack of emigration than to mortality. Thus, while a number of studies suggest that generalist predators are important natural enemies of forest insects, it is almost never possible to say whether and how they regulate their prey. As for the gypsy moth, much more detailed information, collected over long periods, and with suitably designed experiments to prove cause and effect, are needed.

At high prey densities, C. sycophanta can be an effective gypsy moth natural enemy, but it does not respond fast enough to increasing pest populations to bring about consistent control. In many ways this beetle has the population dynamics of a specialist parasitoid more than it does that of a typical predator. This is probably because its life cycle is so closely synchronized with that of the gypsy moth that it needs no other prey, at least when this pest is abundant. Certainly its abundance cycle is closely related to that of the gypsy moth. In one very important way, however, C. sycophanta is different from parasitoids—adults live for several years. Thus beetles do not necessarily die or need to migrate if prey populations are low. Adults are known to be able to survive on fruits in the laboratory (Vasic, 1972) and probably obtain carbohydrates from plants in the field when prey are not available. The length of time adults can live in the field, as well as their behaviour during times of low prey density are not known. Knowing what C. sycophanta does between outbreaks and what its dispersal abilities are under a variety of conditions will be critical in helping to determine why its impact is sporadic and delayed.

It is also important to know how this predator interacts with the gypsy moth at various population densities and with other mortality agents. For example, Capinera and Barbosa (1975) showed that adult beetles not only feed on gypsy moth larvae infected with nuclear polyhedrosis virus with no ill effects, but even excrete viable virus particles in their faeces. But the possibility that the predator might be an important disseminator of disease agents has not been addressed in any field studies. Effectiveness of C. sycophanta might be improved by transporting beetles from one site to another, but the effective implimentation of this and other plans may not be possible until some very important data are available.

With such information in hand, it may be possible to increase intervals between outbreaks by manipulating generalist predators, and to decrease the severity of outbreaks when they do occur by manipulating *C. sycophanta*. While there are mortality factors separate from predators that are important, especially at high densities, an understanding of predators will be critical for understanding and therefore managing populations of the gypsy moth.

References

BESS, H. A. (1961). Population ecology of the gypsy moth *Porthetria dispar* L. (Lepidoptera: Lymantriidae). *Connecticut Agricultural Experiment Station Bulletin* **645**, 43 pp.

BURGESS, A. F. (1911). *Calosoma sycophanta*: its life history, behavior, and successful colonization in New England. *United States Department of Agriculture Bureau of Entomology Bulletin* **101**, 94 pp.

CAMPBELL, R. W. (1975). The gypsy moth and its natural enemies. *United States Department of Agriculture Agricultural Information Bulletin* **381**, 27 pp.

CAMPBELL, R. W. (1976). Comparative analysis of numerically stable and violently fluctuating gypsy moth populations. *Environmental Entomology* **5**, 1218–1224.

CAMPBELL, R. W., HUBBARD, D. L. and SLOAN, R. J. (1975). Location of gypsy moth pupae and subsequent pupal survival in sparse, stable populations. *Environmental Entomology* **4**, 597–600.

CAMPBELL, R. W. and SLOAN, R. J. (1976). Influence of behavioral evolution on gypsy moth pupal survival in sparse populations. *Environmental Entomology* **5**, 1211–1217.

CAMPBELL, R. W. and SLOAN, R. J. (1977). Natural regulation of innocuous gypsy moth populations. *Environmental Entomology* **6**, 315–322.

CAPINERA, J. L. and BARBOSA, P. (1975). Transmission of nuclear-polyhedrosis virus to gypsy moth larvae by *Calosoma sycophanta*. *Annals of the Entomological Society of America* **68**, 593–594.

COLLINS, C. W. and HOLBROOK, J. E. R. (1929). Trapping *Calosoma* beetles. *Journal of Economic Entomology* **22**, 562–569.

DOANE, C. C. and SCHAEFFER, P. W. (1971). Field observations on the flight activity of *Calosoma sycophanta* (Coleoptera: Carabidae). *Annals of the Entomological Society of America* **64**, 528.

FURUTA, K. (1982). Natural control of a *Lymantria dispar* population at low density levels in Hokkaido (Japan). *Zeitschrift für Angewandte Entomologie* **93**, 513–522.

FURUTA, K. (1983). Behavioural response of the Japanese paper wasp (*Polistes jadwigae* Torre-Dalla, Hymenoptera, Vespidae) to the gypsy moth (*Lymantria dispar* L., Lepidoptera, Lymantriidae). *Applied Entomology and Zoology*, **18**, 464–474.

HIGASHIURA, Y. (1980). Analysis of factors affecting bird predation on gypsy moth

egg masses by using Holling's disc-equation. *Researches in Population Ecology*
22, 147–162.

HIGASHIURA, Y. (1989). Survival of eggs in the gypsy moth *Lymantria dispar*. I.
Predation by birds. *Journal of Animal Ecology* **58**, 403–412.

HOLMES, R. L., SCHULTZ, J. C. and NOTHNAGLE, P. (1979). Bird predation on
forest insects: an exclosure experiment. *Science* **206**, 462–463.

LEONARD, D. E. (1970). Feeding rhythm in larvae of the gypsy moth. *Journal of
Economic Entomology* **63**, 1454–1457.

MASON, T. L. and TICEHURST, M. (1984). Predation of *Cryptorhiopalum ruficorne*
(Coleoptera: Dermestidae) on egg masses of the gypsy moth, *Lymantria dispar*
(Lepidoptera: Lymantriidae). *Canadian Entomologist* **116**, 1675–1677.

NOTHNAGLE, P. J. and SCHULTZ, J. C. (1987). What is a forest pest? In: *Insect
Outbreaks*, pp. 59–80 (eds P. Barbosa and J. C. Schultz). New York: Academic
Press.

PRICE, P. W. (1987). The role of natural enemies in insect populations. In: *Insect
Outbreaks*, pp. 287–312 (eds P. Barbosa and J. C. Schultz). New York:
Academic Press.

SEMEVSKIY, F. N. (1973). Studies of the dynamics of the numbers of the gipsy moth
Porthetria dispar L. (Lepidoptera, Lymantriidae) at low population-density
levels. *Entomological Review* **52**, 25–29.

SMITH, H. R. and LAUTENSCHLAGER, R. A. (1981). Gypsy moth predation. In:
The Gypsy Moth: Research Toward Integrated Pest Management, pp. 96–125
(eds C. C. Doane and M. L. McManus). *United States Department of Agriculture
Technical Bulletin* **1584**.

VASIC, K. (1972). *A Biological Method of Control of* Lymantria dispar *L. and* Diprion
pini, *L.* Belgrade: Institute of Forestry and Wood Industry.

WESELOH, R. M. (1985a). Changes in population size, dispersal behavior, and
reproduction of *Calosoma sycophanta* (Coleoptera: Carabidae), associated with
changes in gypsy moth, *Lymantria dispar* (Lepidoptera: Lymantriidae),
abundance. *Environmental Entomology* **14**, 370–377.

WESELOH, R. M. (1985b). Predation by *Calosoma sycophanta* L. (Coleoptera:
Carabidae): Evidence for a large impact on gypsy moth, *Lymantria dispar* L.
(Lepidoptera: Lymantriidae), pupae. *Canadian Entomologist* **117**, 1117–1126.

WESELOH, R. M. (1985c). Dispersal, survival, and population abundance of gypsy
moth, *Lymantria dispar* (Lepidoptera: Lymantriidae), larvae determined by
releases and mark-recapture studies. *Annals of the Entomological Society of
America* **78**, 728–735.

WESELOH, R. M. (1987). Accuracy of gypsy moth (Lepidoptera: Lymantriidae)
population estimates based on counts of larvae in artificial resting sites. *Annals
of the Entomological Society of America* **80**, 361–366.

WESELOH, R. M. (1988). Effects of microhabitat, time of day, and weather on
predation of gypsy moth larvae. *Oecologia* **77**, 250–254.

WESELOH, R. M. (1990a). Estimation of predation rates of gypsy moth larvae by
exposure of tethered caterpillars. *Environmental Entomology*, (in press).

WESELOH, R. M. (1990b). Simulation of predation by ants based on direct observations of attacks on gypsy moth larvae. *Canadian Entomologist* **121**, 1069–1076.

WHELAN, C. J., HOLMES, R. T. and SMITH, H. R. (1989). Bird predation on gypsy moth (Lepidoptera: Lymantriidae) larvae: An aviary study. *Environmental Entomology* **18**: 43–45.

22
Site Factors, Predators and Pine Beauty Moth Mortality

PATRICK J. WALSH

Forestry Commission, Northern Research Station, Roslin, Midlothian, UK

Introduction

Panolis flammea (D. and S.) (Lepidoptera: Noctuidae) is an indigenous species in Scotland, normally feeding on Scots pine (*Pinus sylvestris*). Defoliation of Scots pine has never been recorded in the United Kingdom, but in continental Europe *P. flammea* is a serious pest of Scots pine with defoliation recorded and attributed to it as far back as 1449 in Nurnberg (Watt and Leather, 1988).

Lodgepole pine (*Pinus contorta*), a conifer native to North America, has been planted widely in north and north-east Scotland since the late 1950s, 75% of which, in Scotland is planted on poor upland soils. Twenty-one percent of land planted with conifers in the north and north-east is under lodgepole pine (Watt and Leather 1988).

Panolis flammea was first recorded on lodgepole pine in 1973, but the first outbreak, in the Rimsdale block of Naver forest, Highland region, did not occur until 1976 (Stoakley, 1979). Since 1976, outbreaks have occurred in 32 sites and 12 of these have suffered a second outbreak, making a total of 44 outbreaks. *Panolis flammea* is now considered to be the most serious pest of *P. contorta* in Scotland.

Stoakley (1977) observed that outbreaks of *P. flammea* developed in plantations grown on deep peat. Poor soil types have often been associated with insect outbreaks (e.g. Hanski and Parviainen, 1985; Rhoades, 1985; Wallner, 1987).

Investigations to date have examined the role of pine species, lodgepole pine seed origin and monoterpene profile, in the role of adult oviposition site and larval survival, pupation substrate, humidity and temperature in pupal survival (see Watt and Leather, 1988, and references therein). Although these studies have indicated

Population Dynamics of Forest Insects
© Intercept Ltd, PO Box 716, Andover, Hampshire, SP10 1YG, UK

factors which may affect *P. flammea* survival none of them have explained why the populations reach outbreak levels. Watt (1989) examined larval survival on *P. contorta* and *P. sylvestris*, and in the absence of predators, found *P. sylvestris* to be a better host food plant. Watt concluded that the occurrence of outbreaks on *P. contorta* grown in deep peat, was not due to it being a better host plant than *P. sylvestris*, but a result of the lack of predators or parasitoids normally associated with *P. sylvestris*.

This study compares the communities of predatory arthropods and small mammals in two forests with differing histories of *P. flammea* outbreaks. These are linked to pupal and larval survival on *P. sylvestris* and *P. contorta* in single species stands and mixtures.

Methods

Two forests were studied, the Elchies block of Craigellachie Forest, Grampian region, where only one defoliating outbreak occurred in 1978 (successfully controlled in 1979) and no further outbreaks have occurred since, and the North Dalchork block of Shin Forest, Dornoch Forest District which has had two defoliating outbreaks in a seven-year period and often requires control measures.

Predatory ground beetles were pitfall trapped in both forests and predatory arthropods in the canopy were sampled by mist blowing groups of ten trees in *P. contorta*, *P. sylvestris* and *P. sylvestris*/*P. contorta* stands. The arthropods falling from the canopy were collected in 10 cones of collecting area 0.2 m² each, giving an effective sampling area of 2 m². Small mammals were sampled by pitfall trapping with 20 plastic bowls sunk into the ground in a 4 x 5 grid with 10 m spacing between traps. These were monitored in the three stand types from 31 August 1988 to 1 May 1989, in both forests.

Fertile eggs of *P. flammea* were introduced on to the trees in three treatments, in the three stand types, in both forests. In treatment 1, 20 eggs, oviposited on needles, were used in each replicate. The base of the needles were protected by card and tied together with plastic coated garden ties. These were then tied to branches in the upper canopy and protected by organza sleeves closed at both ends (full predator exclosure). There were 10 trees/stand type, two sleeves/tree. In treatment 2, 40 eggs per sleeve were introduced onto branches. The end of the sleeve nearest the trunk was held open by a wire hoop which was stabilized by tying lengths of wire to neighbouring branches (vertebrate exclosure allowing access by invertebrates to the eggs and larvae). Forty eggs per sleeve were used for this treatment because a calibration experiment demonstrated that 20% of the larvae can migrate from the sleeve during the course of the season. Again, there were 10 trees/stand type, two sleeves/tree. The sleeves, from both treatments, were collected after the fourth instar and the survivors counted. Progression through the instars was monitored by having sleeves with larvae in them on low branches, on which bud burst had occurred. In treatment 3, 200 eggs were placed in the upper canopy of the trees and the head capsules were collected in pots lined with muslin. Four pots per tree constituted an effective collecting area of 0.1 m². This is the same method used by Higashiura (1987), but the area of collection was larger in the present study. The muslin liners were collected after pupation had taken place. The head capsules were sorted and counted in the laboratory. There were 10 trees/stand type in both forests (unprotected introductions).

During late September/early October 1988, *P. flammea* pupae were introduced into the three stand types in both forests. Pupae were placed in perforated plastic cups which had been filled with peat. The top of the cups were covered by a 'cage' of 1.25 cm wire mesh. The pupa was placed in a slight depression, in the peat, at the centre of the cup and then covered by peat. The cups were placed in the ground with the tops slightly below the soil surface. The exposed wire was covered with needle litter. This mimicked the position in which healthy pupae are found during collection, but only allowed access to invertebrates.

Pupae were also placed under 1.25 cm wire mesh 'tables'. These 'tables' were rectangles of wire, measuring 17.5 x 9 cm in area, standing on wire legs, giving an approximate clearance of 5 cm between the litter and the top of the 'table'. This allowed free access to small mammals and arthropods, but excluded birds. The pupae were placed in a depression in the soil, beneath the centre of the table, and the litter firmly replaced. Unprotected pupae were also placed in slight depressions made in the soil and covered in the same way. In all treatments, 'cages', 'tables' and unprotected, the pupae were placed at 1, 2 and 3/m². However, for the purposes of this paper only measures of overall disappearance between treatments and tree species are used. One hundred and twenty pupae were used in each treatment, in each stand type, in both forests, a total of 360 pupae/stand type or 1080 pupae/forest.

A random selection of 300 pupae from those collected for predation trials were examined using the Geimsa staining technique used by the Institute of Virology (Oxford). Using phase/contrast microscopy, this permitted an assessment of the number of pupae contaminated with *P. flammea* nuclear polyhedrosis virus (pfNPV), bacteria or fungi. Only one pupa from these was found to be affected in any way. Fungal mycelia were found in this single specimen. The reason for the apparent health of the population may be due to selection which took place during collection, any apparently unhealthy pupae were discarded, and only pupae which had managed to penetrate to below the soil/litter interface were used. Assessment of overall winter mortality took place during March 1989.

Results

The larval mortality results were analysed using analyses of variance for the closed and open sleeves data and G-tests with Williams Correction (G_{adj}) for the head capsule data. *Table 1* shows the results for closed and open-ended sleeves. No differences were found either within or between forests in any of the treatments. The head capsule data are presented in *Table 2*. Again no differences were found either within or between forests.

The pupal mortality data were also analysed using G_{adj} (*Table 3*). No differences were found in any of the treatments within forests except for the unprotected pupae in the *P. sylvestris*/*P. contorta* compartment in Elchies where mortality was lower than in either of the single species compartments. Between-forest differences were

Table 1.   Average percent mortality of *P. flammea* larvae in closed and open sleeves in the different stand types in two forests (Elchies and North Dalchork). LP, *Pinus contorta;* SP, *Pinus sylvestris;* SP/LP, *Pinus sylvestris/Pinus contorta* mixtures; NS, not significant; d.f. (within forests) = 2,57; d.f. (between forests) = 1,114

	Closed sleeves		Open sleeves	
	Elchies	N. Dalchork	Elchies	N. Dalchork
% mortality				
LP	40.4	37.5	91.7	90.7
SP	43.6	34.7	93.9	92.2
SP/LP	36.1	36.6	95.7	92.6
F within forests	0.7 NS	2.4 NS	1.7 NS	0.2 NS
F between forests		3.6 NS		3.3 NS

Table 2.   Percent mortality of the unprotected *P. flammea* larvae on different tree species for Elchies and North Dalchork; G_{adj}, G-test using Williams Correction for within and between forests; NS, not significant

	Elchies	N. Dalchork
% mortality		
LP	95.5	92.5
SP	97.0	94.0
SP/LP	98.5	95.5
G_{adj} *(within forests)*		
LP x SP	0.19 NS	0.12 NS
SP x SP/LP	2.01 NS	0.66 NS
LP x SP/LP	2.74 NS	1.31 NS
G_{adj} *(between forests)*		
LP x LP	0.50 NS	
SP x SP	0.66 NS	
SP/LP x SP/LP	2.01 NS	

highly significant for all treatments ($P < 0.001$). The analysis of the small mammal numbers included only the common shrew, *Sorex araneus*. *Sorex minutus* is not capable of predating arthropods beneath the litter (Dickman, 1988), and *Clethrionomys glareolus* is confined mainly to areas where *P. flammea* pupae would not ordinarily survive (Walsh, unpublished data). *Table 4* demonstrates no significant differences in numbers between species within forests, but a highly significant difference ($P < 0.001$) between forests.

Numbers of predaceous arthropods in the canopy showed significant differences within Elchies ($P < 0.01$) with the mixed species compartment having the highest population and lodgepole pine having the lowest. No significant differ-

Table 3. Percent mortality of introduced *P. flammea* pupae in Elchies and N. Dalchork and the results of G_{adj}-tests; NS, not significant; * $P<0.05$, ** $P<0.01$, *** $P<0.001$.

	Unprotected		Tables		Cages	
	Elchies	N. Dalchork	Elchies	N. Dalchork	Elchies	N. Dalchork
LP	98.0	55.0	93.0	50.8	95.8	29.0
SP	97.5	59.0	95.0	55.0	91.7	38.0
SP/LP	90.8	63.0	95.0	43.0	92.5	30.8
G_{adj} (within forests)						
LP x SP	0.19 ns	1.29 ns	0.30 ns	0.41 ns	1.78 ns	1.01 ns
SP/LP x SP	4.96*	1.38 ns	0.00 ns	1.38 ns	0.06 ns	1.47 ns
SP/LP x LP	6.95**	0.28 ns	0.31 ns	3.25 ns	1.21 ns	0.08 ns
G_{adj} (between forests)						
LP x LP	74.53 ***		58.51 ***		130.20 ***	
SP x SP	59.92 ***		56.61 ***		81.59 ***	
SP/LP x SP/LP	130.03***		84.59***		106.58***	

Table 4. Average number of *Sorex araneus* caught in each trap over 245 trap nights in both forests and the results of analyses of variance for within and between forest differences; NS, not significant; *** $P<0.001$; d.f. (within forests) = 2,57; d.f. (between forests) = 1,114

	Elchies	N. Dalchork
No./trap		
LP	2.3	0.95
SP	2.0	0.35
SP/LP	3.4	0.60
F within forests	1.21 ns	2.34 ns
F between forests	23.02***	

ences were found between species within N. Dalchork. The difference between forests was highly significant (see *Table 5*).

The ground beetle (Carabidae) communities in both forests exhibited marked differences also. Eighteen species were trapped in Elchies but only six species in N. Dalchork. Significant differences have been found between soil types in Elchies with iron pan soils having greater diversity and numbers, than on deep peat planted with the same tree species (Walsh, Day, Leather and Smith, in prep.). Elchies also had greater numbers of the species found in both forests. To date, it has been determined that only three species of carabid prey on *P. flammea* pupae in the soil, these are, *Carabus glabratus, C. problematicus* and *C. violaceous*. Of these three *C. problematicus* has not been found in N. Dalchork and the remaining two species are present in significantly lower numbers than in Elchies (Walsh, unpublished data).

Table 5. Mean number m⁻² of arboreal arthropods in the three stand types in Elchies and N. Dalchork and the analyses of variance for within and between forest differences; NS, not significant, ** $P<0.01$; *** $P<0.001$; d.f. (within forests) = 2,27; d.f. (between forests) = 1,54

	Elchies	N. Dalchork
No. m⁻²		
LP	13	9
SP	31	15
SP/LP	37	14
F within forests	5.49**	0.45 NS
F between forests		15.58**

Discussion

We have here then, two forests with not only different outbreak histories but also different communities of predatory arthropods and small mammals. The reasons for this may be found in the location of the forests. Elchies is a soil mosaic, i.e. a mixture of surface water gleys, iron pan soils and deep peats. The forest is surrounded by arable land, rough pasture and grouse moor. There is a substantial amount of *P. sylvestris* planted throughout the forest and the *P. sylvestris/P. contorta* compartments are planted in intimate mixtures. North Dalchork has greater than 70% *P. contorta* with very little *P. sylvestris* remaining. The *P. sylvestris/P. contorta* compartments are small, in the region of 8–10 ha, and tend to have species clumped together. This is probably due to 'beating up', the replacement of young trees that die during the first few years of growth. The whole forest is planted on deep peat and is surrounded by unproductive moorland. The opportunity for extant predatory arthropod and small mammal communities to expand is limited by the availability of suitable habitats within N. Dalchork.

P. *contorta* dries the peat and improves the soil over a number of years (Pyatt, 1987). This may allow the gradual migration of arthropods and small mammals into these 'improved' areas. Gravesen and Toft (1987) have shown that predatory arthropods migrate from adjacent areas where they are already established into large agricultural monocultures. This may be one of the reasons why only one outbreak has occurred in Elchies and they continue in N. Dalchork. Small mammals are very active dispersers (Hanski, 1986) but they must have somewhere into which they can migrate. The fact that a community of *S. araneus* exists within N. Dalchork but at significantly lower levels than in Elchies is also an indication of the limitations to population expansion.

Why did *P. flammea* reach outbreak levels in Elchies only once? It is postulated here that a self-maintaining predatory arthropod and small mammal community (Parmenter and MacMahon, 1988) existed outwith the deep peat areas in Elchies. The wet areas were gradually dried out by *P. contorta* and a deep litter layer laid down. During the early stages of this process the sites would be suitable for *P. flammea*, a highly vagile species, to colonize and overwinter successfully. A lag in the colonization of these areas by ground-dwelling predators would allow greater pupal survival and allow the pest population to build up to a point where it would escape the 'predator ravine'. After control, by insecticide, it is proposed that the predatory community and *P. flammea* population recolonized, diversified and expanded simultaneously due to the improved habitats, thus holding the pest population below outbreak levels. The long period spent by *P. flammea* as a sedentary pupa (>120 days, and exclosure is temperature dependent) predisposes it to a high level of mortality due to predation where predator populations are large. The high mortality rate in Elchies compared with N. Dalchork, in all treatments, can be readily explained by random encounters by predators due to the very high populations existing there and the lengthy period the pupae spend in the soil. The anomalies in the data, e.g. lower pupal mortality in *P. sylvestris/ P. contorta* stands of unprotected pupae, may be accounted for by the fact that an animal which enters a restricted area, such as an exclosure, is more likely to encounter anything hidden there. Mortality was still over 90%. The lack of significant differences due to arthropod predation of larvae, where significant differences in predatory arthropod populations occurs within and between forests, may be due to the fact that over 80% of mortality occurs in the first three instars. Since the analysis here was based on survival to fifth instar these differences would not show up here. This still does not alter the fact that overall mortality (unprotected introductions) does not show any significant difference within or between forests, strongly indicating that pupal survival is the most important process in the build-up of populations to outbreak levels. The process of improving habitat is occurring in N. Dalchork, but at a much slower rate (Pyatt, pers. comm.). Large contiguous areas of peat will remain wet for much longer periods, than discrete patches, such as those found in Elchies. Tree growth and canopy closure are also impaired by the prevailing site conditions. The ability for beneficial arthropods and mammals to increase population numbers, in N. Dalchork, is very much reduced due to the unavailability of suitable areas, into which they can expand. This suggests that outbreaks will continue to occur and populations cycle at a much greater rate, and amplitude, than in Elchies.

Acknowledgements

This study was financed and supported by the Department of Education for Northern Ireland and the Forestry Commission. It was jointly supervised by Dr K. R. Day, University of Ulster and Dr S. R. Leather, Forestry Commission, Northern Research Station.

References

DICKMAN, C. R. (1988). Body size, prey size, and community structure in insectivorous mammals. *Ecology* **69**, 569–580.

GRAVESEN, E. and TOFT, S. (1987). Grass fields as reservoirs for polyphagous predators (Arthropoda) of aphids (Homoptera; Aphididae). *Journal of Applied Entomology* **104**, 461–473.

HANSKI, I. (1986). Population dynamics of shrews on small islands accord with the equilibrium model. *Biological Journal of the Linnean Society* **28**, 23–26.

HANSKI, I. and PARVIAINEN, P. (1985). Cocoon predation by small mammals, and sawfly population dynamics. *Oikos* **45**, 125–136.

HIGASHIURA, Y. (1987). Larval densities and a life-table for the gypsy moth, *Lymantria dispar,* estimated using the head capsule collection method. *Ecological Entomology* **12**, 25–30.

PARMENTER, R. R. and MacMAHON, J. A. (1988). Factors influencing species composition and population sizes in a ground beetle community (Carabidae): predation by rodents. *Oikos* **52**, 350–356.

PYATT, D. G. (1987). Afforestation of blanket peatland - soil effects. *Forestry and British Timber* (March), 15–16.

RHOADES, D. F. (1985). Offensive-defensive interactions between herbivores and plants: their relevance in herbivore population dynamics and ecological theory. *American Naturalist* **125**, 205–238.

STOAKLEY, J. T. (1977). A severe outbreak of the pine beauty moth on lodgepole pine in Sutherland. *Scottish Forestry* **31**, 113–125.

STOAKLEY, J. T. (1979). The pine beauty moth - Its distribution, life cycle and importance as a pest in Scottish forests. In: *Control of Pine Beauty Moth in Scotland,* pp. 7–17 (eds A. V. Holden and D. Bevan). Edinburgh: Forestry Commission Occasional Paper 4.

WALLNER, W. E. (1987). Factors affecting insect population dynamics: differences between outbreak and non-outbreak species. *Annual Review of Entomology* **32**, 317–340.

WATT, A. D. (1989). The growth and survival of *Panolis flammea* in the absence of predators on Scots pine and lodgepole pine. *Ecological Entomology* **14**, 225–234.

WATT, A. D. and LEATHER, S. R. (1988). The pine beauty in Scottish lodgepole pine plantations. In: *Dynamics of Forest Insect Populations,* pp. 243–266 (ed. A. A. Berryman). New York: Plenum Press.

23
Small Mammal Predation and the Population Dynamics of *Neodiprion Sertifer*

ILKKA HANSKI

Department of Zoology, University of Helsinki, P. Rautatiekatu 13, SF-00100 Helsinki, Finland

Introduction

Pine sawflies, like other forest insects, have two types of natural enemies. Specialist enemies, which are typically parasitoids, have synchronous dynamics with their host species. Generalist enemies include most arthropod and vertebrate predators, and they have relatively stable populations, buffered against fluctuations in the population size of pine sawflies by the availability of alternative prey species. The bulk of the theoretical literature on the dynamics of insect populations and their natural enemies deals with specialist parasitoids, with conveniently simple life cycles for modelling (Hassell, 1978). It is however possible that generalist predators play an equally or greater role than specialist enemies in the control of many forest insects (Hamilton and Cook, 1940; Buckner, 1966; Holmes, Schultz and Nothnagle, 1979), including pine sawflies (Hanski and Parviainen, 1985; Hanski, 1987). Murdoch, Chesson and Chesson (1985) suggest that, contrary to the conventional wisdom of biological control of insect pests, control of prey populations by generalist predators, if it occurs, is facilitated by polyphagy, consumption of large numbers of prey individuals by single predators, and by relatively stable predator populations.

Population Dynamics of Forest Insects
© Intercept Ltd, PO Box 716, Andover, Hampshire, SP10 1YG, UK

Pine sawflies have three groups of generalist predators: arthropods preying on their eggs, larvae and cocoons; birds preying mostly on larvae; and small mammals preying on cocoons in the forest floor. This paper is largely limited to small mammals as predators, partly because of personal interest and the abundance of information available on them, partly because they seem to possess the greatest potential to control sawfly populations (they closely match the list of attributes that Murdoch, Chesson and Chesson, 1985, suggested to enhance the control of prey populations by generalist predators). *Neodiprion sertifer* (Geoff.), the most serious outbreak species in northern Europe (Hanski, 1987), provides most of the examples, but studies on other sawfly species will also be cited.

This paper is structured as follows, with the aim of making three main points. I start with the predators' point of view, and ask how rewarding food items the sawfly cocoons are to small mammals. The first point to be made is that, generally, relatively large food items, such as insect pupae, are the most profitable ones for shrews, the main group of small mammal predators, and as sawfly cocoons are additionally very suitable for transport and short-term food caching, it is not surprising that they are a favourite prey type for the predators. The prey's means of defending itself against the cocoon predators are limited, in contrast to the behavioural and chemical adaptations that the sawfly larvae possess against arthropod and bird predators. I shall next discuss the characteristic numbers of small mammals in forests and the level of predation inflicted by them on sawfly cocoons. Contrary to Holling's (1959) frequently cited results on small mammal predators on *N. sertifer* cocoons, the rate of predation seems to be generally inversely, rather than directly, density dependent, which is my second main point and a conclusion that seems to exclude the possibility of successful control of prey populations at a low equilibrium level. I suggest that such control may none the less occur in prey metapopulations, systems of local populations connected by dispersal. Sawfly populations may frequently go extinct due to heavy predation at fertile sites with large numbers of small mammals, but local populations may persist at poor-quality sites with generally small numbers of predators. The latter populations may be controlled by small mammal predators, often perhaps due to dispersal of predators from good sites when the sawfly populations start to increase.

The predators' point of view

SAWFLY COCOONS— A PREFERRED FOOD TYPE

The most significant vertebrate predators of sawfly cocoons are shrews and other small mammals. Shrews are highly opportunistic predators with a wide diet (Hanski, 1990), but most species prefer relatively large prey items, roughly of the size of sawfly cocoons (Platt and Blackley, 1973; Barnard and Brown, 1981; Dickman, 1988; Hanski, 1990).

Shrews may prefer relatively large prey items because they are often, though not necessarily always, the most profitable ones, and because they may be easier to detect than small ones. One large prey item may be equivalent to 10 or more small ones in energy content, and concentrating on large prey may often reduce total search time. Large prey items, such as insect pupae, can be easily cached, and the food caches may critically help the shrews to survive short-term fluctuations in food availability (below). Optimal foraging theory (Stephens and Krebs, 1986) suggests that the availability of alternative prey should not significantly affect the use of the preferred prey. The important point here is that shrews are expected to search for and use sawfly cocoons whenever available, regardless of the presence of other types of prey. The shrews' functional response to sawfly cocoons should rise steeply to its asymptotic value.

FOOD CACHING

Shrews and other insectivorous mammals often find their food in places where it would be risky to stay for long periods, for instance, the open forest floor. The food item is picked up and transported to a spot where the shrew is safe from its predators and competitors (McLeod, 1966).

A simple extension of the habit of transporting cocoons before consumption is short-term food caching. Food caches have several important functions in the ecology of shrews and omnivorous small mammals: they amplify the predators' functional response; they reduce the risk of starvation in short-term food shortages and thereby contribute to the stability of the predator populations (Hanski, 1985b, 1990); and they promote a numerical response to a prey type that is otherwise effectively seasonal (Buckner, 1958).

The sawfly's responses to predation

Sawfly larvae exhibit defensive displays when disturbed by their predators and parasitoids. Such group defence responses have been shown to reduce predation by the pentatomid bug *Podisus modestus* (Tostowaryk, 1971) and by parasitoids (Prop, 1960; Tripp, 1962).

Larvae could escape predation by small mammals in the forest floor by spinning cocoons in trees. Montgomery and Wallner (1988) suggest that the reason why the gypsy moth larvae in the Ukraine stay in trees while in North America the larvae aggregate in the forest litter is the much higher density of small mammals in the Ukraine than in North America. Regarding pine sawflies, Lyons (1963) found that 3% of *N. sertifer* cocoons were in fact located in trees, and in central Europe the first-generation larvae of *Diprion pini* regularly spin cocoons in the canopy (Schwenke, 1982). There is thus ample intraspecific variation in the location where the larvae spin their cocoons. However, tree trunks and the foliage are not necessarily any safer places for cocoons than is the forest floor. Tits have been observed to prey on *Diprion pallidum* cocoons in trees (Kurir, 1977), and Hanski and Parviainen (1985) showed experimentally, by glueing *N. sertifer* cocoons at different heights in trees, that predation by birds in trees was even heavier (85%) than predation by small mammals in the forest floor (70%).

Most larvae spin cocoons shallowly below the forest litter, at a depth of 5 cm or less (Schoenfelder *et al.*, 1978; Hanski and Parviainen, 1985), where they are easily detected by small mammals (Holling, 1955; Buckner, 1958) using olfactory stimuli (Holling, 1958). It is not clear why larvae do not spin the cocoon at a greater depth. The message from this section is, in brief, that sawflies have no effective means of defence against mammalian cocoon predators.

Rate of predation by small mammals

How many cocoons are consumed by small mammals naturally depends on their numbers. Shrews, the most significant predators, have relatively stable popula- tions (Skarén, 1972; Pankakoski, 1979), and their density typically peaks in August–September, just in time to attack the freshly spun cocoons of *N. sertifer*. Numbers of small mammals vary from place to place, but commonly the pooled density of insectivorous and omnivorous species ranges from 20 to 100 individuals per ha in northern Europe. Assuming that there are 50 individuals per ha, each consuming 100 cocoons per day, which is a conservative estimate (Buckner, 1964), small mammals would consume 150 000 cocoons in 30 days, the pupal period of *N. sertifer*. Field studies have shown that small mammals frequently

Table 1. Rate of predation by shrews and omnivorous small mammals on sawfly cocoons. The period over which predation rate has been estimated varies from one to several months, but generally most predation occurs in 1–2 months in late summer and early autumn. Some studies report several estimates of predation rate for several study plots or years, in which case an average value is used here.

Predators	Cocoon density (1000/ha)	Percent predation	Reference
Small mammals	<1* Exp	99	Hanski and Otronen (1985)
Small mammals	<1* Exp	72	Buckner (1958)
Small mammals	1* Exp	68	Hanski and Parviainen (1985)
Shrews	30	95	Olofsson (1987)
Shrews	43	95	Buckner (1964)
Shrews	150	55	McLeod (1966)
Small mammals	206	71	Kolomiets *et al.* (1979)
Shrews	1400	45	Buckner (1964)
Small mammals	c. 1500	50	Obrtel *et al.* (1978)
Small mammals	1700 Exp	45	Schoenfelder *et al.* (1978)
Small mammals	15 000	20	Griffiths (1959)

*The density estimate does not take into account the naturally occurring cocoons; the true density is likely to be between 1000 and 10 000 cocoons per ha.
Exp: Results from experiments in which cocoons were planted in the forest floor.

consume some 80% or more of cocoons in endemic sawfly populations (*Table 1*), when the density of cocoons in coniferous forests in Fennoscandia is less than 10 000/ha (Juutinen, 1967; Larsson and Tenow, 1980; Hanski, 1987; Olofsson, 1987). In contrast, during severe outbreaks, when there are several million cocoons per ha, only a small fraction is destroyed by small mammals. These figures clearly indicate the potential of small mammals in controlling sawfly populations at a low level. The not novel but perpetually important question is how successful they are in doing so, and in preventing sawfly populations from increasing to outbreak levels?

Functional and numerical responses

Holling's (1959) classical field experiments on three species of small mammal (*Sorex*, *Blarina* and *Peromyscus*) preying on the cocoons of *N. sertifer* demonstrated Type 3 functional responses, consistent with his laboratory observations on *Peromyscus*. It is not obvious why the functional responses should be of Type 3, because sawfly cocoons as a highly preferred food type (previous section) are expected to be consumed whenever encountered (according to the optimal foraging theory). Even more surprising is that Holling's (1959) results on *Peromyscus* and especially on *Sorex* revealed a clear numerical response, which is not expected, because *N. sertifer* cocoons are generally available only for a short period in late summer (Buckner, 1964; Hanski, 1987). The explanation of the observed numerical responses probably is that Holling (1959) worked in pine plantations with little alternative food for insectivorous and omnivorous small mammals and during a *N. sertifer* outbreak, when large numbers of cocoons probably stayed in prolonged diapause during most of the year (Hanski, 1988) and allowed an increase in small mammal numbers. Whatever the reasons for the observed functional and numerical responses, Holling's (1959) results provide the first and so far the only thorough

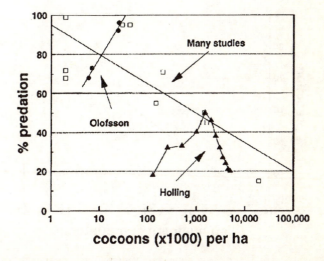

Figure 1.   Rate of predation on sawfly cocoons (mostly *Neodiprion sertifer*) as a function of cocoon density in three sets of data, Holling's (1959) study (black triangles), Olofsson's (1987) study (black dots), and a compilation of several studies (open squares; from *Table 1*). The inverse density-dependence in the latter set of data is highly significant ($t = -3.52$, $P<0.007$).

examples of functional and numerical responses in small mammals to increasing numbers of pine sawflies.

Combining the functional and numerical responses gives the rate of predation, which in Holling's (1959) study first increased, then declined with increasing prey density (*Figure 1*). Within the range of prey densities where predation is directly density dependent, it may contribute or suffice to control the prey population at a low equilibrium level (Holling, 1959). Unfortunately, there are not many other studies which have examined density dependence in the rate of predation inflicted by insectivorous mammals on sawflies. Olofsson (1987), studying different mortality factors operating in a *N. sertifer* population found that only predation by small mammals, probably largely by shrews, was directly density dependent (*Figure 1*). However, Olofsson (1987) had data for four years only, and the observed variation in prey density was small. Ives (1976) working with the larch sawfly *Pristiphora erichsonii* found cocoon predation by small mammals to vary widely, up to an extremely high level of 95% at the outbreak density of nearly 2 million cocoons per ha, but overall the rate of predation was density independent.

Data in *Table 1* allow us to examine general trends in density dependence in the rate of predation. The picture which emerges is clearly different from the one described by Holling (1959); in these data the rate of predation is highest at the lowest prey density, declining monotonically with increasing prey numbers (*Figure 1*). These data are heterogeneous, as they have been collected by several individuals working with different species at different localities, yet the trend is convincing. And the result is not entirely unexpected in view of the shrews'

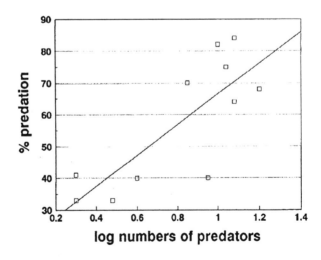

Figure 2.   Rate of predation by small mammals on *Neodiprion sertifer* cocoons in 11 study plots during one month as a function of the numbers of predatory small mammals trapped from the same plots. The study plots represent barren pine forests and bogs with pine. The slope of the regression line is significantly different from zero, $t = 3.95$, $P<0.003$ (from Hanski and Parviainen, 1985).

preference for food items such as sawfly cocoons (see previous section), and the expected lack of an important numerical response in most situations. Note that the level of predation in Olofsson's (1987) study is consistent with the general pattern; the positive density dependence found by him looks like a chance result. Holling's (1959) results for high prey densities, above 1 million cocoons per ha, are also consistent with the general pattern. The striking difference is in the rate of predation at lower prey densities, which was directly density dependent in Holling's (1959) study but is clearly inversely density dependent in the pooled results for many studies (*Figure 1*).

What might be the cause of this critical difference? As mentioned above, Holling (1959) worked in pine plantations with little alternative food for insectivorous small mammals. The density of small mammals was consequently extremely low in the study plots with the lowest density of sawflies, at fewer than 5 individuals per ha (Holling, 1959). These few individuals could hardly be expected to have destroyed the greater part of cocoons within the entire study plots. In contrast, in more natural situations, when the density of small mammals is typically 50/ha or more, due to the presence of alternative food sources, the rate of predation on the preferred prey type of sawfly cocoons can be extremely high even at the lowest densities. This point is demonstrated by results on predation rates in 11 study plots with varying numbers of predators (*Figure 2*). The density of cocoons was low in all plots, certainly less than 10 000/ha, and all the plots represented barren types of pine forest. Where the numbers of predators was small, the rate of predation was comparable to that found by Holling (1959) at low prey density, around 30%, while where the predators were abundant, the rate of predation was high, around 80%, fitting well the general pattern in *Figure 1*.

Control of sawfly populations by small mammals at the metapopulation level

Control of a prey population at low density requires the operation of one or more directly density-dependent factors; for example, density-dependent predation. In this perspective one may be tempted to dismiss predation by shrews and omnivorous mammals as unimportant when no density dependence has been detected (Ives, 1976) or when predation is inversely density dependent (*Figure 1*).

Outbreaks of pine sawflies (Juutinen, 1967; Juutinen and Varama, 1983) and many other forest insects (Berryman, 1988) develop in time and space, starting and spreading from restricted 'outbreak foci' (Campbell and Sloan, 1977). The spatial element in outbreaks suggests that questions about the dynamics of sawfly populations should generally be examined not in the context of single local populations but in the context of metapopulations, systems of many local populations connected by dispersal. The results reviewed in the previous section suggest the following scenario, in which small mammals might play a significant role in the control of sawfly metapopulations.

Let us assume that there are two kinds of sites for local populations (in reality, there is a continuum of different kinds of sites). At good sites the populations of small mammals are large and relatively stable, due to the abundance of many kinds of prey, while at poor sites small mammals are scarce ('poor' and 'good' are thus defined by small and large numbers of predators). Several studies have reported a positive correlation between the density of shrews and the density of their prey populations (Judin, 1962; Butterfield, Coulson and Wanless, 1981; Hanski, 1990). At good sites, predation on sawfly cocoons is extremely high in low sawfly densities (*Table 1, Figure 1*), and the local sawfly populations may go frequently extinct due to heavy predation. At poor sites, the rate of predation is generally relatively low because of small numbers of predators, but may increase with increasing prey density, as found by Holling (1959); often perhaps due to movements of small mammal predators from adjacent good sites. Sawfly populations may thus be controlled by small mammals at a low equilibrium density at poor sites, especially if located close to some good sites. Such control may however fail in years when the conditions are exceptionally favourable for the sawfly but unfavourable for small mammals. In such years sawfly populations at (some) poor sites may escape from the control of local predators, and a regional (metapopulation) outbreak may develop, as more and more local populations at all kinds of sites are pushed beyond the threshold density by dispersal of the sawfly from the outbreak area. The outbreak is likely to be ended by a combination of deteriorating food resources, viral diseases and an increasing level of parasitism (Hanski, 1987).

The above scenario may be encapsulated in metapopulation models with alternative stable equilibria (Hanski, 1985a, 1991), corresponding to regional rarity and to regional outbreaks. Note that both poor and good sites are required for this metapopulation scenario to work. Poor sites, where the prey is unlikely to go extinct, prevent the extinction of the entire metapopulation, while dispersal of predators from the good sites may be a critical ingredient in the control of the prey at the poor sites.

The dynamics of *Neodiprion sertifer* in northern Europe is a likely example of this model. In most years the species appears to be entirely absent from large areas (M. Varama, personal communication). It is impossible to be entirely sure about negative observations, but because the larvae are gregarious, they are relatively easy to detect. When there are no outbreaks, small populations have been found on marginal habitats, e.g. small islands in lakes (M. Varama, personal communication).

Discussion

The metapopulation model of forest insect dynamics makes a number of predictions that may be compared with field observations:

1. Rate of predation at low prey density should be directly density dependent in poor-quality sites with generally small numbers of predators, but inversely density-dependent in good-quality sites with large numbers of predators. I suggest that the conflicting trends in density-dependence reported in *Figure 1* are due to differences in habitat quality.

2. Outbreaks should start from poor-quality forest stands, such as pine forests on

bogs and sandy soils. This pattern is well documented for pine sawflies (e.g. Juutinen, 1967; McLeod, 1970; Kolomiets, Stadnitskii and Vorontzov, 1979; Hanski, 1987).

3. Outbreaks should start in years when the environmental conditions are favourable for the sawfly but unfavourable for small mammals. Dry and hot summers is one example of such conditions: sawfly larvae develop rapidly and have little diseases, but small mammals suffer because of general scarcity of food (Pankakoski, 1984; Myllymäki, 1969). It has been frequently found that sawfly outbreaks are preceded by a series of dry and hot summers (Kolomiets, Stadnitskii and Vorontzov, 1979; Larsson and Tenow, 1984).

4. Assuming that the control of sawflies by small mammals at the poor sites is largely due to movements of the predators from adjacent good sites, the metapopulation model predicts that outbreaks are more frequent in regions with extensive tracts of poor habitat than in regions with a higher proportion of good habitat. Observations that outbreaks are largely restricted to regions with infertile soils (Larsson and Tenow, 1984) are consistent with this prediction.

Hanski (1987) has examined several mechanisms other than small mammal predation which could explain the above and other patterns in pine sawfly population dynamics and outbreaks. The conclusion was that the plant defence and cocoon predation hypotheses explain more observations about sawfly dynamics than the microparasite (virus) and macroparasite (parasitoid) hypotheses. However, experiments are needed that would allow us more conclusively to discriminate between the competing hypotheses. Experiments on forest insects and their mammalian predators may either attempt to change the suitability of the environment for the predators (e.g. Buckner, 1966) or to manipulate directly the numbers of predators (e.g. Campbell and Sloan, 1977; Warren, 1970). Unfortunately, due to the logistical problems in conducting relevant experiments with vertebrate predators, no good experimental evidence yet exists for small mammal predation on pine sawflies.

Conclusions

The three main points of this paper are as follows. First, shrews and omnivorous small mammals tend to prefer relatively large food items, such as insect pupae, which are therefore eagerly searched for even when scarce and when alternative food is available. Secondly, as a result of the preference for sawfly cocoons, rate of predation by small mammals is directly density dependent at low sawfly densities only in poor-quality sites, where the numbers of small mammals are generally small, but may increase when the cocoon density significantly increases. In more fertile sites predation is inversely density dependent at all densities of sawflies. Thirdly, sawfly populations in good sites may frequently go extinct due to heavy predation, but local populations persisting in poor sites prevent regional extinction and may function as foci for subsequent outbreaks. Dispersal of predators from adjacent good sites may significantly contribute to the control of the prey at poor sites.

References

BARNARD, C. J. and BROWN, C. A. J. (1981). Prey size selection and competition in the common shrew (*Sorex araneus* L.). *Behavioural Ecology and Sociobiology* **8**, 239–243.

BERRYMAN, A. A. (1988). *Dynamics of Forest Insect Populations: Patterns, Causes, Implications*. New York: Plenum.

BUCKNER, C. H. (1958). Mammalian predators of the larch sawfly in eastern Manitoba. *Proceedings of the International Congress of Entomology, Montreal* **4**, 353–361.

BUCKNER, C. H. (1964). Metabolism, food consumption, and feeding behaviour in four species of shrews. *Canadian Journal of Zoology* **42**, 259–279.

BUCKNER, C. H. (1966). The role of vertebrate predators in the biological control of forest insects. *Annual Review of Entomology* **11**, 449–470.

BUTTERFIELD, J., COULSON, J. C. and WANLESS, G. (1981). Studies on the distribution, food, breeding biology and relative abundance of the pygmy and common shrews (*Sorex minutus* and *S. araneus*) in upland areas of northern England. *Journal of Zoology, London* **195**, 169–180.

CAMPBELL, R. W. and SLOAN, R. J. (1977). Release of gypsy moth populations from innocuous levels. *Environmental Entomology* **6**, 323–330.

DICKMAN, C. R. (1988). Body size, prey size, and community structure in insectivorous mammals. *Ecology* **69**, 569–580.

GRIFFITHS, K. J. (1959). Observations on the European pine sawfly, *Neodiprion sertifer* (Geoff.) and its parasites in southern Ontario. *Canadian Entomologist* **91**, 501–512.

HAMILTON, W. J. Jr. and COOK, D. B. (1940). Small mammals and the forest. *Journal of Forestry* **38**, 468–473.

HANSKI, I. (1985a). Single-species spatial dynamics may contribute to long-term rarity and commonness. *Ecology* **66**, 335–343.

HANSKI, I. (1985b). What does a shrew do in an energy crisis? In: *Behavioural Ecology*, pp. 247–252 (eds R. M. Sibly and R. H. Smith). Oxford: Blackwell.

HANSKI, I. (1987). Pine sawfly population dynamics: patterns, processes, problems. *Oikos* **50**, 327–335.

HANSKI, I. (1988). Four kinds of extra long diapause in insects: a review of theory and observations. *Annales Zoologici Fennici* **25**, 37–53.

HANSKI, I. (1990). Insectivorous mammals. In: *Natural Enemies, The Population Biology of Predators, Parasites and Diseases* (ed. M. J. Crawley). Oxford: Blackwell (in press).

HANSKI, I. (1991). Single-species metapopulation models. In: *Metapopulation Dynamics* (eds M. Gilpin and I. Hanski). New York: Academic Press (in press).

HANSKI, I. and OTRONEN, M. (1985). Food quality induced variance in larval performance: comparison between rare and common pine-feeding sawflies (Diprionidae). *Oikos* **44**, 165–174.

HANSKI, I. and PARVIAINEN, P. (1985). Cocoon predation by small mammals, and pine sawfly population dynamics. *Oikos* **45**, 125–136.

HASSELL, M. P. (1978). *The Dynamics of Arthropod Predator-Prey Systems*. Princeton NJ: Princeton University Press.

HOLLING, C. S. (1955). The selection of certain small mammals of dead, parasited and healthy prepupae of the European pine sawfly, *Neodiprion sertifer* (Geoff.). *Canadian Journal of Zoology* 33, 404–419.

HOLLING, C. S. (1958). Sensory stimuli involved in the location and selection of sawfly cocoons by small mammals. *Canadian Journal of Zoology* 36, 633–653.

HOLLING, C. S. (1959). The components of predation as revealed by a study of small-mammal predation of the European pine sawfly. *Canadian Entomologist* 91, 293–330.

HOLMES, R. T., SCHULTZ, J. C. and NOTHNAGLE, P. (1979). Bird predation on forest insects: an enclosure experiment. *Science* 206, 462–463.

IVES, W. G. H. (1976). The dynamics of Larch sawfly (Hymenoptera: Tenthredinidae) populations in southeastern Manitoba. *Canadian Entomologist* 108, 701–730.

JUDIN, B. S. (1962). Ecology of shrews (genera *Sorex*) of West Siberia. In: *The Problems of Ecology, Zoogeography and Systematics of Animals*, pp. 33–134 (ed. by A. I. Cherepanov). Novosibirsk.

JUUTINEN, P. (1967). Zür Bionomie und zum Vorkommen der roten Kiefernbuschhornblättwespe (*Neodiprion sertifer* Geoffr.) in Finnland in den Jahren 1959–65. *Communicationes Instituti Forestalis Fenniae* 63, 1–129.

JUUTINEN, P. and VARAMA, M. (1983). Ruskean mäntypistiäisen esiintyminen vuosina 1980–1983. *Metsäntutkimuslaitoksen tiedonantoja* 99.

KOLOMIETS, N. G., STADNITSKII, G. V. and VORONTZOV, A. I. (1979). *The European Pine Sawfly*. New Delhi: Amerind Publishing.

KURIR, A. (1977). Observations on the bionomy of *Diprion pallidum* (Hymenoptera: Diprionidae) during the gradation in Carinthia 1971–1972. *Zeitschrift für Angewandte Entomologie* 84, 155–163.

LARSSON, S. and TENOW, O. (1980). Needle-eating insects and grazing dynamics in a mature Scots pine forest in Central Sweden. In: *Structure and Function of Northern Coniferous Forests—An Ecosystem Study* (ed. T. Persson). *Ecological Bulletin (Stockholm)* 32, 269–306.

LARSSON, S. and TENOW, O. (1984). Areal distribution of a *Neodiprion sertifer* (Hym., Diprionidae) outbreak on Scots pine as related to stand condition. *Holarctic Ecology* 7, 81–90.

LYONS, L. A. (1963). The European pine sawfly *Neodiprion sertifer* (Geoff.) (Hymenoptera: Diprionidae). A review with emphasis on studies in Ontario. *Proceedings of the Entomological Society of Ontario* 94, 5–37.

MCLEOD, J. M. (1966). The spatial distribution of cocoons of *Neodiprion swainei* Middleton in a jack pine stand. I. Cartographic analysis of cocoon distribution, with special reference to production of small mammals. *Canadian Entomologist* 98, 430–447.

MCLEOD, J. M. (1970). The epidemiology of the Swaine jack-pine sawfly, *Neodiprion swainei* Midd. *Forestry Chronicle* 46, 126–133.

MONTGOMERY, M. E. and WALLNER, W. E. (1988). The gypsy moth: a westward migrant. In: *Dynamics of Forest Insect Populations: Patterns, Causes, Implications*, pp. 353–375 (ed. A. A. Berryman). New York: Plenum.

MURDOCH, W. W., CHESSON, J. and CHESSON, P. L. (1985). Biological control in theory and practice. *American Naturalist* 125, 344–366.

MYLLYMÄKI, A. (1969). Productivity of a free-living population of the field vole,

Migrotus agrestis (L.). In: *Energy Flow Through Small Mammal Populations*, pp. 255–265 (eds K. Petrusewicz and L. Ryszkowski). Warsawa: International Biological Program.

OBRTEL, R., ZEJDA, J. and HOLISOVA, V. (1978). Impact of small rodent predation on an overcrowded population of *Diprion pini* during winter. *Folia Zoologia* **27**, 97–110.

OLOFSSON, E. (1987). Mortality factors in a population of *Neodiprion sertifer* (Hymenoptera: Diprionidae). *Oikos* **48**, 297–303.

PANKAKOSKI, E. (1979). The influence of weather on the activity of the common shrew *Sorex araneus*. *Acta Theriologica* **24**, 522–526.

PANKAKOSKI, E. (1984). Relationships between some meteorological factors and population dynamics of *Sorex araneus* L. in southern Finland. *Acta Zoologica Fennica* **173**, 287–289.

PLATT, W. J. and BLACKLEY, N. R. (1973). Short-term effects of shrew predation upon invertebrate prey. *Proceedings of the Iowa Academy of Science* **80**, 60–66.

PROP, N. (1960). Protection against birds and parasites in some species of tenthredinid larvae. *Archives Néerlandaises de Zoology* **13**, 380–447.

SCHOENFELDER, T. W., HOUSEWEART, M. W., THOMPSON, L. C., KULMAN, H. M. and MARTIN, F. B. (1978). Insect and mammal predation of yellow-headed spruce sawfly cocoons (Hymenoptera, Tenthredinidae). *Environmental Entomology* **7**, 711–713.

SCHWENKE, W. (1982). *Die Forstschädlinge Europas. Haultflügler und Zweiflügler*. Hamburg: Paul Parey.

SKARÉN, U. (1972). Fluctuations in small mammal populations in mossy forests of Kuhmo, eastern Finland, during eleven years. *Annales Zoologici Fennici* **9**, 147–151.

STEPHENS, D. W. and KREBS, J. R. (1986). *Foraging Theory*. Princeton NJ: Princeton University Press.

TOSTOWARYK, W. (1971). Relationship between parasitism and predation of diprionid sawflies. *Annals of the Entomological Society of America* **64**, 1424–1427.

TRIPP, H. A. (1962). The relationship of *Spathimeigenia spinigera* Townsend (Diptera: Tachinidae) to its host, *Neodiprion swainei* Midd. (Hymenoptera: Diprionidae). *Canadian Entomologist* **94**, 809–818.

WARREN, G. L. (1970). Introduction of the masked shrew to improve control of forest insects in Newfoundland. *Proceedings of the Tall Timbers Conference*. Florida: Tallahassee.

24

Are Parasitoids of Significance in Endemic Populations of Forest Defoliators? Some Experimental Observations from Gypsy Moth, *Lymantria dispar* (Lepidoptera: Lymantriidae)

N. J. MILLS

CAB International Institute of Biological Control, Silwood Park, Ascot, Berks, SL5 7TA, UK

Introduction

Forests often cover extensive areas and are only harvested at infrequent intervals. Forest trees, therefore, are at an extreme in the continuum of cropping system environments both in terms of their spatial scale and temporal stability. Thus forests support a greater diversity of insect herbivores than other environments (Strong, Lawton and Southwood, 1984). Despite this, rather more than 90% of forest insects never attain pest status and remain at stable endemic densities (e.g. Mason, 1987; Schultz, 1983). Even many pest insects remain at endemic levels of abundance for considerable periods between outbreaks and may only erupt when natural constraints within the life system fail due to unpredictable disturbance.

Population Dynamics of Forest Insects
© Intercept Ltd, PO Box 716, Andover, Hampshire, SP10 1YG, UK

While a variety of trophic interactions operate to constrain population growth in insect herbivores (e.g. Lawton and McNeill, 1979), parasitism is of particular interest in the case of forest defoliators. Pschorn-Walcher (1977) pointed out that the parasitoid complexes of forest insects are more mature than those of other insects. The greater diversity of parasitoids supported by forest defoliators has been achieved by a greater level of niche diversification and specialization within the parasitoid guilds. One important dimension for niche separation is host density and several examples of high- and low-density parasitoids are known from the forest insect literature (*Table 1*). The low-density parasitoids tend not to be represented during outbreaks of their hosts and are unlikely to be involved in the collapse of host outbreaks. The nature of parasitism and the high host-searching efficiency of these parasitoids, however, point to their potential for regulation of defoliator populations at endemic densities.

Table 1.   Examples of low and high host density parasitoids of forest defoliators.

Host	Parasitoid species Low host density/high host density	Reference
Operophtera brumata	*Agrypon flaveolatum/ Cyzenis albicans*	Embree (1966)
Zeiraphera diniana	*Sympeisis punctifrons/ Phytodietus griseanae*	Delucchi (1977)
Choristoneura fumiferana	*Synetaeris tenuifemur/ Apanteles fumiferanae*	Miller and Renault (1977)
Neodiprion sertifer	*Lophyroplectus luteator/ Lamachus eques*	Pschorn-Walcher (1973)
Neodiprion swainei	Larval parasitoids/ cocoon parasitoids	Price (1973)

The gypsy moth is a notorious defoliator of hardwood forests and has been of particular concern in the United States since its accidental introduction from Europe in 1868 (Doane and McManus, 1981). As an exotic pest it has been the target of an extensive biological control programme in the US (e.g. Dahlsten and Mills, 1990) with the importation, release and establishment of a number of natural enemies from its native Europe. Unfortunately, these natural enemies have not been able to prevent the further defoliation of hardwoods in the North-eastern States and gypsy moth continues to expand its range in North America. It is worth noting at this point that all introduced natural enemies were collected from gypsy moth outbreaks in Europe and did not include any low host density specialists. Thus an alternative strategy for gypsy moth biological control in North America would be to introduce any potential low-density parasitoids from Europe with a view to achieving increased intervals of endemic pest densities and the prevention of the development of outbreaks. This paper describes the results of a survey in western Europe for low-density larval parasitoids of the gypsy moth and the potential for further biological control of the gypsy moth in North America.

Methods

Experimental cohorts of gypsy moth larvae were used to expose this host to parasitism in the southern Alsace in eastern France. Exposures were carried out at the same site over a series of successive years using individual trees in a relatively small area of approximately 0.5 ha. This region is composed of scrub oak (*Quercus robur*) and has never experienced a gypsy moth outbreak. Background population levels are very low and were estimated at 1.13 males per trap (s.d. 1.26, $n=31$) from pheromone trapping in 1981.

The host exposure and recovery methods were described by Mills, Fischer and Glanz (1986) and consisted of placing foliage-reared second instar larvae on small isolated oaks for periods of approximately three weeks. From three to six successive cohorts were exposed at weekly intervals from mid-May to mid-June, each year from 1980 to 1988, to span the activity period of larval parasitoids. Each cohort comprised 30 or 40 000 second instar larvae split between three to five trees, within a 10 m radius, such that smaller trees received fewer larvae and no trees could be substantially defoliated. On recovery the cohorts were totally recollected and reared on foliage to collect and identify emerging parasitoids.

Results

SUCCESS OF HOST EXPOSURE

A summary of the total numbers of larvae exposed each year, the number of successive cohorts used and the total recovery of exposed larvae are presented in *Table 2*. The gypsy moth larvae suffered substantial losses within the first two days

Table 2. A summary of the gypsy moth larval cohorts exposed to parasitism in the Alsace

Year	000s exposed	No. of cohorts	000s recovered
1980	90	3	1.2
1981	90	3	1.2
1982	90	3	0.7
1983	30	1	1.4
1984	120	3	7.0
1985	120	3	5.2
1986	160	4	11.3
1987	240	6	27.7
1988	240	6	17.0

of field exposure due to failure of establishment on the experimental trees. Counts of the number of larvae observed per five minute search on six trees in 1982 showed that from 38 to 60% of larvae were lost during the period between 1 and 48 h after exposure. Sticky traps placed under experimental trees indicated that many larvae dropped from the trees during this period but few appeared to successfully disperse to neighbouring trees.

LARVAL PARASITOID SPECIES REARED

A total of four braconid, two ichnuemonid and seven tachinid species were reared from the exposed gypsy moth larvae between 1980 and 1988. The complex included both generalist and specific parasitoids and varied in composition between years (*Table 3*). *Ceranthia samarensis*, *Compsilura concinnata* and *Parasetigena silvestris* occurred each year while *Blondelia nigripes*, *Ceromyia* sp. and *Exorista larvarum* were present only during one year. In the latter case, the parasitoids were represented by several individuals and so formed a true component of the complex in individual years but were incidental parasitoids in the sense that they were not consistent each season. Hymenopteran parasitoid species richness increased notably in 1986, coinciding with years of warmer and drier weather in late May and early June, but in general the tachinid parasitoids were more frequent.

Table 3.   The parasitoid complex of the exposed gypsy moth larval cohorts, giving the number of parasitoids obtained from exposed hosts each year. An asterisk (*) indicates years in which *P. silvestris* abundance was not adequately monitored.

Species	1980	1981	1982	1983	1984	1985	1986	1987	1988
Braconidae									
Apanteles liparidis							5	21	
A. melanoscelus						46	298	30	19
A. porthetriae	2		120	4	5	34	65	15	19
Ichneumonidae									
Hyposoter tricoloripes					4		6		
Phobocampe disparis					2		20		4
No. Hymenoptera spp.	1	1	1	1	3	2	5	3	3
Tachinidae									
Blepharipa pratensis	55	9	23	9					
Blondelia nigripes								14	
Ceranthia samarensis	4	14	23	510	1788	1379	3012	356	555
Compsilura concinnata	12	39	81	84	186	78	156	63	69
Ceromyia sp.							8		
Exorista larvarum		3							
Parasetigena silvestris	41	37	17	19	509	92	*	*	29
Zenillia libatrix	18							2	
No. tachinid spp.	5	5	4	4	3	3	3	4	5
Total parasitoid spp.	6	6	5	5	6	5	8	7	8

Amongst the more constant parasitoids of the experimental cohorts, the relative frequency changed with time (*Table 4*). During the first three years a greater range of more generalist tachinids, such as *Compsilura concinnata*, *Exorista larvarum* and *Zenillia libatrix*, were present in the complex and *C. concinnata* and the braconid, *Apanteles porthetriae*, were most frequent. From 1983, however, the more specific tachinid, *Ceranthia samarensis*, dominated the complex and, in attacking and killing the younger host larval instars, reduced the abundance of hosts available to other tachinid parasitoids.

Table 4. The relative frequency of the most constant parasitoids attacking exposed cohorts of gypsy moth larvae, expressed as a percentage of the total number of parasitoids obtained from exposed hosts each year. An asterisk (*) indicates years in which *P. silvestris* abundance was not adequately monitored.

Species	1980	1981	1982	1983	1984	1985	1986	1987	1988
A. porthetriae	2	54	0	1	<1	<1	2	3	3
C. samarensis	3	6	16	79	72	85	85	72	79
C. concinnata	9	18	56	15	7	5	4	13	10
P. silvestris	31	17	12	3	20	6	*	*	4

The impact of parasitism on the experimental cohorts varied both with the sequence of successive cohorts within a year and for comparable cohorts between years. The first cohort to be exposed each year frequently escaped extensive parasitism due to their presence in the field in advance of parasitoid activity. In contrast, the later cohorts often experienced much higher levels of parasitism, as shown in *Table 5*. The data are very variable due to changes in the overall level of parasitism from year to year but do indicate an increasing trend in parasitism of cohorts recollected from the beginning of June through to a peak in early July. This could result from extended instar durations in response to ageing and toughening of foliage as the season progresses, although no data are available to support this hypothesis. Mean levels of parasitism for cohorts recollected between 15 June and 11 July, most closely approximating natural gypsy moth phenology in the region,

Table 5. Parasitism of exposed gypsy moth in relation to relative time of the season.

Recollection date		No. of recollections	Percentage parasitism (± S.D.)
June	1–7	1	0
	8–14	2	6.0 ± 6.0
	15–21	6	12.4 ± 13.5
	22–28	11	19.3 ± 17.3
July	29–5	9	17.9 ± 16.5
	6–12	6	39.9 ± 20.2
	13–19	4	21.4 ± 12.0

are presented in *Figure 1*. Parasitism increased rapidly from 1980 to 1984, largely due to the single species, *Ceranthia samarensis*. Parasitism declined partially in 1985 and substantially in 1987 coinciding with years of unusually cold and wet weather during June, which appears to have affected the activity of all parasitoids but particularly *C. samarensis*. Parasitism also appears to have declined since 1984 following the use of experimental cohorts containing a greater number of individuals and a longer series of successive exposures (*Table 2*). The numbers of hosts recovered from these cohorts, which more closely represents the numbers of hosts available for parasitoid attack, also increased suggesting that the parasitoid potential within the experimental site may have been saturated.

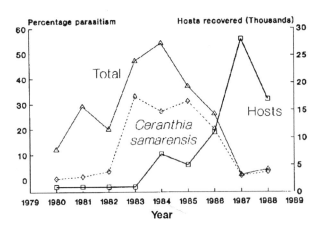

Figure 1.   The percentage parasitism of exposed gypsy moth larval cohorts, recollected between 15 June and 11 July, in relation to the number of hosts recovered. Hosts recovered (□), parasitism by all parasitoids (Δ) and by *Ceranthia samarensis* (◊).

Discussion and conclusions

COMPARISON OF OUTBREAK AND EXPERIMENTAL LOW-DENSITY PARASITISM

A comparison of the parasitoid complex of typical outbreak populations of gypsy moth in Europe (from Fuester *et al.*, 1983) with that of the experimental non-outbreak cohorts in the Alsace (*Table 6*), indicates some important differences. First, and most significantly, *Ceranthia samarensis* is not recorded from outbreaks of gypsy moth and yet dominates the experimental exposures. This suggests that *C. samarensis* is a low host density specialist and a parasitoid that is unable to persist during the development of host outbreaks. The only other record of gypsy moth parasitism by this tachinid is from Austria (Fuester *et al.*, 1983) following the collapse of an outbreak, when host populations had returned to low-density levels.

The relative importance of the other gypsy moth parasitoids also differs between outbreak and non-outbreak host densities. Typical outbreak parasitoids include the tachinids *Blepharipa pratensis* and *Parasetigena silvestris* which attack older larvae and are specific univoltine species (Weseloh, 1974; Godwin and O'Dell, 1981). These parasitoids are replaced by *Compsilura concinnata* and *Apanteles porthetriae* in the experimental non-outbreak populations, although secondarily to *Ceranthia samarensis*. Interestingly, these non-outbreak parasitoids attack earlier host larval instars, a situation that was also noted by Price (1973) for parasitism of *Neodiprion swainei*. The two former species are less specific and are unsynchronized, requiring alternate hosts for persistence throughout the year (Burgess and Crossman, 1929; Ticehurst *et al.*, 1978). In contrast, *C. samarensis* is a much more specific parasitoid and is better synchronized (only a small partial second generation annually) (Mills and Nealis, 1990).

Table 6. A comparison of the most frequent parasitoids in outbreak and non-outbreak populations of the gypsy moth in Europe

	Outbreak	Non-outbreak
Rank order of abundance	*Blepharipa pratensis*	*Ceranthia samarensis*
	Parasetigena silvestris	*Compsilura concinnata*
	Apanteles spp.	*Apanteles porthetriae*
	Hyposoter/Phobocampe spp.	*Parasetigena silvestris*
Source	Fuester *et al.* (1983)	Exposures in Alsace

GYPSY MOTH DYNAMICS AND BIOLOGICAL CONTROL

Gypsy moth populations are eruptive and spreading in North America (Montgomery and Wallner, 1988), exhibit cyclical outbreaks in southern and eastern Europe (Montgomery and Wallner, 1988) but remain at endemic levels with only infrequent outbreaks in central, northern and western Europe (e.g. Wellenstein, 1978). The use of experimental cohorts in regions of low natural host densities in northern USA (this volume, Chapter 25) indicate that no low host density specialist parasitoids occur in North America. The distribution of *Ceranthia samarensis* in Europe appears to be confined to just those regions where gypsy moth does not exhibit outbreak cycles (Herting, 1984; Mihalyi, 1986). It is possible, therefore, that *C. samarensis*, in acting as a specialized low host density parasitoid, is one of the most important factors preventing gypsy moth from producing more frequent outbreaks in central, northern and western Europe.

Ceranthia samarensis has many characteristics which indicate its potential as a biological control agent (Mills and Nealis, 1990). Introductions are currently underway in Ontario, Canada, and it will be interesting to see if this novel control agent can be established in North America and bring about a decline in the frequency of outbreak eruptions.

PARASITOID REGULATION OF ENDEMIC HOST DENSITIES

Ceranthia samarensis provides a further example of the diversity and maturity of the parasitoid complexes of forest insect defoliators (Pschorn-Walcher, 1977). Specific synchronized parasitoids that specialize in attack of low density host populations provide another mechanism, in addition to host plant defences, for the maintenance of endemic population equilibria in phytophagous insects. These two mechanisms are likely to operate simultaneously and may be synergistic in many cases, with the result that the majority of forest defoliators are rare.

The frequency of low density specialist parasitism remains poorly known due to the practical difficulties of working with endemic insect populations (e.g. Miller and Renault, 1976). The use of experimental cohorts to expose hosts to parasitism in regions where natural populations are endemic provides a very valuable tool to examine such parasitoid–host relationships (Price, 1972) and for the collection of low-density parasitoids for use in biological control (Bartlett and van den Bosch, 1964).

Low-density parasitoids have often been used in biological control, where the host has not been of pest status in its region of origin. However, biological control has been less successful when the target is also a pest in its native region and parasitoids have been collected during host outbreaks. At least in the case of forest defoliators, host exposure techniques provide an alternative approach to the selection of biological control agents and provide further opportunities for the control of some notorious pests, such as the gypsy moth.

Acknowledgements

It is a pleasure to thank the many summer students who have assisted in this work and in particular to Peter Fischer and Heather Dewar who have provided much input into the project. Dr B. Herting kindly identified the Tachinidae, Dr J. Papp the Braconidae and Dr K. Horstmann the Ichneumonidae.

References

BARTLETT, B. R. and VAN DEN BOSCH, R. (1964). Foreign exploration for beneficial organisms. In: *Biological control of insect pests and weeds*, pp. 283–304 (ed. P. DeBach). London: Chapman and Hall.

BURGESS, A. F. and CROSSMAN, S. S. (1929). Imported insect enemies of the gypsy moth and the brown tail moth. Washington DC: *United States Department of Agriculture Technical Bulletin* 86.

DAHLSTEN, D. L. and MILLS, N. J. (1990). Biological control of forest insects. In: *Principles and Application of Biological Control* (ed. T. W. Fisher). Berkeley: University of California Press (in press).

DELUCCHI, V. (1982). Parasitoids and hyperparasitoids of *Zeiraphera diniana* (Lep. Tortricidae) and their role in population control in outbreak areas. *Entomophaga* 27, 77–92.

DOANE, C. C. and MCMANUS, M. L. (eds) (1981). The gypsy moth: research toward integrated pest management. Washington DC: *United States Department of Agriculture Forest Service, Technical Bulletin* 1584.

EMBREE, D. G. (1966). The role of introduced parasites in the control of the winter moth in Nova Scotia. *Canadian Entomologist* **98**, 1159–1168.

FUESTER, R. W., DREA, J. J., GRUBER, F., HOYER, H. and MERCARDIER, G. (1983). Larval parasites and other natural enemies of *Lymantria dispar* (Lepidoptera: Lymantriidae) in Burgenland, Austria and Wurzburg, Germany. *Environmental Entomology* **12**, 724–737.

GODWIN, P. A. and O'DELL, P. A. (1981). Intensive laboratory and field evaluations of individual species: *Blepharipa pratensis* (Meigen) (Diptera: Tachinidae). In: *The Gypsy Moth: Research Toward Integrated Pest Management*, pp. 375–394 (eds C. C. Doane and M. L. McManus). Washington DC: United States Department of Agriculture Forest Service, Technical Bulletin 1584.

HERTING, B. (1984). Catalogue of Palearctic Tachinidae (Diptera). *Stuttgarter Beitrage zur Naturkunde, Series A.*, No. 369, 228 pp.

LAWTON, J. H. and MCNEILL, S. (1979). Between the devil and the deep blue sea: on the problem of being a herbivore. In: *Population Dynamics*, pp. 223–244 (eds R. M. Anderson, B. D. Turner and L. R. Taylor). Oxford: Blackwell.

MASON, R. R. (1987). Nonoutbreak species of forest Lepidoptera. In: *Insect Outbreaks*, pp. 31–57 (eds P. Barbosa and J. C. Schultz). New York: Academic Press.

MIHALYI, F. (1986). Tachinidae – Rhinophoridae. *Fauna Hungariae 161*, vol. XV, Part 14–15, Diptera II.

MILLER, C. A. and RENAULT, T. R. (1976). Incidence of parasitoids attacking endemic spruce budworm (Lepidoptera: Tortricidae) populations in New Brunswick. *Canadian Entomologist* **108**, 1045–1052.

MILLS, N. J., FISCHER, P. and GLANZ, W.-D. (1986). Host exposure: a technique for the study of gypsy moth larval parasitoids under non-outbreak conditions. *Proceedings of the 18th IUFRO World Congress, Division 2*, Vol. II, pp. 777–785.

MILLS, N. J. and NEALIS, V. G. (1990). European field collections and Canadian releases of *Ceranthia samarensis* (Diptera: Tachinidae); a parasitoid of gypsy moth. *Bulletin of Entomological Research*, (in press).

MONTGOMERY, M. E. and WALLNER, W. E. (1988). The gypsy moth: a westward migrant. In: *Dynamics of Forest Insect Populations*, pp. 354–375 (ed. A. A. Berryman). New York: Plenum Press.

PRICE, P. W. (1972). Methods of sampling and analysis for predictive results in the introduction of entomophagous insects. *Entomophaga* **17**, 211–222.

PRICE, P. W. (1973). Parasitoid strategies and community organisation. *Environmental Entomology* **2**, 623–626.

PSCHORN-WALCHER, H. (1973). Die Parasiten der gesellig lebenden Kiefern-Buchshornblattwespen als Beispiel für Koexistenz und Konkurrenz in multiplen Parasit-Wirt- Komplexen. *Verhanden Deutsch Zoologische Geselschaft* **66**, 136–145.

PSCHORN-WALCHER, H. (1977). Biological control of forest insects. *Annual Review of Entomology* **22**, 1–22.

SCHULTZ, J. C. (1983).   Impact of variable plant defensive chemistry on susceptibility of insects to natural enemies. *ACS Symposium Series* **208**, 37–54.

STRONG, D. R., LAWTON, J. H. and SOUTHWOOD, T. R. E. (1984). *Insects on
Plants: Community Patterns and Mechanisms.* Oxford: Blackwell.

TICEHURST, M. R., FUSCO, R. A., KLING, R. P. and UNGER, J. (1978).
Observations on parasites of gypsy moth in first cycle infestations in Pennsyl-
vannia from 1974–1977. *Environmental Entomology* **7**, 355–358.

WELLENSTEIN, G. (1978).   *Lymantria dispar* L., Schwammspinner.   In: *Die
Forstschaedlinge Europas. 3 Band. Schmetterlinge*, pp. 335–349 (ed. W.
Schwenke). Hamburg: Paul Parey.

WESELOH, R. M. (1974).   Host-related microhabitat preferences of the gypsy moth
parasitoid, *Parasetigena silvestris. Environmental Entomology* **3**, 363–364.

25
Experimental Manipulation of Gypsy Moth Density to Assess Impact of Natural Enemies

J. S. ELKINTON*, J. R. GOULD*, C. S. FERGUSON*, A. M. LIEBHOLD**, AND W. E. WALLNER†

*Department of Entomology, University of Massachusetts, Amherst, MA 01003 USA
**USDA Forest Service, Northeastern Forest Experimental Station, PO Box 4360, Morgantown, WV 26505, USA
†USDA Forest Service, Northeastern Forest Experimental Station, 51 Mill Pond Rd., Hamden, Connecticut 06514, USA

Introduction

Varley and Gradwell (1960,1968) established a methodology for quantifying mortality and analysing the dynamics of insect populations. These methods included key factor analysis and detection of density- dependent mortality. They were developed for the analysis of long-term data on changes in density and sources of mortality of populations of insects with discrete generations. Various authors have subsequently addressed some of the limitations of these methods or the practical difficulties that typically arise when researchers attempt to apply them. Some of the criticisms are statistical in nature. For instance, Varley and Gradwell's method of detection of density dependence is susceptible to spurious positive correlations (Kuno, 1973; Royama, 1977; Gaston and Lawton, 1987) and a variety of alternative techniques have been suggested (e.g. Bulmer, 1975; Pollard, Lakhani and Rothery, 1987). Another limitation concerns selection of the correct spatial scale on which to conduct such research. Hassell, Southwood and Reader (1987) and Heads and Lawton (1983) have shown that some studies may fail to detect density dependence because they are conducted on an inappropriate spatial scale. Furthermore, correlation does not prove causation. For example, analyses might indicate density-dependent mortality due to a certain parasitoid, but it is entirely possible that both host and parasitoid are responding to some other underlying factor such as overwintering survival which produces synchronous changes in the density of both species.

Other difficulties are more practical in nature. Many forest insects are difficult or impossible to sample, particularly at low population density. Furthermore, most of the analytical techniques require that samples be obtained from the same population over a period of many generations (at least 15). For univoltine insects this requires an unusual level of commitment by individual researchers, as well as by the agencies that fund their work. For all of these reasons, various researchers have advocated experimental manipulation of populations as an alternative to quantifying mortality and density in natural, unmanipulated populations (Murdoch, 1970; Hassell, 1987; Murdoch and Reeve, 1987).

For gypsy moth, *Lymantria dispar* L., sampling problems are especially severe. The larvae are mobile and seek daytime resting locations in the litter or under bark flaps. The proportion of larvae that exhibit such behaviour varies with population density. Consequently, techniques for sampling larvae per unit of foliage, an approach that is used for many forest defoliators, are rarely applied to gypsy moth. Current methods for sampling gypsy moth larvae include counts of frass particles (Liebhold and Elkinton, 1988a,b) or head capsules (Higashiura, 1987). These procedures work well only in populations of high-to-moderate density. Counting larvae under burlap bands (Weseloh, 1987; Wallner, Zarnoch and Devito, 1989) is the only feasible method of enumerating larval populations at very low density, but it is affected by factors that influence larval behaviour and its adequacy as a population index has not been fully evaluated. Finally, many gypsy moth populations remain for many years at densities so low that no method is feasible for quantifying larval density or determining cause of mortality.

Because of these difficulties, long-term studies of gypsy moth dynamics in North America have been lacking despite an intense research effort on this insect. Campbell (1967) conducted analyses of population data collected in New York and Connecticut and concluded that the mortality most highly correlated with generational change was that which occurred during the late larval stage. Further studies indicated that predation by small mammals, particularly *Peromyscus leucopus*, was the principal cause of mortality during this stage in low-density populations (Campbell and Sloan 1977a,b, 1978; Campbell, Sloan and Biazak, 1977). None of these analyses indicated that parasitoids played a significant role in gypsy moth population dynamics. However, most of these studies were based on fairly limited samples.

Here we describe experiments that involve creating populations of gypsy moths of different densities on 1 ha plots in forest stands where the natural populations are extremely sparse. The purpose of these experiments was to see if parasitoids or other natural enemies would cause spatially density-dependent mortality in these artificial populations. These studies were conducted on Cape Cod in eastern

Massachusetts and in Cadwell State Forest in western Massachusetts. Additional experiments were conducted the following year to see if spatially density-dependent mortality would translate into between-generation (temporal) density dependence.

Methods

Egg masses were collected from high-density populations a few weeks before hatch and were rinsed with a 10% formalin solution to kill any viable particles of nuclear polyhedrosis virus, which can be a major source of mortality among neonates from high-density populations. The eggs were then weighed and distributed into small-screen packets which were attached to the boles of trees. After hatching, neonates crawled through the screen mesh and ascended the trees to begin feeding. The eggs were placed in the field in early May and the timing of hatch was monitored. In each case we were able to synchronize the hatch with that of naturally occurring egg masses and with bud break. By measuring the number of neonates emerging from such eggs held in the laboratory we could estimate the total number of neonates emerging. By varying the number of packets or the weight of eggs per packet, we were able to create populations of gypsy moths of any desired density. Details of these experiments have been described in Gould, Elkinton and Wallner (1990), Liebhold and Elkinton (1989a) and Ferguson, Elkinton and Wallner (unpublished).

In western Massachusetts in 1987, we created populations on eight 1 ha plots. We created four densities with two plots at each density ranging from *ca.* 45,000 to 1.1 million neonates per ha. The former density is typical of a moderately low-density population from which we would expect no noticeable defoliation and the latter density is typical of a moderately high-density population from which we might expect some defoliation (Wilson and Talerico, 1981). Egg mass counts were conducted on these sites within 15 m diameter circles (Kolodny-Hirsch, 1986) before we released neonates in 1987 and in the following years. Beginning in 1985 we also made egg mass counts in naturally occurring unmanipulated populations in eight stands near the experimental sites (within 10 km). These counts indicated that the densities in these surrounding populations were at levels less than 3000 neonates per ha during this period. No outbreak populations or defoliation from gypsy moth had been observed in the surrounding region since 1981.

In 1988 we returned to the same eight experimental plots that we had augmented the previous year in Cadwell State Forest. We also selected two new plots that we had not manipulated in 1987. In all of these plots we released 500 000 neonates, a density equivalent to the second highest density we had created in 1987. This number was so much larger than the numbers surviving in the plots from the previous generation, that the resulting number of neonates in all plots was essentially the same. We quantified mortality and changes in density as we had done the previous year. The purpose of this follow-up study was to see if between-generation differences in parasitism would result from the differences in density of adult parasitoids that we had produced the previous year.

On Cape Cod, Massachusetts in 1987 we created two populations in excess of 1 million neonates (Liebhold and Elkinton, 1989a). One of these populations was created with eggs collected in the field, the other with eggs obtained from females that were mated with irradiated males as part of the so-called F1 sterile male programme (Mastro, Schwalbe and O'Dell, 1981). The eggs produced by such females hatch and mature into healthy larvae, but the adults that arise from them are sterile. A large research and development program exists to release such sterilized eggs in new infestations of gypsy moth (Mastro, Schwalbe and O'Dell 1981). The idea behind this programme is that males arising from such eggs will mate with naturally occurring females, the eggs they produce will be sterile and the infestation will be thereby eradicated. One purpose of our Cape Cod study was to see if there was a difference in terms of subsequent mortality between the sterilized and unsterilized larvae.

In each population we measured weekly changes in density using frass traps (Liebhold and Elkinton, 1988a,b) and, near the end of the larval stage when densities were very low, with counts under burlap bands. Mortality from parasitoids and disease was measured by collecting and rearing *ca.* 100 gypsy moth larvae and pupae per plot per week. Mortality caused by each agent was expressed in terms of *k*-values (Varley and Gradwell, 1960) using procedures derived from Royama (1981).

Results

DETECTION OF SPATIAL DENSITY DEPENDENCE

In 1987 the population densities of gypsy moths declined dramatically in all eight plots in western Massachusetts, especially in the high density plots (*Figure 1*). In the two lowest density plots, there was less than a ten-fold difference between the number of neonates released and the numbers entering the next generation (*Figure 1a*). In contrast, the numbers in the highest density plots declined by about four orders of magnitude (*Figure 1d*) to densities that were lower than in the plots where we had released a much smaller number of neonates. Total mortality for the generation was significantly density dependent as indicated by regression of total *K* against log initial density (slope = 1.86, *P* = 0.012). In contrast, most naturally occurring populations in the surrounding stands increased in density (*Figure 2*). During the mid-larval stages, the tachinid *Compsilura concinnata* (Meigen) caused higher mortality than any other agent including residual mortality, which is a catch-all category that includes losses from dispersal and predation. Furthermore, *C. concinnata* was the only agent that exhibited strongly density dependent mortality (*Figure 3*) and was largely responsible for the density dependent decline of the population. Another tachinid *Parasetigena silvestris* (Robineau-Desvoidy), which is a univoltine specialist on gypsy moth, did not show overall density dependence in the mid-larval (*Figure 3*) or late-larval periods (Gould, Elkinton and Wallner, 1990). However, analyses of weekly counts of the large macrotype eggs laid by this agent on the integument of the host indicated that *P. silvestris* attacks were initially inversely density dependent but then switched to being

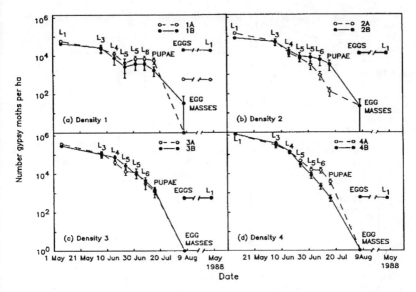

Figure 1. Density of gypsy moths during the summer of 1987 and surviving to the next generation; L_1-L_6 = instars 1–6. Each pair of plots received neonates ranging in density from 45 000 (a) to 1.1 million (d) per ha. (Reprinted from Gould, Elkinton and Wallner, 1990; courtesy of *Journal of Animal Ecology*)

positively density dependent near the end of the larval stage (*Figure 4*). We think that the most likely cause of this pattern is that *P. silvestris* adults were attracted in greater numbers to the higher density plots but the initial difference in host density was so much greater than the difference in *P. silvestris* density, that

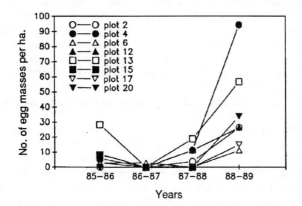

Figure 2. Estimates of gypsy moth egg mass density in naturally-occurring populations on eight sites near (within 10 km) of the experimentally created populations.

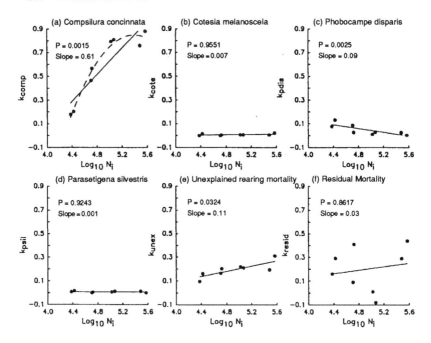

Figure 3.   Regressions of the k-values for six mortality agents acting during the mid-larval stages (instars 3–4) on $\log_{10} N_i$ (the density at the beginning of the period). Solid lines indicate linear trends and dashed lines indicate quadratic trends (if present). (Reprinted from Gould, Elkinton and Wallner, 1990; courtesy of *Journal of Animal Ecology*)

inversely density dependent attack rates resulted. By the end of the larval stage, however, the differences in host density among plots had been greatly reduced due to the action of *C. concinnata*, so that the higher numbers of *P. silvestris* in the higher density plots resulted in positively density-dependent attacks.

We suspect that only differences among plots in parasitoid aggregation could account for the substantial difference in parasitism we observed between high and low density plots. It is possible, however, that some other change in parasitoid behaviour, such as learning or host-switching, resulted in a Type 3 functional response and accounted for at least part of our results.

The largest rate of mortality (k-value) for the entire generation occurred during the pupal stages as indicated by the steep decline in the survivorship curves evident in *Figure 1*. Losses from parasitoids during this period were negligible. Experiments with deployed pupae indicated that most of this mortality was caused by small mammal predation which was, however, *inversely* density dependent (Gould, Elkinton and Wallner, 1990; Elkinton *et al.*, 1989). These results are consistent with the findings of Campbell and Sloan (1977a,b; 1978) who postulated that inverse density dependence in predation by small mammals at intermediate gypsy moth densities was the key to the usually rapid transition from low

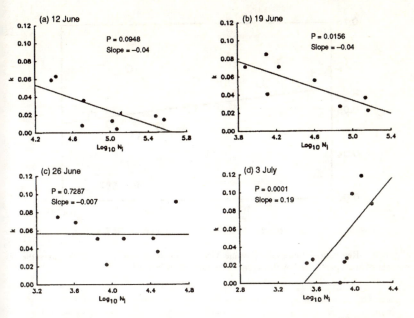

Figure 4. Regressions of the k-values of the oviposition rate of *P. silvestris* on four weekly sample occasions on $\log_{10} N_i$ (the density of larvae prior to the sample occasion). (Reprinted from Gould, Elkinton and Wallner, 1990; courtesy of *Journal of Animal Ecology*.)

density to the outbreak phase of the gypsy moth population system. Loss rates during early instars, including the first instar dispersal phase, were comparatively minor (*Figure 1*), a finding that corresponds to what we have observed in naturally occurring populations (Elkinton *et al.*, 1989).

Similar results occurred on Cape Cod (Liebhold and Elkinton, 1989a). The densities of both of the experimentally elevated populations collapsed to levels that were comparable to surrounding, naturally occurring populations. Furthermore, parasitism by *C. concinnata* and *P. silvestris* was higher in these populations than that recorded in the 8 ha area immediately surrounding the 1 ha experimental plots and in three nearby populations of naturally occurring gypsy moths. Indeed, the level of parasitism caused by *C. concinnata* in the experimentally elevated populations was higher than we had ever observed on any of these sites over the preceding four years, during which the naturally occurring populations fluctuated over a large range of densities (Elkinton *et al.*, 1989). Some of these densities exceeded those which we have created in any of our experimental populations. Thus it seems that it is the contrast in density with the surrounding populations rather than the absolute density in the artificial population that leads to enhanced parasitism. There were also differences in parasitism observed between the two experimental populations (sterilized versus unsterilized larvae). These differences were probably explained by the slower developmental rate of the sterilized larvae and are discussed in Liebhold and Elkinton (1989a).

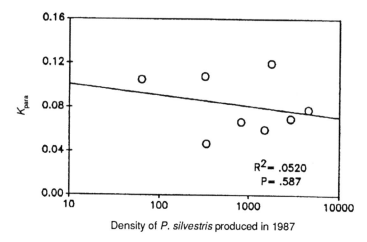

Figure 5.   Regression of mortality (*k*-values) from *P. silvestris* observed in 1988 against the number of hosts parasitized by this agent in the same plots in 1987.

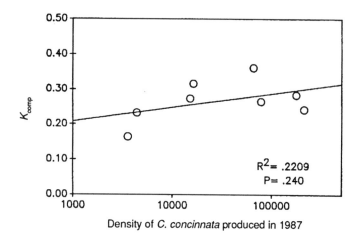

Figure 6.   Regression of mortality (*k*-values) from *C. concinnata* observed in 1988 against the number of hosts parasitized by this agent in the same plots in 1987.

DETECTION OF BETWEEN-GENERATION RESPONSES

In 1988 we attempted to find out whether the spatially density-dependent mortality observed in 1987 would translate into temporal density dependence. In other words, would there be a between-generation numerical response of parasitoids in these populations? We knew from the 1987 results that, compared to the low-density plots, a higher proportion, and thus a much higher number of hosts, were parasitized by *C. concinnata* in the high-density plots. The same was true, to lesser extent, for *P. silvestris*. Most theoretical treatments of host–parasitoid interactions for oligophagous parasitoids such as *P. silvestris* assume that the level of parasitism in a population depends on the number of adult parasitoids produced during the previous host generation. For generalists such as *C. concinnata*, which is multivoltine and depends on alternate hosts, no such link to the previous host generation would be expected. Our experimental strategy to look for such between-generation responses was to create populations of 500 000 neonates on the same eight 1 ha plots in western Massachusetts that we used in 1987, as well as two new sites that had received no gypsy moth eggs the previous year. As in 1987, all the surrounding populations had densities that were less than 10 000 neonates per ha, although we did observe a density increase in these background populations between the two years (*Figure 2*). We hypothesized that we would see higher parasitism caused by the specialist *P. silvestris*, but not the generalist *C. concinnata*, on sites that had received the higher gypsy moth densities the previous year.

The detailed results of the 1988 experiments will be presented in Ferguson, Elkinton and Wallner (unpublished). To quantify between-generation responses we regressed percent parasitism (expressed as a k-value) against the estimated $\log_{10}$ number of adult parasitoids of each species produced on each site the previous year. A significant positive correlation would be evidence for such a between-generation response. Contrary to our expectations, we saw absolutely no evidence for a between-generation response for *P. silvestris* (*Figure 5.*). There was at least an indication of a positive correlation for *C. concinnata*, but the relationship was not statistically significant (*Figure 6*). Furthermore, compared to the eight plots that received gypsy moths the previous year, the levels of parasitism observed in the two new plots in 1988 were no different for *P. silvestris* but lower for *C. concinnata*. Overall survival of gypsy moths was higher in 1988 than in 1987 in the two plots that received equivalent densities of gypsy moths (*ca.* 500 000 neonates) in both years (plots 3A,B; *Figure 1c*). Overall survival of gypsy moths in the surrounding naturally occurring populations was also greater in 1988 than 1987 (*Figure 2*). Parasitism by *C. concinnata* was lower in 1988 compared to 1987, possibly due to the decreased contrast between the experimental and surrounding populations. On the other hand, parasitism by *P. silvestris* was higher than in 1987, possibly due to the increases in abundance that one might expect due to increases in the surrounding host populations.

Discussion

The findings of these experiments indicate that tachinid parasitoids, particularly *C. concinnata*, have a heretofore unrecognized ability to locate and decimate high density populations of gypsy moth on a hectare sized spatial scale. Results from other studies suggest that our findings are not a unique occurrence. Populations of gypsy moth created by researchers at the University of Vermont (using F1-sterilized larvae) have shown high levels of mortality caused by *C. concinnata* (T. O'Dell, personal communication). Some of these experiments were done on a larger spatial scale (25 ha). Several other studies have attempted to create high-density populations by releasing egg masses, but these efforts failed because the populations declined precipitously for reasons that were never determined (T. O'Dell, M. McManus, unpublished data). Our results suggest an explanation for these observations.

Our findings illustrate the important distinction that exists between spatial and temporal density dependence. This distinction has been discussed in the recent ecological literature (e.g. Dempster and Pollard, 1986; Hassell, 1987). Most researchers agree that temporal density dependence is required for a mortality agent to act as a regulatory factor on a host population. Theoretical studies of Nicholson–Bailey host–parasitoid models suggest that parasitoid aggregation to patches of high host density contributes to the stability of host populations (Hassell and May, 1973), but similar investigations involving Lotka-Volterra equations have lead to the opposite conclusion (Murdoch and Stewart-Oaten, 1989). All of these studies involve specialist parasitoids whose numbers are determined by the number of hosts parasitized in the previous generation. Simulations conducted by Hassell (1986) and Latto and Hassell (1988) indicate that, under some conditions, spatially density-dependent mortality translates into temporal density dependence even in the absence of any between-generation numerical response. However, it is not clear whether these models capture the essence of our population system. We found no evidence for any between-generation response by *P. silvestris*, even though it specializes on gypsy moth and apparently plays an important regulatory role in gypsy moth populations in eastern Europe (Sisojevic, 1975; Montgomery and Wallner, 1988; Bogenschutz, Maier and Trzebitzky, 1989). Possibly some mortality factor affecting this parasitoid in North America prevents consistent between-generation numerical responses. Alternatively, perhaps the response occurred but the adults dispersed in the spring before attacking gypsy moth on our plots. With *C. concinnata* our failure to detect a between-generation numerical response is not surprising given requirement of this multivoltine tachinid for alternate hosts.

Our findings also illustrate the importance of spatial scale in experimental studies and raise the question as to what role these agents play in maintaining the apparent stability of low-density populations of gypsy moth. Perhaps, if we manipulated populations on a much larger scale, we would see between-generation numerical responses of *P. silvestris* because there would be less dispersal away from the populations. The level of parasitism caused by *C. concinnata* in the populations we created on Cape Cod exceeded that which we had ever observed in any naturally occurring population over the previous four years of intensive monitoring. It seems likely that if densities rise synchronously over large areas, the spatial response of these parasitoids would be nullified and they would be unable to suppress an increase in gypsy moth populations. Such regional synchronous trends are characteristic of the gypsy moth population system in North America (Liebhold and Elkinton, 1989b). On the other hand, within the general regional trend, there is a lot of spatial heterogeneity in gypsy moth density on the scale of a few hectares. For instance, certain stands known as 'focal areas' or epicentres experience high-density gypsy moth populations more frequently than surrounding 'resistant stands' (Houston and Valentine, 1977; Wallner, 1987). We have often observed outbreak populations of gypsy moths coexisting with low density populations within a distance of a few 100 m. It seems likely that aggregation of parasitoids may act to reduce densities in such patches of high density to levels comparable to that in surrounding, lower-density populations. In this sense it seems that such spatial responses would contribute to the stability (or apparent stability) of low-density populations even in the absence of any between-generation numerical responses. Whether or not parasitoid aggregation could, by itself, stabilize a population system without a between-generation numerical response requires further theoretical investigation.

Acknowledgements

We thank J. Boettner, C. Boettner, P. Grinberg and our summer student employees for assistance with the field work. We thank C. Schwalbe for use of laboratory facilities, D. Arganbright, D. Moran, the Massachusetts District Commission and the Massachusetts Air National Guard for use of the research plots. This research was supported by a Cooperative Agreement (23-142) with the USDA Forest Service and by a grant (87-CRCR-1-2498) from the USDA Competitive Research Grants Program.

References

BOGENSCHUTZ, H., MAIER, K. and TRZEBITZKY, C. (1989). Gypsy moth outbreak and control in southwest Germany, 1984–86. In: *The Lymantriidae: Comparisons of Features of New and Old World Tussock Moths*, pp. 89–99 (ed. W. E. Wallner). Washington DC: United States Department of Agriculture, Forest Service General Technical Report NE- 123.

BULMER, M. G. (1975). The statistical analysis of density dependence. *Biometrics* **31**, 901–911.

CAMPBELL, R. W. (1967). The analysis of numerical change in gypsy moth populations. *Forest Science Monographs* **15**, 1–33.

CAMPBELL, R. W. and SLOAN, R. J. (1977a). Natural regulation of innocuous gypsy moth populations. *Environmental Entomology* **6**, 315–322.

CAMPBELL, R. W. and SLOAN, R. J. (1977b). Release of gypsy moth populations from innocuous levels. *Environmental Entomology* **6**, 323–330.

CAMPBELL, R. W. and SLOAN, R. J. (1978). Numerical bimodality among North American gypsy moth populations. *Environmental Entomology* **7**, 641–646.

CAMPBELL, R. W., SLOAN, R. J. and BIAZAK, C. E. (1977). Sources of mortality among late instar gypsy moth larvae in sparse populations. *Environmental Entomology* **6**, 865–871.

DEMPSTER, J. P. and POLLARD, E. (1986). Spatial heterogeneity, stochasticity and the detection of density dependence in animal populations. *Oikos* **46**, 413–416.

ELKINTON, J. S., GOULD, J. R., LIEBHOLD, A. M., SMITH, H. R. and WALLNER, W. E. (1989). Are gypsy moth populations in North America regulated at low density? In: *The Lymantriidae: Comparisons of Features of New and Old World*

Tussock Moths, pp. 233–249 (ed. W. E. Wallner). Washington DC: United States Department of Agriculture, Forest Service General Technical Report NE-123.

GASTON, K. J. and LAWTON, J. H. (1987). A test of statistical techniques for detecting density dependence in sequential censuses of animal populations. *Oecologia* 74, 404–410.

GOULD, J. R., ELKINTON, J. S. and WALLNER, W. E. (1990). Density dependent suppression of experimentally created gypsy moth, *Lymantria dispar* (Lepidoptera: Lymantriidae), populations by natural enemies. *Journal of Animal Ecology* 59, 213–233.

HASSELL, M. P. (1986). Detecting density dependence. *Trends in Ecology and Evolution* 1, 90–93.

HASSELL, M. P. (1987). Detecting regulation in patchily distributed animal populations. *Journal of Animal Ecology* 56, 705–713.

HASSELL, M. P. and MAY, R. M. (1973). Stability in insect host-parasite models. *Journal of Animal Ecology* 42, 692–726.

HASSELL, M. P., SOUTHWOOD, T. R. E. and READER, P. M. (1987). The dynamics of the viburnum whitefly (*Aleurotrachelus jelinekii*): a case study of population regulation. *Journal of Animal Ecology* 56, 283–300.

HEADS, P. A. and LAWTON, J. H. (1983). Studies on the natural enemy complex of the holly leaf miner: the effects of scale on the detection of aggregative responses and the implications for biological control. *Oikos* 40, 267–276.

HIGASHIURA, Y. (1987). Larval densities and a life-table for the gypsy moth, *Lymantria dispar*, estimated using the head-capsule collection method. *Ecological Entomology* 12, 25–30.

HOUSTON, D. R. and VALENTINE, H. T. (1977). Comparing and predicting forest stand susceptibility to gypsy moth. *Canadian Journal of Forestry Research* 7, 447–461.

KOLODNY-HIRSCH, D. M. (1986). Evaluation of methods for sampling gypsy moth (Lepidoptera:Lymantriidae) egg mass populations and development of sequential sampling plans. *Environmental Entomology* 15, 122–127.

KUNO, E. (1973). Statistical characteristics of the density-independent population fluctuation and the evaluation of density-dependence and regulation in animal populations. *Researches in Population Ecology* 15, 99–120.

LATTO, J. and HASSELL, M. P. (1988). Generalist predators and the importance of spatial density dependence. *Oecologia* 77, 375–377.

LIEBHOLD, A. M. and ELKINTON, J. S. (1988a). Techniques for estimating the density of late instar gypsy moth (Lepidoptera: Lymantriidae), populations using frass-drop and frass production measurement. *Environmental Entomology* 17, 381–384.

LIEBHOLD, A. M. and ELKINTON, J. S. (1988b). Estimating the density of larval gypsy moth *Lymantria dispar* (Lepidoptera: Lymantriidae), using frass-drop and frass production measurement. Sources of variation and sample size. *Environmental Entomology* 17, 385–390.

LIEBHOLD, A. M. and ELKINTON, J. S. (1989a). Elevated parasitism in artificially augmented gypsy moth, *Lymantria dispar* (Lepidoptera: Lymantriidae), populations. *Environmental Entomology* 18, 986–995.

LIEBHOLD, A. M. and ELKINTON, J. S. (1989b). Characterizing spatial patterns of gypsy moth regional defoliation. *Forest Science* **35**, 557–568.

MASTRO, V. C., SCHWALBE, C. P. and O'DELL, T. M. (1981). Sterile-male technique. pp. 669-679 In: *The Gypsy Moth: Research toward Integrated Pest Management*. (eds C. C. Doane and M. L. McManus). Washington DC: United States Department of Agriculture, Forest Service Technical Bulletin 1584.

MONTGOMERY, M. E. and WALLNER, W. E. (1988). The gypsy moth, a westward migrant. In: *Dynamics of Forest Insect Populations,* pp. 353–375 (ed. A. A. Berryman). New York: Plenum Press.

MURDOCH, W. W. (1970). Population regulation and population inertia. *Ecology* **51**, 497–502.

MURDOCH, W. W. and REEVE, J. D. (1987). Aggregation of parasitoids and the detection of density dependence in field populations. *Oikos* **50**, 137–141.

MURDOCH, W. W. and STEWART-OATEN, A. (1989). Aggregation by parasitoids and predators: effects on equilibrium and stability. *The American Naturalist* **134**, 288–310.

ROYAMA, T. (1977). Population persistence and density-dependence. *Ecological Monographs* **47**, 1–35.

ROYAMA, T. (1981). Evaluation of mortality factors in insect life table analysis. *Ecological Monographs* **5**, 495–505.

POLLARD, E., LAKHANI, K. H. and ROTHERY, P. (1987). The detection of density-dependence from a series of annual censuses. *Ecology* **68**, 2046–2055.

SISOJEVIC, P. (1975). Population dynamics of tachinid parasites of the gypsy moth (*Lymantria dispar* L.) during a gradation period (in Serbo-Croatian), *Plant Protection* **206**, 97–170.

VARLEY, G. C. and GRADWELL, G. R. (1960). Key factors in population studies. *Journal of Animal Ecology* **29**, 399–401.

VARLEY, G. C. and GRADWELL, G. R. (1968). Population models for the winter moth. In: *Insect Abundance* pp. 132-142 (ed. T. R. E. Southwood). Symposium of the Royal Entomological Society of London No. 4.

WILSON, R. W. and TALERICO, R. L. (1981). Egg mass density/defoliation relationships. In: *The Gypsy Moth: Research Toward Integrated Pest Management,* pp. 49–50 (eds C. C. Doane and M. L. McManus). Washington DC: United States Department of Agriculture, Forest Service Technical Bulletin 1584.

WALLNER, W. E. (1987). Factors affecting insect population dynamics: differences between outbreak and non-outbreak species. *Annual Review of Entomology* **32**, 317–340.

WALLNER, W. E., ZARNACK, S. and DEVITO, A. (1989). Regression estimators for late-instar gypsy moth larvae at low population density. *Forest Science* **35**, 789–800.

WESELOH, R. M. (1987). Accuracy of gypsy moth (Lepidoptera: Lymantriidae) population estimates based on counts of larvae in artificial resting sites. *Annals of Entomological Society of America* **80**, 361–366.

26
Interaction of Parasitism and Predation in the Decline of Winter Moth in Canada

JENS ROLAND

Boreal Institute for Northern Studies and Department of Botany, University of Alberta, Edmonton, Alberta, T6G 2E9, Canada

Introduction

The most common evidence used to judge the success of an introduced biological control agent is the strong correlation in time between introduction of the natural enemy and the decline of the pest population. To understand host population dynamics, however, it is essential to determine the impact of the natural enemy relative to other mortality factors acting on the host insect population. This is particularly important since insect populations are notorious for exhibiting large changes in abundance, and the pest population can decline regardless of any specific management action. Thus, both the experimental study of mortality factors on insect populations (e.g. Schonfelder *et al.*, 1978; Campbell and Torgersen, 1983; Roland, Hannon and Smith, 1986), and the use of life tables and mortality-factor analysis (Morris, 1957; Varley and Gradwell, 1968) are necessary to partition the effect of different mortality agents during population decline.

One of the most dramatic successes in biological control of forest insect pests has been the introduction of *Cyzenis albicans* (Fall.) (Diptera: Tachinidae) and *Agrypon flaveolatum* (Grav.) (Hymenoptera: Ichneumonidae) for the control of the winter moth *Operophtera brumata* (L.) (Lepidoptera: Geometridae) in oak forests in Nova Scotia, Canada (Embree, 1965, 1966, 1971; Embree and Sisojevic, 1965). The impact of introduced parasitoids was surprising since native populations of winter moth at Wytham Wood, England, were hardly affected by parasitoids, including *Cyzenis albicans* (Varley and Gradwell, 1968; Hassell

Population Dynamics of Forest Insects
© Intercept Ltd, PO Box 716, Andover, Hampshire, SP10 1YG, UK

1980). Among biotic factors affecting populations in Britain, predation on winter moth pupae in the soil by invertebrate predators was the principal regulating factor (Varley and Gradwell, 1968). A more recent introduction of winter moth to the west coast of North America (Gillespie *et al.*, 1978; Embree and Otvos, 1984; Kimberling, Miller and Penrose, 1986; Roland 1988) has provided a third opportunity to examine the impact of parasitoids on winter moth populations. In British Columbia, the two parasitoid species considered responsible for winter moth control in Nova Scotia, were introduced from 1979 through 1981 (Williamson, 1981; Embree and Otvos, 1984). Since then, host populations have declined to approximately one-tenth peak density (Roland, 1989). From the perspective of pest management, the parasitoids introduced for control of winter moth have been very effective in causing population decline.

In this chapter, I examine evidence for population decline of winter moth caused by introduced parasitoids (from life-table data collected in British Columbia from 1982 to 1989), and re-examine published data for populations in Nova Scotia from 1954 to 1961 (Embree, 1965). In each case, the effect of introduced parasites is separated from the effect of other mortality stages. By the use of exclusion experiments, I also examine the relative impact of parasitism and predation during winter moth population decline, in the British Columbia population.

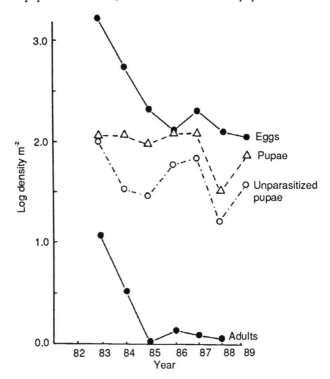

Figure 1. Log-transformed density of four life-stages of the winter moth at Mt. Tolmie Park, Victoria, British Columbia 1983–89.

Temporal pattern of winter moth mortality

Density of winter moth at Mt. Tolmie Park, Victoria, British Columbia, was censused (Roland, 1986a) at four life stages from 1982 to1989 (*Figure 1*). The stages sampled were: egg potential (from female density and mean fecundity), pre-pupae dropping to pupate, unparasitized pupae, and adult females climbing trees to oviposit. From the density estimates, mortality (*k*-values [Varley, Gradwell and Hassell, 1973]) were calculated. Egg and larval mortality (*k*-larv) was estimated from the difference between egg potential and the density of fully fed fifth instar larvae dropping to pupate. Parasitism (*k*-par) was estimated for each tree from

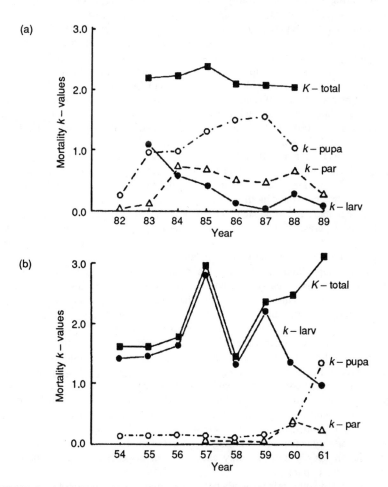

Figure 2. *k*-values for winter moth at (a) Mt. Tolmie Park, Victoria, British Columbia from 1982 through 1989, and (b) at Oak Hill (site 1–2) Nova Scotia from 1954-1961 (from Embree, 1965, 1966, after Roland, 1988).

examination of pupae collected in peat-filled trays set under each of 16 oak trees. Winter moth larvae ingest *Cyzenis* eggs while feeding, but are not killed until after pupation. Pupal mortality (*k*-pupa) was estimated from the known density of unparasitized pupae entering the soil in spring (from pupation trays) to the density of adults emerging in the fall. In addition, mortality of the parasitoid in the soil was estimated from the density of parasitized hosts entering the soil in late spring and the density of adult parasites in emergence traps the following spring.

Parasitism from introduced parasites (almost exclusively *Cyzenis albicans*) rose from zero before introduction, to about 80% in 1984 (*k*-par increasing from 0.0 to 0.70) and has subsequently declined to 47% in 1989 (*k*-par = 0.28) (*Figure 2a*). During this period, mortality of the pupal stage of the winter moth increased from about 25% (*k*-pupa = 0.12) up to 98 percent (*k*-pupa = 1.66) in 1985, and has remained above 90 percent (*k*-pupa >1.00) since then (*Figure 2a*). Mortality of pupae, therefore, increased more (increase of *k*-pupa = 1.54) than did parasitism by introduced parasites (increase = 0.70). It is of interest that pupal mortality increased only after parasitoid introduction; a pattern discussed below.

In Nova Scotia, parasitoids were introduced in 1955 (Embree, 1971). At sites where the contribution of each of several mortality factors can be estimated from published data (Embree, 1965) parasitism rose to a peak of 61% in 1960 (*k*-par increase from 0.00 to 0.41). Parasitism then declined to 50% (*k*-par = 0.30) in 1961 (*Figure 2b*). During the same interval, pupal mortality in the soil increased from about 37% (*k*-pupa = 0.20) to 94% (*k*-pupa = 1.20; *Figure 2b*), an increase in *k*-pupa of 1.00. As in British Columbia, pupal mortality increased more than did parasitism from introduced parasitoids, and did so only after parasitoid introduction. Decline of winter moth in North America is therefore associated with an increase in mortality of pupae in the soil, greater than that due to parasitism from introduced parasites. In Britain, pupal mortality was much higher than was parasitism (Varley and Gradwell, 1968; Hassell, 1978), and was the principal regulating factor for winter moth populations.

In both populations in Canada, pupal mortality rose to about 93–98%, four to five years after parasites were introduced. The increase coincided with the decline of moth numbers. This pattern suggests that there is a strong interaction between parasitism by *Cyzenis* and subsequent mortality of unparasitized pupae in the soil (parasitism by *A. flaveolatum* was never observed above 2% in British Columbia). Two possible mechanisms for the interaction have been suggested previously (Roland, 1988): (1) winter moth are in the soil only five months of the year, compared to 10 months for *Cyzenis*. Build-up of predator numbers may have been precluded until there were abundant prey virtually year-round (i.e. after *Cyzenis* introduction); (2) parasitism and predation are not acting independently of each other (independence is assumed in the use of *k*-values), and predation rises in the presence of parasitism due to this interaction. A third explanation for the rise in pupal mortality is that agents killing pupae in the soil have a minor effect at high population density and can only exert a major effect after populations have declined due to parasitism.

The first hypothesis is presently being evaluated. The remainder of this chapter presents data on the latter two.

Independence of parasitism and other pupal mortality

Preliminary experiments, where pupae were planted in the soil (technique described by Buckner, 1959), indicated that among pupal mortality factors, predation was by far the most important in British Columbia (*Figure 3*). Three experiments were conducted to determine: (1) the impact of different types of predators; (2) the timing of losses due to predation; and (3) potential interaction between parasitism by *Cyzenis* and subsequent predation.

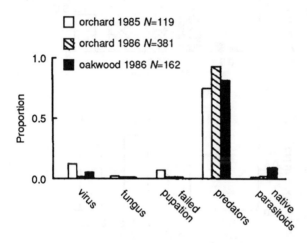

Figure 3. Cause of death for pupae planted in the soil in 1985 and 1986. Total pupal mortality in both years was greater than 90% (from Roland, 1988).

TYPE OF PREDATOR

Sets of 10 pupae were buried 1–2 cm below soil and litter in one of five treatments: (1) no exclusion cage; (2) 11 mm mesh exclosure; (3) 5 mm mesh exclosure; (4) 1.5 mm mesh exclosure; and (5) 0.5 mm mesh exclosure. Each treatment was replicated twice at each of six sites on the Saanich Penninsula, Vancouver Island (described in Roland, 1986a) for a total of 600 pupae. Pupae were planted at the end of May when *O. brumata* pupate, and were recovered at the end of October when adults begin to emerge. Proportions preyed on (arcsin-transformed) in each replicate were analysed with a two-way ANOVA with site and cage type as factors. Predation was affected significantly by cage type ($F = 8.16$, d.f. $=4$, $P<0.001$) with pupae in the 0.5 mm mesh cages suffering less predation than in any of the other cage types (Scheffe's method for multiple comparisons, Sokal and Rohlf, 1981). There was no cage effect among those with mesh size greater than 0.5 mm. Although other predators are present, their impact is apparently masked by the dominant impact of small predators. Overall losses were about 30–40% among all sites in cages with mesh size greater than 0.5 mm, and 11% in the 0.5 mm cage (*Figure 4a*). At the principal study site, Mt. Tolmie, losses were about 70%, except in the 0.5 mm mesh cages where there were no losses. There was no interaction

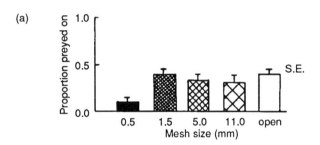

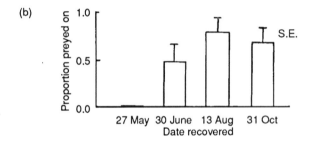

Figure 4.   Proportional loss of winter moth pupae due to predation (a) in cages of differing mesh size at all sites, and (b) when uncaged and recovered on each of three dates after their initial planting (27 May).

between site and cage type ($F = 1.68$, d.f. $= 20$, $P = 0.10$) indicating that predators between 0.5 and 1.5 mm in size in their smallest dimension dominate predation losses at all sites. In Britain, the most important predators of winter moth pupae were larvae of the staphylinid beetle *Philonthus decorus* Gr. (Coleoptera: Staphylinidae) (Frank, 1967; East, 1974; Kowolski, 1976). Staphylinids are abundant at the Mt. Tolmie site especially *Staphylinus aeneocephalus* DeGeer and *Ocypus melanarius* Heer (Roland, 1986b); both introduced species. By their small size, these beetles are implicated as possible major predators.

TIMING OF PREDATION

To determine the timing of pupal losses in 1987, 18 sets of ten pupae each were buried on 27 May, 1987 at the Mt. Tolmie site. Five sets were recovered on 30 June (34 days exposure); five sets recovered on 13 August (79 days); and eight sets of pupae were recovered on 31 October (157 days). By the date of first recovery (30

June), 45% of pupae had been preyed on (*Figure 4b*). By 13 August, 70% were lost, but this was not significantly different from the level on 30 June. There was no further increase in predation to the end of October ($F = 1.47$, d.f. = 2, $P = 0.27$) (*Figure 4b*). Despite the small number of replicates in each time interval, the bulk of the predation losses occurred by mid-summer. In Britain, most predation also occurred by August (East, 1974).

Pit-fall trap data from April 1983 to October 1985, in the British Columbia population, indicate that beetle larvae were abundant each year from April to July (Roland, 1986b); adults were abundant into the winter. Beetle abundance and phenology, therefore, support the suggestion that beetle larvae cause most losses due to predation.

In Nova Scotia, losses of pupae were attributed to small mammals, although winter moth remains were never found in stomach contents (Embree, 1965). Predation by beetle adults and larvae could have caused some of the increased pupal mortality in the Nova Scotia population, but this was not examined.

INTERACTION OF PARASITISM AND PREDATION

This experiment examined the effect of: (1) presence/absence of introduced parasitoids, (2) presence/absence of predatory beetles; and (3) the magnitude of any interaction between them, to determine whether their action is independent of each other. In 1987 and 1988, parasitoids were excluded from a section of each of 10 oak trees by the use of 1.5 mm mesh fly screening. Larvae from these sections of trees were collected and allowed to pupate in soil collected from under each tree. Larvae were also collected from the uncaged part of each tree, providing samples with natural but unknown levels of parasitism. Sets of the two types of resulting pupae were planted under a number of trees. Half of each type was exposed to predation, and half were protected from predation with 0.5 mm mesh cages. For each tree, therefore, four treatments with 10 pupae each were set out (i.e. presence/ absence of parasitoids was crossed with presence/absence of predators). In 1987, a total of 400 pupae were planted under the 10 trees; in 1988, 1000 pupae were planted under 25 trees (loss of some replicates reduced the number to 800 pupae under 20 trees in the analysis). Pupae were buried on 27 May, and were recovered on 31 October when the fate of each pupa was recorded. Fates which pupae suffered were: healthy, parasitized by *Cyzenis*, depredated, infected by pathogens, parasitized by native parasitoids.

For each tree, predation rate was recorded in the absence of parasitism by examining pupal remains and the number of pupae missing. Parasitism for each replicate was not known at the beginning of the experiments because opening cocoons could affect the probability of being preyed on. Parasitism was recorded at the end of the experiment in those sets where predators were excluded; these pupae were inside 0.5 mesh cages and so all remains could be examined for an unbiased estimate. From each of the above two treatments (par and pred), expected total mortality from predation and parasitism was calculated for each tree (1- [(1par)*(1-pred)]), under two assumptions: (1) that the two agents act independently of each other; (2) that parasitism was, on average the same in the caged and uncaged treatments. For example, if parasitism was 0.50 and predation was 0.50, the total expected mortality from both agents would be 0.75 if they act independently. For each tree, the expected total mortality from parasitism and predation was compared to the observed total in the set of pupae exposed to both parasitoids and predators.

In both years, observed total mortality from parasitism and predation was greater than was expected if the two factors had acted independently of each other (*Figure 5*). Because *Cyzenis albicans* attack larvae, not pupae, the parasitism rate of pupae could not change after being planted, therefore, the increased *total* mortality, above that expected, is the result of increased predation on unparasitized pupae. It should be noted that other incidental mortality from pathogens and native parasitoids is not included in the observed total mortality.

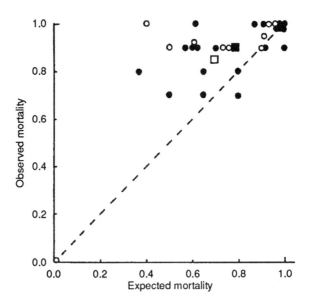

Figure 5. Observed total mortality from parasitism and predation when both agents were present plotted against total mortality expected if parasitism and predation were acting independently of each other. (O) for individual trees in 1987 and (□) for over all values in 1987; closed symbols are for 1988 data.

A direct evaluation of differential predation on the two types of pupae has not been made. However, two sets of data support the above conclusion: (1) Number of winter moth *not* preyed on at the end of the experiment was significanly affected by presence/absence of predators (1988 data: paired t-test, $t = 4.35$, d.f. = 19, P <0.001); numbers of *Cyzenis* not preyed on was unaffected by cage treatment ($t = 1.43$, d.f. = 19, $P = 0.17$). (2) life-table data confirm that in each year, in British Columbia, *Cyzenis* survive better in the soil than do winter moth (*Figure 6*) despite being in the soil twice as long. Most mortality in the soil was due to predation (*Figure 3*) and *Cyzenis* suffer lower mortality indicating that predators do not take many parasitized pupae. Data from Nova Scotia, for the one year in which pupal survival was estimated for both host and parasite (Embree, 1965, 1966), also showed higher survival for the parasite (*Figure 6*).

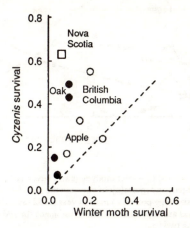

Figure 6.   Survival of *Cyzenis albicans* relative to survival of winter moth in the oakwood (●) and apple orchard (O) in British Columbia, and an oakwood in Nova Scotia (□). Data for Nova Scotia from Embree, 1965 and 1966.

Temporal interaction between parasitism and predation

When each stage-specific mortality is plotted against winter moth population size over the eight years of censuses, it is clear that each is most effective over a different host density range (*Figure 7*). Egg and larval mortality were most important at population densities 1000–2000 eggs per m² and declined as competition among larvae declined (pathogens play a minor role in winter moth populations; Wigley, 1976; Cunningham, Tonks and Kaupp, 1981). Parasitism by introduced parasites peaked when populations had declined to densities of 200–300/m². Predation is currently most important, at densities of 100–150/m². These patterns suggest that predators were precluded from becoming a major factor until winter moth populations declined from the effect of parasitism. There appears, therefore, to be a temporal interaction between parasitism and predation. Such a pattern may also be true for populations in Nova Scotia, but the cause of rising pupal mortality observed there can only be speculated upon.

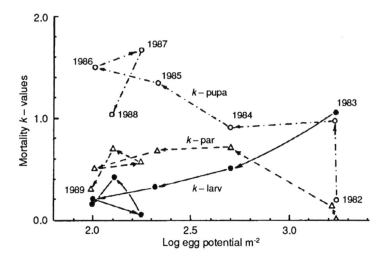

Figure 7.   Mortality (k-values) of winter moth eggs and larvae (k-larv), parasitism from introduced parasitoids (k-par) and mortality of pupae in the soil (k-pred) as a function of winter moth population size (egg potential) in each year (1982–89). Egg potential in 1982 was not measured, but was assumed to be similar to estimates for 1983 based on the same levels of defoliation in the two years.

Discussion

It is unlikely that the success of a biological control programme is due to any single factor introduced to the system. In the case of the winter moth in Canada, it appears that there are strong interactions between parasitism from introduced parasitoids, and other mortality agents already present in the system, especially predation. The combined effect of these, and their interaction is, I suggest, reponsible for the observed successful control of winter moth (not merely the addition of mortality from introduced parasitoids). Predation rises more than does parasitism during the period of population decline for introduced winter moth populations on both the east and west coasts of Canada. The large rise in predation, following parasitoid introduction, is suggested to result from both a within-generation interaction between introduced parasitoids and predators (preferential predation), and an among-generation interaction manifest by their respective impact at different host density.

Several sets of data suggest that the within-generation interaction is the result of small beetle larvae preying preferentially on unparasitized host pupae. The mechanism causing this difference is as yet unknown and has not been evaluated explicitly. Other studies on interaction of mortality agents have implicated difference in location of parasitized and unparasitized individuals, either within host colonies (Tostowaryk, 1971) or within plants (Frazer and van den Bosch, 1973) as a mediating factor. No difference was observed in depth of pupation in the soil for parasitized and unparasitized winter moth larvae (Roland, 1986b) so predators probably encounter the two types at a rate proportional to their abundance. One potential mechanism causing differential predation, is that the fly puparium is simply too hard and smooth for the mandibles of a small predator successfully to grasp and puncture compared to the thinner, more sutured cuticle of the winter moth pupa. This mechanism remains to be tested.

In an attempt at biological control, the pattern of heavier predation on the unparasitized individuals complements the impact of the introduced parasitoid. If the pattern had been the opposite (heavier predation on parasitized hosts), not only would the added mortality from parasitism be nullified (Morris, 1965; Campbell and Torgersen, 1983), but the natural enemy might disappear from the system. Of the few studies where interaction between parasitism and predation has been evaluated, the pattern is for insect predators to take parasitized hosts preferentially (Tostowaryk, 1971; Frazer and van den Bosch, 1973; Jones, 1987). The current study is the only one of which I am aware where unparasitized individuals suffer higher predation by invertebrate predators. I suggest that part of the dramatic success of biological control of winter moth is attributable to this pattern of interaction.

Different mortality factors may have their major impact over different ranges of host density. For example, for gypsy moth *Lymantria dispar* L. (Lepidoptera: Lymantriidae) pathogens exert their greatest effect at very high host abundance, parasites are most effective at moderate density, and predators only have a large impact at low prey abundance (Campbell, 1975; Campbell and Sloan, 1977). Predation on gypsy moth only becomes important after populations are reduced by other agents (Campbell and Sloan, 1977; Gould, Elkinton and Wallner, 1990). A similar pattern appears to hold for winter moth populations in both British Columbia and Nova Scotia. Decline from very high density likely results from the negative effect of chronic severe defoliation of trees on winter moth fecundity (Roland and Myers, 1987) and survival. Further decline has resulted from avian predators (Roland, Hannon and Smith, 1986) and from parasitism, to the point where pupal mortality has become the dominant factor.

Parasitism by *C. albicans* at Mt. Tolmie Park, has subsequently declined from a peak of 80% to 47%. The reduction in parasitism by *Cyzenis* may reflect its indirect mode of attack (oviposition on damaged foliage and subsequent ingestion of eggs by the feeding host larva). Although it can cause high rates of parasitism when hosts are ubiquitous, *Cyzenis* may be inefficient at low density when leaf damage is less likely to be caused by winter moth, and more likely caused by other insects (Embree, 1971). Predators inflict their greatest effect once population density declines below $100-150/m^2$. Because they appear to prey preferentially on unparasitized pupae, the usable prey base is reduced even further by the presence of parasitoids, and their impact further enhanced.

There is no evidence yet that winter moth populations in British Columbia are held at low density by predators or by parasitoids. The magnitude of predation is, however, much higher than is parasitism, and the incidence of spatial density-dependent mortality of winter moth is much more evident for predation than it is for parasitism (Roland, 1989). This pattern was also true for populations in Britain (Kowolski, 1976; Hassell, 1980). As winter moth populations in Canada have declined, the pattern increasingly resembles that observed in Britain, where parasitism plays a minor role, and predation on pupae a major regulating role (Varley, Gradwell and Hassell, 1973).

From the perspective of forest pest management, the introduction of *C. albicans* for control of winter moth has been an unequivocal success. I suggest, however, that from the perspective of understanding the dynamics of the system, much of the success is attributable to factors already present in the system, combined with a fortuitous pattern of interaction of these agents with the introduced parasitoids.

Acknowledgements

Helpful suggestions on an earlier version of this work were provided by A. A. Berryman, D. G. Embree, S. J. Hannon, M. P. Hassell, R. Kowolski, J. H. Myers, I. S. Otvos and J. K. Waage and two anonymous referees. Field studies were assisted by M. A. Smith, T. Danyk and D. Morewood. Staphylinid beetles were identified by A. Smetana. Financial support was provided by NSERC operating grants to J. H. Myers (University of British Columbia) and W. G. Evans (University of Alberta), and by a Canadian Forestry Service Human Resources Grant and a Boreal Institute (Alberta) Grant to the author.

References

BUCKNER, C. H. (1959). The assessment of larch sawfly cocoon predation by small mammals. *Canadian Entomologist* **91**, 275–282.

CAMPBELL, R. W. (1975). *The Gypsy Moth and its Natural Enemies*. Washington DC: United States Department of Agriculture Information Bulletin 381. 27 pp.

CAMPBELL, R. W. and SLOAN, R. J. (1977). Natural regulation of innocuous gypsy moth populations. *Environmental Entomology* **6**, 315–322.

CAMPBELL, R. W. and TORGERSEN, T. R. (1983). Compensatory mortality in defoliator population dynamics. *Environmental Entomology* **12**, 630–632 .

CUNNINGHAM, J. C., TONKS, N. V. and KAUPP, W. J. (1981). Viruses to control winter moth *Operophtera brumata* (Lepidoptera: Geometridae). *Journal of the Entomological Society of British Columbia* **78**, 17–24 .

EAST, R. (1974). Predation on the soil dwelling stages of the winter moth at Wytham Wood, Berkshire. *Journal of Animal Ecology* **43**, 611–626 .

EMBREE, D. G. (1965). The population dynamics of the winter moth in Nova Scotia, 1954–1962. *Memoirs of the Entomological Society of Canada* **46**, 1–57 .

EMBREE, D. G. (1966). The role of introduced parasites in the control of the winter moth in Nova Scotia. *Canadian Entomologist* **98**, 1159–1168 .

EMBREE, D. G. (1971). The biological control of the winter moth in eastern Canada by introduced parasites. In: *Biological Control*, pp. 217–226 (ed. C. B. Huffaker). New York: Plenum Press.

EMBREE, D. G. and OTVOS, I. S. (1984). *Operophtera brumata* (L.) winter moth (Lepidoptera: Geometridae). In: *Biological Control Programmes Against Insects and Weeds in Canada 1969–1980*, pp. 353–357 (eds J . S. Kelleher and M. A. Hulme). Slough: Commonwealth Agricultural Bureaux.

EMBREE, D. G. and SISOJEVIC, P. (1965). The bionomics and population density of *Cyzenis albicans* (Fall.) (Tachinidae: Diptera) in Nova Scotia. *Canadian Entomologist* **97**, 631–639 .

FRANK, J . H . (1967). The insect predators of the pupal stage of the winter moth, *Operophtera brumata* (L.) (Lepidoptera: Hydriomenidae). *Journal of Animal Ecology* **36**, 375–389 .

FRAZER, B. D. and VAN DEN BOSCH, R. (1973) . Biological control of the walnut aphid in California: the interrelationship of the aphid and its parasite. *Environmental Entomology* **2**, 561–568 .

GILLESPIE, D. R., FINLAYSON, T., TONKS, N. V. and ROSS, D. A. (1978) . Occurrence of the winter moth, *Operophtera brumata* (Lepidoptera: Geometridae), on southern Vancouver Island, British Columbia. *Canadian Entomologist* **110**, 223–224.

GOULD, J. R., ELKINTON, J. S. and WALLNER, W. E. (1990). Density dependent suppression of experimentally created gypsy moth, *Lymantria dispar* (Lepidoptera: Lymantriidae) populations by natural enemies. *Journal of Animal Ecology* **59**, 213–233.

HANSKI, I. and PARVIAINEN, P. (1985). Cocoon predation by small mammals, and pine sawfly population dynamics. *Oikos* **45**, 125–136.

HASSELL, M. P. (1978). *The Dynamics of Arthropod Predator-Prey Systems*. Princeton, NJ: Princeton University Press.

HASSELL, M. P. (1980). Foraging strategies, population models and biological control: a case study. *Journal of Animal Ecology* **49**, 603–628 .

JONES, R. E. (1987). Ants, parasitoids, and the cabbage butterfly *Pieris rapae*. *Journal of Animal Ecology* **56**, 739–749 .

KIMBERLING, D. N., MILLER, J. C. and PENROSE, R. L. (1986). Distribution and parasitism of winter moth, *Operophtera brumata* (Lepidoptera: Geometridae) in western Oregon. *Environmental Entomology* **15**, 1042–1046.

KOWOLSKI, R. (1976). Biology of *Philonthus decorus* (Coleoptera, Staphylinidae) in relation to its role as a predator of winter moth pupae [*Operophtera brumata* (Lepidoptera:Geometridae)]. *Pedobiologia* **16**, 233–242.

MORRIS, R. F. (1957). The interpretation of mortality data in studies on population dynamics. *Canadian Entomologist* **89**, 49–69.

MORRIS, R. F. (1965). Contemporaneous mortality factors in population dynamics. *Canadian Entomologist* **97**, 1173–1184.

ROLAND, J. (1986a). Parasitism of winter moth in British Columbia during build-up of its parasitoid *Cyzenis albicans*: attack rate on oak *vs* apple. *Journal of Animal Ecology* **55**, 215–234.

ROLAND, J. (1986b). Success and failure of *Cyzenis albicans* in controlling its host the winter moth. PhD thesis, University of British Columbia, Vancouver.

ROLAND, J. (1988). Decline of winter moth populations in North America: direct versus indirect effect of introduced parasites. *Journal of Animal Ecology* **57**, 523–531.

ROLAND, J. (1989). Parasitoid aggregation: chemical ecology and population dynamics. In: *Critical Issues in Biological Control*, pp. 185–211 (eds M. Mackauer, L. E. Ehler and J. Roland). Andover: Intercept.

ROLAND, J. and MYERS, J. H. (1987). Improved insect performance from host-plant defoliation: winter moth on oak and apple. *Ecological Entomology* 12: 409–414.

ROLAND, J., HANNON, S. J. and SMITH, M. A. (1986). Foraging pattern of pine siskins and its influence on winter moth survival in an apple orchard. *Oecologia* 69, 47–52.

SOKAL, R. R. and ROHLF, F. J. (1981). *Biometry*. San Francisco: W. H. Freeman. 859 pp.

SCHONFELDER, T. W., HOUSEWEART, M. W., THOMPSON, L. C., KULMAN, H. M. and MARTIN, F. B. (1978). Insect and mammal predation of yellow-headed spruce sawfly cocoons (Hymenoptera: Tenthredinidae). *Environmental Entomologist* 7, 711–713.

STAMP, N. E. (1982). Behavior of parasitized, aposematic caterpillars: advantageous to the parasitoid or host? *American Naturalist* 118, 715–725.

TOSTOWARYK, W. (1971). Relationship between predation and parasitism of diprionid sawflies. *Annals of the Entomological Society of America* 64, 1424–1427.

VARLEY, G. C. and GRADWELL, G. R. (1968). Population models for the winter moth. In: *Insect Abundance*, pp. 132–142 (ed. T. R. E. Southwood). Oxford: Blackwell Scientific.

VARLEY, G. C., GRADWELL, G. R. and HASSELL, M. P. (1973). *Insect Population Ecology*. Berkeley: University of California Press.

WIGLEY, P. J. (1976). The epizootiology of a nuclear polyhedrosis virus of the winter moth, *Operophtera brumata* L. at Wistman's Wood, Dartmoor. PhD Thesis, University of Oxford.

WILLIAMSON, G. D. (1981). *Insect Liberations in Canada. 1979. Parasites and Predators*. Agriculture Canada, Research Branch.

27

Host Plant-mediated Impacts of a Baculovirus on Gypsy Moth Populations

JACK C. SCHULTZ*, MICHAEL A. FOSTER* AND MICHAEL E. MONTGOMERY**

*Department of Entomology, Penn State University, University Park, PA 16802, USA
**USDA Forest Service, 51 Mill Pond Road, Hamden, CT 06514, USA

Introduction

The gypsy moth (GM), *Lymantria dispar* L., is an irruptive pest of north temperate deciduous forests. In North America, GM feeding and populations are centered on oaks (*Quercus* spp.) and a few other favoured tree species. Many studies have shown that growth of individual larvae and fecundity of females can be influenced strongly by the foliage on which individuals feed (e.g. Barbosa and Capinera 1977; Hough and Pimentel 1978). As a result, caterpillar performance and population growth can vary dramatically among forest stands (Campbell, 1983; Kleiner, 1989).

The observation that GM growth and fecundity decline over the course of outbreaks in the absence of starvation (Bess, Spurr and Littlefield, 1947; Campbell, 1983) has led to several studies showing that the food quality of oaks and other species declines during intensive defoliation (Wallner and Walton, 1979; Schultz and Baldwin, 1982; Rossiter, Schultz and Baldwin, 1988). Taken together, these studies support the view that GM population dynamics may be influenced by host plant quality (Baker and Cline, 1936; Behre, 1939; Schultz and Baldwin, 1982; Rossiter, Schultz and Baldwin, 1988).

Population Dynamics of Forest Insects
© Intercept Ltd, PO Box 716, Andover, Hampshire, SP10 1YG, UK

There are many other candidate regulatory factors in the GM system. Parasites, predators, weather, 'physiological disease', and pathogens have been suggested to play roles in either the release of populations from innocuous levels, or in their decline (see Campbell, 1983). There is as yet no definitive way to judge the relative importance of the many potential factors involved (see ch. 20, this volume). However, several authors have concluded that naturally occurring populations of a baculovirus, the gypsy moth nuclear polyhedrosis virus (GMNPV), play an important role in the decline of GM populations from outbreak levels (Campbell, 1983; Wallner and McManus, 1983; Woods and Elkinton, 1987).

Worldwide, tree species preferred by the GM are characterized by leaf chemistries comprised almost exclusively of phenolics (Barbosa and Krischik, 1987). Hence, recent studies of the impact of leaf quality on GM growth have focused on this class of allelochemicals. Rossiter, Schultz and Baldwin (1988) showed that GM pupal mass and fecundity vary inversely with the range of phenolic concentrations found in a stand of red oaks (*Q. rubra* L.). In their study, the highest phenolic concentrations occurred in trees subjected to manual defoliation rates of 50% or more; female pupal mass on these trees was reduced by as much as 30%. This suggested that induced changes in leaf phenolics may have a negative impact on GM populations via reduced fecundity.

At the same time, Rossiter, Schultz and Baldwin (1988) reported a positive relationship between red oak condensed tannins and gypsy moth fecundity. Hydrolysable tannins were negatively correlated with fecundity but the regressions were not significant. It appears that the mixture of phenolics in oak leaves may have diverse activities in an insect like the gypsy moth.

Phenolics, including the tannins, are widely known to possess antimicrobial activity (Zucker, 1983), including inhibitory effects on viruses (Cadman, 1960; Mink, Huisman and Sarsena, 1966; Keating and Yendol, 1987). Recently, we have shown that oak phenolics, including tannins, reduce susceptibility of the GM larva to GMNPV (Keating, Yendol and Schultz, 1988; Keating, Hunter and Schultz, 1990). We have shown that this inhibition derives from leaf chemicals actually present in the larval gut when viral polyhedral inclusion bodies (PIBs) are ingested with leaf material; there is no apparent generalized physiological change in larval susceptibility (Keating, 1988; Hunter and Schultz, unpublished data). Hence, GM larvae preferentially ingest tree leaves which can reduce their growth and fecundity, but may also increase their resistance to a common, powerful disease.

In this paper, we describe a study of variation in larval susceptibility to GMNPV among eight tree stands and relate this to the phenolic contents of foliage in those stands. From these results, we derive a regression relationship between leaf constituents and LD_{50} in controlled bioassays. We then compare the relationship between chemistry and mortality due to virus with that between chemistry and fecundity as a way of determining the relative importance of leaf traits on these two aspects of GM life. Finally, we discuss the likely consequences of this foliage/GMNPV/larva interaction for GM population dynamics, with brief reference to published models of forest pest populations.

A study of interstand variation in leaves and susceptibility

MATERIALS AND METHODS

Eight stands at four locations in central Pennsylvania, USA, were selected for study. At each location, we located a ridge-top stand and an adjacent lowland stand for study, because of the frequent assertion that ridge-top sites are prone to more frequent GM outbreaks than are adjacent valleys (Houston and Valentine, 1977). Leaf samples were obtained from 10 red oak trees in each stand during the first week of June 1988. At this time most larvae were entering the third instar at most sites, and it is at this point that the majority of larvae should consume viral PIBs deposited on leaves by dying, infected neonates (Woods and Elkinton, 1987).

Leaves were returned to the laboratory on dry ice, and samples from each tree were analysed separately. After leaf discs were removed for bioassays (below), the remaining leaf material was lyophilized, ground to a fine powder in a UDY air mill, washed in diethyl ether, and extracted in 70% acetone for phenolic analyses (Rossiter, Schultz and Baldwin, 1988; Hagerman, 1988). Analyses carried out are as described in Rossiter, Schultz and Baldwin (1988), and included hydrolysable tannins, proanthocyanidin condensed tannins, total Folin-reactive phenolics, and astringency or protein binding capacity (haemoglobin). Condensed tannin units were derived from a wattle tannin standard (Leon Monnier, Inc., single batch, unnumbered) using a standard curve, and are expressed as wattle tannin equivalent dry leaf mass (*%WTE*). The standard for hydrolysable tannins, total phenolics and astringency was tannic acid (Sigma, Inc., batch #T-0125) and values are expressed as tannic acid equivalent dry leaf mass (*%TAE*).

Discs 10 mm in diameter were punched from fresh leaves and used in bioassays of GMNPV activity. Larvae were obtained from egg masses provided by the APHIS, USDA Methods Development Laboratory (Otis, MA, USA). GM nuclear polyhedrosis virus ('Hamden Isolate') was produced and purified as described by Magnoler (1974). Bioassay methods have been described previously in Keating and Yendol (1987). Briefly, larvae were reared through second instar on an artificial diet, removed from food for 24 h upon moulting to third instar, and then allowed 24 h to consume a single 10 mm disc of hostplant foliage treated with one of 3 doses of GMNPV PIBs; controls received an equal amount of distilled water (zero dose). Five larvae were used per tree per dose (i.e. 20 larvae/tree, 200/stand of 10 trees for a total of 1600 larvae). Following inoculation, larvae were returned to artificial diet and held individually for a 17-day incubation period during which they were checked daily for mortality. Since larvae were treated identically before and after inoculation, differences in mortality among larvae inoculated with the same viral dose but on different leaf discs can be attributed to the direct effect of foliage on virus or the infection process.

LD_{50} values were calculated for each stand using Probit analysis, and linear regressions were employed to ascertain relationships between the chemical components of the leaves from each stand and the LD_{50} derived for each stand (SAS Institute).

RESULTS

Calculated LD_{50}s for GMNPV on leaf discs varied significantly among stands (analysis of variance, $P< 0.05$). Stand LD_{50}s on the most inhibitory foliage were about double those from the least inhibitory stand (*Figures 1–3*). The chemical measures most closely correlated with LD_{50} values on red oak discs were hydrolyzable tannins ($r^2 = 0.72$, *Figure 1*) and astringency ($r^2 = 0.54$, *Figure 2*), both of

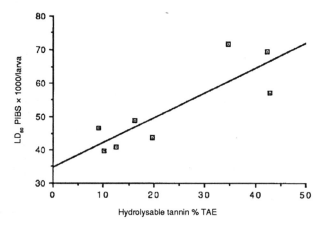

Figure 1. Relationship between mean hydrolyzable tannin concentrations of leaves from 8 red oak stands in Pennsylvania (10 trees/stand) and LD_{50} values calculated for larvae administered GMNPV on leaf discs from the same trees. Regression: $LD_{50} = 34.8 + 0.75$(H Tannin) ($r^2 = 0.72$, $P<0.01$).

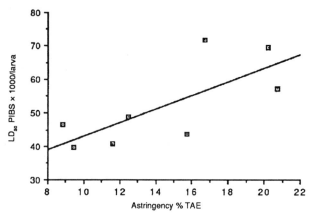

Figure 2. Relationship between mean astringency (binding to hemoglobin) of leaf extracts from eight red oak stands (10 trees/stand) and LD_{50} values calculated for larvae administered GMNPV on leaf discs from the same tress. Regression: $LD_{50} = 23.2 + 2.0$(Astringency) ($r^2 = 0.54$, $P<0.01$).

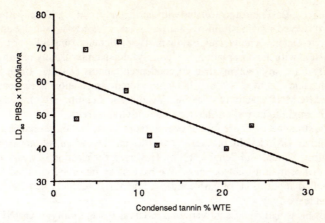

Figure 3. Relationship between mean condensed tannin concentrations of leaves from 8 red oak stands (10 trees/stand) and LD_{50} values calculated for larvae administered GMNPV on leaf discs from the same trees. Regression: $LD_{50} = 63.3 - 0.98$(C tannin) ($r^2=0.34$, $P<0.05$).

which were positively related to LD_{50}. Hydrolysable tannin contents ranged from 9% *TAE* to 43% *TAE*, and protein binding measures ranged from just under 9% *TAE* to over 20% *TAE*. Condensed tannin content was negatively related to LD_{50}s on leaf discs ($r^2 = 0.34$, *Figure 3*). Condensed tannin concentrations ranged from 2.7% *WTE* to 23.4% *WTE*.

DISCUSSION

As found previously (Keating and Yendol, 1987; Keating, Yendol and Schultz, 1988; Keating, Hunter and Schultz, 1989), high hydrolysable tannin contents and high astringency were associated with reduced effectiveness of the GMNPV. Almost three quarters of the among-stand variance in LD_{50}s could be explained by hydrolyzable tannin concentrations in red oak. Astringency was a slightly weaker predictor of LD_{50}s, as has been found previously (Keating, Yendol and Schultz, 1988). These results support the earlier suggestions (Keating, Yendol and Schultz, 1988; Keating, Hunter and Schultz, 1990) that some chemical association, perhaps hydrogen or covalent binding, between phenolics and viral proteins may be responsible for viral inhibition.

A negative relationship was observed between condensed tannin concentrations and LD_{50}s (*Figure 4*). This situation could arise from viral facilitation by condensed tannins, or from negative correlations between hydrolysable and condensed tannin concentrations within oak leaves. We believe that the latter explanation is most reasonable.

In the trees we studied, concentrations of total phenolics, hydrolysable tannins and protein binding were all strongly, positively correlated with each other ($r > + 0.77$). Condensed tannin contents were negatively correlated with hydrolysable tannins ($r = -0.31$) and uncorrelated with astringency ($r = 0.0$). These patterns have been observed previously in several tree species (Baldwin, Schultz and Ward, 1985; Rossiter, Schultz and Baldwin, 1988; Schultz, unpublished), and are consistent with the observation that hydrolyzable tannins usually have a stronger affinity for many proteins than do condensed tannins (Zucker, 1983; Rossiter, Schultz and Baldwin, 1988). If tannin-protein association is responsible for reduced GMNPV activity (Keating, Yendol and Schultz, 1988), hydrolysable tannins ought to be responsible for most antiviral activity.

The strongest evidence for our interpretation is the fact that condensed tannins do inhibit GMNPV in bioassays on artificial diet when administered alone (Keating, Yendol and Schultz, 1988). However, their action is weak compared with that of hydrolyzable tannins. We have never observed viral facilitation in controlled applications of pure condensed tannins, and doubt that they play that role in leaves.

It is clear that significant variation in larval susceptibility to GMNPV exists among stands of red oak, and is caused by interstand variation in leaf traits. Larvae in some stands would have been half as susceptible to virus as in others. If the virus were present, and if it is indeed a significant influence on GM populations, there is the potential for great variation among stands in the population dynamics of the gypsy moth.

Leaf impacts on fecundity versus mortality

Rossiter, Schultz and Baldwin (1988) derived regression relationships between oak leaf phenolic concentrations and female pupal mass for larvae reared on red oaks that were defoliated to various degrees. They found that hydrolysable tannin concentrations and astringency were most closely, and negatively, related to pupal mass. Pupal mass is strongly correlated with fecundity in GM females (Rossiter, Schultz and Baldwin, 1988) so pupal mass could be taken to represent reproductive capacity. Variation in pupal mass in relation to variation in leaf traits may then suggest what the impact of leaf traits on population parameters might be. Their regression equation relating pupal mass (*PM*) to hydrolysable tannins (*HT*) is:

$$PM = 1.98 - 0.01 \ (HT)$$

and is significant at $P = 0.0001$. However, they also performed regressions relating leaf traits to fecundity directly for a subset (15) of the (60) trees in their study. These revealed a negative association between defoliation and fecundity and a positive relationship between condensed tannins and fecundity; no other analyses were significant.

We performed linear and second-order polynomial regressions of the mortality data at each dose (from which LD_{50}s were derived) obtained in the present study on mean stand leaf hydrolysable tannin (*HT*) and condensed tannin (*CT*) concentrations and astringency (*A*) to assess the strength of associations between phenolic measures and viral effectiveness. In general, hydrolysable tannin concentrations and astringency were negatively related to mortality (indicating inhibition of

GMNPV) and condensed tannin measures were positively related to mortality. As one might expect, the amount of variance in mortality 'explained' by chemical measures ranged from 1 to 61%, depending on dose (mortality variance is low at very low and very high doses).

At an intermediate dose (6000 polyhedral inclusion bodies per larva), regressions were:

$$Mortality = 0.68 - 0.01(HT) \; (r^2=0.53)$$
$$Mortality = 0.82 - 0.03(A) \; (r^2=0.34)$$
$$Mortality = 0.16 + 0.01(CT) \; (r^2=0.44).$$

Again, this reflects an apparent positive effect on the virus by condensed tannins, a negative effect by hydrolysable tannins and astringency. Polynomial regressions did not provide significantly better fits to the data.

We have calculated regression relationships between gypsy moth fecundity and these tannin measures from Rossiter, Schultz and Baldwin's (1988) data for comparison with those above. These are:

$$Fecundity = 315 + 12.43(HT) - 0.360(HT)^2 \; (r^2=0.12)$$
$$Fecundity = 445 + 8.7(A) - 0.95(A)^2 \; (r^2=0.18)$$
$$Fecundity = 176 + 57.95(CT) - 3.30(CT)^2 \; (r^2=0.11).$$

Polynomial regressions provide equivalent or better fits than linear in all cases, although no regression provides good explanation of variance in fecundity. The shape of these relationships suggests that both tannin types might have a positive influence on fecundity at low concentrations, switching to negative effects at higher concentrations.

It is evident that the relationship between both types of tannin and fecundity is complex and weak. From a population point of view, these regression relationships seem to suggest that the impact of tannins at low concentrations could be to increase fecundity while decreasing mortality due to virus. At higher concentrations, fecundity may be depressed, but mortality due to virus should continue to be suppressed.

Because condensed tannin concentrations are low whenever hydrolysable tannin concentrations are high in red oak, the nature of tannin effects is likely to depend on the relative concentrations of the two tannin types in these trees. Multiple regressions of fecundity or mortality on hydrolysable and condensed tannins ($Fec = 533 - 2.6HT + 10.7CT$; $Mort = 0.32 - 0.003HT + 0.02CT$) have r^2 of 25% and 44% respectively, but neither represents the non-linear aspects of these interactions. The net effect of tannins cannot be appreciated without understanding these interactions in more detail. Again, since we can show that condensed tannins inhibit (not facilitate) the virus in laboratory tests, it seems likely that their apparent facilitation of viral activity inferred here is a statistical artifact of negative co-correlations among tannin measures (above).

In all of our experiments, every larva ingested viruses; this is unlikely to be true in nature. Hence, the effect of leaf chemistry on mortality rates in populations could be reduced but more variable than we can estimate. The LD_{50} values—

ingested dose required to kill 50% of the larvae in a population—may give a better impression of population effects. The intercepts for both LD_{50} regressions using leaf traits are substantially greater than zero (34 800 and 23 200 PIBs/larva) and well above those calculated on phenol-free diets (200–1000 PIBs/larva). Hence, the susceptibility of larvae or the transmission efficiency of the virus (Anderson and May, 1981) is quite low on leaves, whatever the cause.

We conclude that the ability of the GMNPV to influence GM populations could be influenced strongly by the foliage on which PIBs are ingested. The variation we found in LD_{50}s among stands of red oak in the present study is small (a factor of 2) compared with the differences in LD_{50}s seen when PIBs are ingested on different tree species (a factor of 50 for aspens, *Populus* spp., *versus* oaks; Keating and Yendol, 1987). We speculate that the virus is likely to play a very large regulatory role in stands dominated by trees having leaves which are weakly inhibitory, and may play a smaller role in inhibitory stands. Moreover, induced increases in leaf phenolics (Rossiter, Schultz and Baldwin, 1988) should reduce larval mortality due to virus and perhaps increase fecundity at first. This may facilitate the development of outbreaks, at least in the short term.

Implications for modelling population dynamics

There have been many attempts to model irruptive forest pest populations (Berryman, 1987), using both analytical and simulation approaches. One of the more interesting attempts of the analytical type which incorporates pathogens is that of Anderson and May (1981). Developing a schematic for the insect–pathogen interaction based on assumptions and data about various systems, they suggested that pathogens can generate cyclic population dynamics in their hosts. Indeed, Anderson and May (1980) have identified the gypsy moth as having population dynamics likely to be influenced strongly by pathogens, especially GMNPV.

It is not our intent to re-evaluate completely the Anderson–May models in light of new information about the gypsy moth and the impact of its diet on GMNPV. We do wish to point out several ways in which our findings would influence the outcome of their models, and to indicate departures from their assumptions and possible alterations in the model's structure.

Variation in leaf phenolics should affect four aspects of the Anderson–May models. As we have shown (Rossiter, Schultz and Baldwin, 1988), increasing phenolics should reduce fecundity and hence host population growth rate. However, we have argued that this effect is likely to be small. High or increasing phenolic levels decrease the pathogenicity of the virus (alpha in the Anderson–May model), and hence decrease virus production (lambda) and transmission among hosts. Anderson and May (1981) suggested that pathogens would fail to regulate host populations as pathogenicity was reduced in relation to host reproduction (alpha $<r$). The regression results we have provided here suggest that increasing phenolic concentrations may cause a much more dramatic decrease in pathogenicity than in reproduction. Hence, we suggest that the effect of high constitutive levels of phenolics in tree leaves, or of induced increases in phenolics, will be to reduce any

regulatory influence of GMNPV on larval populations. The net effect should be to destabilize the host population, perhaps leading to outbreak.

There are several important differences between the Anderson–May models and the GM/GMNPV system. The first is that there is no 'recovery' or transition from infected to susceptible class. In addition, GMNPV infection necessarily follows death of larvae. The effect of these differences is to make pathogenicity (their alpha) more influential. Anderson and May (1981) calculated transmission as the product of the number of free-living pathogens (PIBs), the number of susceptible hosts, and the transmission coefficient (rate at which free-living viruses actually infect hosts). In the gypsy moth system, the number of free-living pathogens depends in turn on the mortality already caused by virus. In the Anderson–May model, virus-killed insects are not sources of additional infection. This is clearly not the case in the gypsy moth system: killed larvae are the primary source of free-living PIBs and hence infection.

Leaf phenolic traits ought to reduce the supply of inoculum (by lowering mortality) and efficiency of transmission (by inhibiting infection). As a consequence, we would expect diet quality to have a very important impact on the outcome of models of this type modified to resemble more accurately the gypsy moth system.

Summary and conclusions

We have found that variation in host tree leaf phenolics can influence fecundity and possibly GM population growth, but that the same variation is likely to have a stronger effect on larval resistance to viral disease, and may even increase insect fecundity at low tannin concentrations. Hence, the effect of the host plant's chemistry on GM populations seems likely to be greater via disease mortality than via fecundity and growth. Likely mechanisms involve some interaction between phenolics and viral proteins, although the specific mode of inhibition is unknown.

Substantial variation exists among oak stands in the USA in both leaf chemistry and GM susceptibility to virus. This could produce quite different population trends in various stands, even with the same tree species composition. Variation in population dynamics could be even greater among stands differing in tree species composition when the species present have different phenolic chemistries. We suspect that stand susceptibility to the gypsy moth, and changes in stand susceptibility, may arise from factors which alter leaf chemistry and change GM mortality due to pathogens. The utility of microbial pesticides ought to vary among stands for the same reasons.

The predictive power of population models like those of Anderson and May (1981) is likely to vary dramatically among forest stands, as well as among insect species. In some forests or systems, the pathogen could have a major impact on pest populations but, in others (e.g. with highly inhibitory leaves), models emphasizing the pathogen's influence may not be accurate. The validity of these models depends in part on incorporating hostplant traits. Neither population dynamics nor the nature of the interaction between a forest pest like the gypsy moth and its hosts can be understood without an appreciation of the role of diet-pathogen interactions.

Acknowledgements

We thank M. D. Hunter, K. Kleiner, M. C. Rossiter, W. G. Yendol, and N. A. C. Kidd for comments on the manuscript. E. K. Hollis and V. M. Schultz aided in field and laboratory work. Special thanks to J. L. Colbert of the US Forest Service for help with understanding models and Wind Ridge Farm for permission to sample trees. Research was supported by USDA Forest Service Cooperative Agreement FS-23-155 and USDA grant 88-37251-4046. This paper is a product of the PSU Gypsy Moth Research Center. Dedicated to Virginia Schultz, our main fieldperson.

References

ANDERSON, R. M. and MAY, R. M. (1980). Infectious diseases and population cycles of forest insects. *Science* **210**, 658–661.

ANDERSON, R. M. and MAY, R. M. (1981). The population dynamics of microparasites and their invertebrate hosts. *Philosophical Transactions of the Royal Society of London, Series B* **291**, 452–524.

BAKER, W. L. and CLINE, A. C. (1936). A study of the gypsy moth in the town of Petersham, Mass. in 1935. *Journal of Forestry* **34**, 749–765.

BALDWIN, I. T., SCHULTZ, J. C. and WARD, D. (1985). Patterns and sources of leaf tannin variation in yellow birch (*Betula alleghemiensis*) and sugar maple (*Acer saccharum*). *Journal of Chemical Ecology* **13**, 1069–1077.

BARBOSA, P. and CAPINERA, J. L. (1977). The influence of food on developmental characteristics of the gypsy moth, *Lymantria dispar* (L.). *Canadian Journal of Zoology* **55**, 1424–1429.

BARBOSA, P. and KRISCHIK, V. A. (1987). Influence of alkaloids on feeding preference of eastern deciduous forest trees by the gypsy moth, *Lymantria dispar. American Naturalist* **130**, 53–69.

BEHRE, C. E. (1939). The opportunity for forestry practice in the control of gypsy moth in Massachusetts woodlands. *Journal of Forestry* **37**, 546–551.

BERRYMAN, A. A. (1987). The theory and classification of outbreaks. In: *Insect Outbreaks*, pp. 3–30 (eds P. Barbosa and J.C. Schultz). New York: Academic Press.

BESS, H. A., SPURR, S. H. and LITTLEFIELD, E. W. (1947). Forest site conditions and the gypsy moth. *Harvard Forest Bulletin* **22**.

CADMAN, C. H. (1960). Inhibition of plant virus infection by tannins. In: *Phenolics in Plants in Health and Disease*, pp. 161–164 (ed J. B. Pridham). New York: Pergamon.

CAMPBELL, R. W. (1983). Population dynamics; Historical review. *USDA Forest Service Technical Bulletin* **1584**, 65–86.

HAGERMAN, A. E. (1988). Extraction of tannin from fresh and preserved leaves. *Journal of Chemical Ecology* 14, 453–462.

HOUGH, J. A. and PIMENTEL, D. (1978). Influence of host foliage on development, survival, and fecundity of the gypsy moth. *Environmental Entomology* 7, 97–102.

HOUSTON, D. T. and VALENTINE, H. T. (1977). Comparing and predicting forest stand susceptibility to gypsy moth. *Canadian Journal of Forest Research* 7, 447–461.

KEATING, S. T. (1988). The effect of host plant foliage on the susceptibility of gypsy moth (*Lymantria dispar* L.) larvae to the gypsy moth nuclear polyhedrosis virus. PhD. dissertation, Department of Entomology, Pennsylvania State University, University Park, Pennsylvania, USA.

KEATING, S. T. and YENDOL, W. G. (1987). Influence of selected host plants on gypsy moth (Lepidoptera: Lymantriidae) larval mortality caused by a baculovirus. *Environmental Entomology* 16, 459–462.

KEATING, S. T., HUNTER, M. D. and SCHULTZ, J. C. (1990). Leaf phenolic inhibition of the gypsy moth nuclear polyhedrosis virus: the role of polyhedral inclusion body aggregation. *Journal of Chemical Ecology*, (in press).

KEATING, S. T., YENDOL, W. G. and SCHULTZ, J. C. (1988). Relationship between susceptibility of gypsy moth larvae (Lepidoptera: Lymantriidae) to a baculovirus and host plant foliage constituents. *Environmental Entomology* 17, 952–958.

KLEINER, K. W. (1989). Sources of variation in oak leaf quality as food for the gypsy moth: implications for forest stand susceptibility. Unpublished PhD thesis, Pennsylvania State University.

MAGNOLER, A. (1974). Effects of a cytoplasmic polyhedrosis on larval and postlarval stages of the gypsy moth, *Porthetria dispar*. *Journal of Invertebrate Pathology* 23, 263–274.

MINK, G. I., HUISMAN, O. I. and SAKSENA, K. N. (1966). Oxidative inactivation of tulare apple mosaic virus. *Virology* 29, 437–443.

ROSSITER, M. C., SCHULTZ, J. C. and BALDWIN, I. T. (1988). Relationships among defoliation, red oak phenolics, and gypsy moth growth and reproduction. *Ecology* 69, 267–277.

SCHULTZ, J. C. and BALDWIN, I. T. (1982). Oak leaf quality declines in response to defoliation by gypsy moth larvae. *Science* 217, 149–151.

WALLNER, W. E. and MCMANUS, M. L. (1983). Population dynamics; summary. *USDA Forest Service Technical Bulletin* 1584, 202–203.

WALLNER, W. E. and WALTON, G. S. (1979). Host defoliation: a possible determinant of gypsy moth population quality. *Annals of the Entomological Society of America* 72, 62–67.

WOODS, S. A. and ELKINTON, J. S. (1987). Bimodal patterns of mortality from nuclear polyhedrosis virus in gypsy moth (*Lymantria dispar* L.) populations. *Journal of Invertebrate Pathology* 50, 151–157.

ZUCKER, W. V. (1983). Tannins: does structure determine function? An ecological perspective. *American Naturalist* 121, 335–365.

Population Models and Pest Management

28

A Synoptic Model to Explain Long-term Population Changes in the Large Pine Aphid

NEIL A. C. KIDD

School of Pure and Applied Biology, University of Wales, Cardiff CF1 3TL, UK

Introduction

The large pine aphid, *Cinara pinea* (Mordvilko), is a common parasite of Scots pine (*Pinus sylvestris* L.) throughout the Palaearctic region. It can also be found in those areas of North America where Scots Pine has been introduced and grown on a commercial basis. In common with most aphid species, the insect feeds by extracting phloem sap, in this case from the current year's shoots. As a result, tree growth may be retarded, particularly when population numbers are high. There is some commercial incentive, therefore, for elucidating the factors which influence the abundance of the aphid from year to year.

To this end, *C. pinea* has been the subject of an intensive population study in South Wales since 1977. The results of this study have been discussed in detail elsewhere (Kidd, 1985, 1989) and are only briefly summarized here.

Aphid populations tend to show a single mid-summer peak in abundance each year, with some trees consistently more heavily infested than others. The use of simulation models has revealed that the important factors influencing population numbers are plant quality and natural enemies. Variations in plant quality limit population growth by affecting aphid development, mortality and growth rates, and largely account for variations in peak densities from year to year and tree to tree. Natural enemies influence the number of eggs laid in any year, and thus

Population Dynamics of Forest Insects
© Intercept Ltd, PO Box 716, Andover, Hampshire, SP10 1YG, UK

indirectly affect population levels the following year. Intrinsic density-dependent processes, while shown to occur in this species, were found to be of little significance in determining population numbers in the field.

A number of important questions remain to be answered, however. What, for example, is the nature of 'tree quality' to the aphid? Investigations are currently in progress to test how far variations in plant chemistry affect aphid performance. So far, it has become clear that both amino acids and plant allelochemicals have a role to play (Ch. 17, this volume; Kidd *et al.*, unpublished). Also, where long-term population monitoring has been carried out for 12 years, a 4–5 year cycle in peak numbers has been revealed (*Figure 1*). The causes of this cyclical behaviour have not yet been explored.

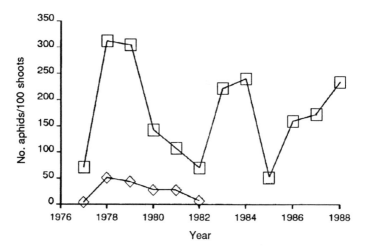

Figure 1.   Average number of C. pinea (□) and its predators (◇) per 100 shoots monitored on one tree between 1977 and 1988.

This paper presents a synoptic analytical model of the pine aphid system, which draws on much of the information gleaned from the more elaborate computer simulation approach (Kidd, 1985, 1989). Despite the model's simplicity, it retains much of the realism and many of the properties of the simulation models. The new model is used here to obtain preliminary insights regarding two possible causes of cyclicity; namely, predation by natural enemies and defensive reactions by the host tree.

The analytical model

Let :

$$S_t = F_t \cdot R_1 \tag{i}$$
$$A_t = S_t \cdot R_2 \tag{ii}$$

$$E_t = f(A_t) \tag{iii}$$
$$F_{t+1} = f(E_t) \tag{iv}$$

where S_t is the peak summer density of aphids in year t, F_t is the peak density of fundatrices in year t, A_t is the peak autumn density in year t, E_t is the density of eggs laid in year t, R_1 is the rate of change in population density between F_t and S_t while R_2 is the rate of change between S_t and A_t.

The function $f(A_t)$ was derived from an equation, relating autumn egg densities to the average density of autumn adults (D_t):

$$E_t = 4.1 \cdot \log D_t + 2.4$$

The derivation of this equation is described in detail elsewhere (Kidd, 1990), but for the purposes of simplicity, autumn peak adult density was here assumed to be equivalent to average aphid density, i.e. $D_t = A_t$. In practice, the two values are unlikely to be very different, as autumn densities in any year varied little from week to week. Thus, in the model:

$$E_t = f(A_t) = 4.1 \log. A_t + 2.4 \tag{v}$$

Approximately 50% of autumn eggs survive to hatch in the spring, this being independent of density (Kidd and Tozer, 1985). Thus, autumn egg densities can be related to fundatrix densities the following year as follows:

$$F_{t+1} = f(E_t) = 0.5 E_t \tag{vi}$$

The rate of change in population numbers between two time periods is determined by a number of factors, some of which operate in a density-independent way, others being density dependent (see Kidd, 1985, 1989).

In the present model, the action of these factors between two time periods is combined and summarized in the single equation :

$$R = M. \exp(-\alpha N) \tag{vii}$$

where R is the net rate of change in population numbers from stage N, α is the coefficient of density-dependence and M sets the maximum rate of change. The larger the value of α, the greater the strength of density dependence acting on the stage.

The equations describing the model now become:

$$S_t = F_t. M_1. exp(-\alpha_1 F_t) \tag{1}$$
$$A_t = S_t. M_2. exp(-\alpha_2 S_t) \tag{2}$$
$$E_t = 4.1. \log A_t + 2.4 \tag{3}$$
$$F_{t+1} = 0.5 E_t \tag{4}$$

with density dependence (α_1 and α_2) acting on both F_t and S_t respectively.

Values of either $M_1 = 25$ or $M_1 = 20$ were used in the simulations, being considered a not unreasonable description of the high population growth rates achieved between the first generation fundatrices of early spring and the high population densities of mid-summer. The action of density-dependent regulating factors is relatively weak during this phase of the life cycle due to a combination of low aphid densities per shoot and high quality foliage (see Kidd, 1989), so a low value of $\alpha_1 = 0.001$ was used.

Between the summer and autumn peaks, however, population numbers generally fall, sometimes dramatically. This is partly due to a reduction in the suitability of the foliage for feeding, which increases the restlessness and mobility of aphids. As population densities are high, interference between aphids is accentuated leading to a multiplying effect on mortality. This is simulated in the model by the simple expedient of making $M_2 < 1$ and increasing the density-dependent feedback effect on R, by making α_2 relatively high, e.g. $\alpha_2 = 0.02$.

With these parameter values, the model produces a pattern of numbers which agrees with the predictions of the computer models when simulating field populations in the absence of natural enemies (*Figure 2a, b*). Peak numbers are constant from year to year in both cases, as yearly differences in tree chemistry are ignored at this stage. Such differences can most easily be incorporated either by making M_1 a uniformly distributed random variable, varying for example between 15 and 25, or by subjecting S_t to a random mortality. This produces a variable peak number from year to year but, given the cyclic nature of longer-term population changes, at least in trees with higher than average population densities, the question arises as to how far such tree-induced changes from year to year are entirely random. Haukioja (1980), for example, has proposed that, where high population densities

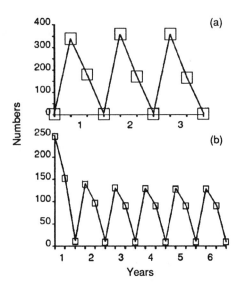

Figure 2.   Examples of aphid population patterns (in the absence of natural enemies) generated by (a) computer simulation, and (b) the analytical model.

of phytophagous tree-dwelling insects induce defensive reactions in their host plants, this may lead to population cycles. The next section explores the possibility that this explanation could account for the 4–5 year period cycles in *C. pinea*.

Modelling aphid-induced plant defences

Where plant defences induced by insect attack are likely to be of short duration, waning in their effects after, say, one year, this can be adequately modelled by incorporating a time delay into the density dependent feedback response. This gives the following modification of equation (5):

$$R = M. \, exp(-\alpha N_{t-1}) \qquad (5)$$

Equations 1 and 2 can now become, respectively:

$$S_t = F_t . \, M_1 . \, exp(-\alpha_1 F_{t-1}) \qquad (6)$$
$$A_t = S_t . \, M_2 . \, exp(-\alpha_2 S_{t-1}) \qquad (7)$$

depending on the stage at which the time delay acts. With the time delay acting on summer peak numbers (i.e. using equation (7)), and the same parameter values as before, the model predicts rapid stabilization of year to year numbers (*Figure 3a*). Thus, short-lived defensive reactions fail, in this case, to induce sustained population cycles.

To mimic the action of prolonged defences, that is, those which may continue to adversely affect the insect population for some years (Haukioja, 1980), a more elaborate model is necessary. In this case I assume that a critical aphid population

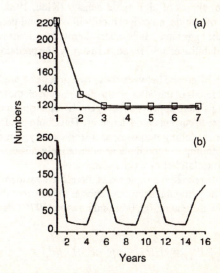

Figure 3. The predicted effects of aphid-induced plant defences on aphid population patterns with the defensive response waning after (a) 1 year, and (b) 3 years.

density triggers a defensive reaction which reduces summer population growth rates (M_1) from 25 to eight and that this effect is sustained over three years. In the fourth year population growth rates return to normal. Incorporating these effects into the model, with a critical threshold density of 100 aphids, produces a five year cycle in aphid abundance (*Figure 3b*). Thus, induced plant defences with long relaxation times could conceivably account for the observed population cycles.

The effects of natural enemies

So far, I have not considered the effects of predators and parasitoids in the model. There is no doubt that these have an important role to play in the population dynamics of *C. pinea*, as inferred from earlier simulation studies, but the interactions have yet to be modelled in detail. Pine aphids are attacked by a large variety of natural enemies, many of which are generalist predators (Kidd, 1989).

Here, I consider only the effects of insect predators, amongst which two basic types of life cycle can be discriminated. First, can be identified those, such as *Phytocoris pini* Kirschbaum (Heteroptera:Miridae) which overwinter on the trees as adults, their reproductive success depending on their ability to capture aphids in spring and early summer. When aphid numbers are low in a particular year, predator reproduction and subsequent survival of offspring is also low, therefore, the number of predators in the following year is determined in large measure by their feeding success in the current year.

Secondly, there are those more loosely coupled predator species, such as syrphids and coccinellids, which as adults actively seek out patches of high prey density for oviposition. On pine, predator oviposition may be carried out preferentially on trees or shoots of high aphid density (Kidd, 1982). Below a certain threshold density of aphids, no oviposition will occur and predators will search elsewhere. After the eggs have hatched, larval growth and survival will depend on the continued availability of aphids and, at maturity, the predators may disperse to new habitats.

In practice, the difference between these two groups of predators may be less pronounced. Individuals of the latter group often remain on the trees to overwinter, while in both groups there appears to be a general redistribution of adults in mid-summer in response to current aphid densities. Thus, both groups of predators can be considered, for modelling purposes, as a single entity, united in the common features of either failing to reproduce or alternatively dispersing from the trees when aphid densities fall below a certain threshold.

In the model, the predator complex was therefore considered in its entirety as one population. The particular submodel chosen to represent this population was the predator model of Beddington, Free and Lawton (1976), described by the following equation:

$$P_{t+1} = c[\{N_t[1 - exp(-aP_t)]\} - \beta P_t] \qquad (8)$$

where P is the predator population, a is the search rate (= area of discovery), c is the conversion efficiency of consumed prey to new predators and β is the minimum prey consumption for reproduction to occur (see Hassell, 1978, for a detailed discussion). A linear relationship is assumed between the *per capita* predator rate of increase and the number of prey consumed during the predator's life. However, realism is sacrificed in a number of important respects. For example, the effects of prey depletion within a time interval are ignored, and the predators have a constant search rate, a. Nevertheless, the predator model is sufficiently useful in the present context to provide some general insights into how predators may influence pine aphid dynamics in the long term.

In the present model the predator rate of increase was related to the peak summer density of aphids, a realistic assumption in that predator reproduction only becomes frequent from around the time of the aphid peak. Thus equation (8) becomes :

$$P_{t+1} = c[\{S_t[1-\exp(-aP_t)]\} - \beta P_t] \qquad (9)$$

With this equation incorporated, the model was able to generate coupled cycles in both predators and prey (*Figure 4a*), the degree of intrinsic density dependence influencing the amplitude of the fluctuations. The cycles showed a similar pattern to those in the field, with a periodicity of 4–5 years (*Figure 1*). Reducing the rate of aphid population growth in the spring (M_1), however, reduces the average population level and removes the tendency for the populations to cycle (*Figure 4b*). This is consistent with observations from the field populations and simulation

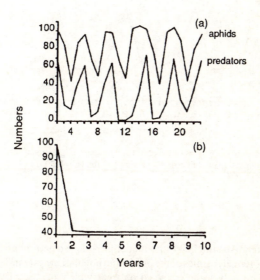

Figure 4. Predicted dynamics of the aphid and predator populations in the absence of aphid-induced plant defences, where the rate of aphid population growth in spring is (a) relatively high ($M_1 = 20$), and (b) relatively low ($M_1 = 10$).

studies. Trees with consistently poorer aphid growth rates and survival have lower population growth rates in early summer, and support smaller populations which do not tend to cycle (Kidd, 1985, 1989).

Combining the effects of predators and plant defences

So far, it appears that both predators and plant defences are each separately capable of causing cycles with a period of approximately five years. But what is likely to be the effect on the pine aphid population if both factors operate simultaneously? This is highly pertinent to the field populations as any induced defence is unlikely to occur in the absence of predation. When both the predation and long-lasting plant defence functions are included in the model together, extinction of the predator population is the usual outcome over a wide range of parameter value combinations. However, a stable five year cycle is possible (*Figure 5*), but, under a very restricted set of conditions, i.e. when intrinsic density dependence is weak (for example, with α_1 and $\alpha_2 = 0.0001$).

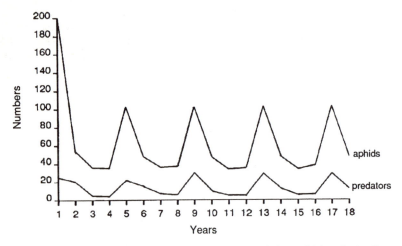

Figure 5.   Predicted dynamics of the aphid and predator populations, with long-lasting (3 years) aphid-induced plant defences also acting (α_1 and $\alpha_2 = 0.0001$)

Discussion

Despite its relative simplicity, the model provides some important new insights into the long-term dynamics of the pine aphid population system. The model shows that cycles are only to be expected on trees which are 'nutritionally' superior and consistently support higher populations of aphids. Both predators and induced defences are shown to be theoretically capable of generating cycles of the periodicity and amplitude found in field populations. Indeed, the predator function does not have to be very complicated to do this. Whether the predator complex of the aphid conforms to the simple predation model used here is, of course, open to doubt; conclusions about their impact on the aphid population therefore have to remain tentative at this stage. However, the model is available for testing with more realistic predator functions at a later stage. Nevertheless one sign of encouragement at present is that the model predicts the same synchrony between aphid and predator population numbers seen in the field (compare *Figures 1*, *4a* and *5*).

While aphid-induced plant 'defences' are shown to be theoretically capable of causing population cycles, it is as yet unclear whether *C. pinea* can elicit such effects in Scots pine. Indeed, the whole question of whether tree aphids can commonly duplicate the kinds of defensive chemical changes brought about by leaf-chewing insects needs to be resolved. One problem is that few aphid species have had this aspect of their ecology studied in adequate detail. Amongst those which have, the evidence is equivocal. The sycamore aphid, *Drepanosiphum platanoidis* (Schr.) does not appear to induce chemical changes of detriment to the aphid (Wellings and Dixon, 1987), but there is some evidence that the lime aphid, *Eucallipterus tiliae* L. can do so, at least when feeding has been sustained at very high population densities for some time (Kidd, 1977; Barlow and Dixon, 1980).

Perhaps the best evidence so far, comes from studies carried out on another closely related *Cinara* species, *C. pini* L., feeding on the bark of Scots Pine (ch. 17, this volume). This aphid lives in highly aggregated colonies on twigs and branches and induces localized chemical and physical changes in the bark, to the detriment of the aphid colonies. As yet, however, we have no information on whether similar chemical or physical defensive reactions can be induced by *C. pinea*. The little indirect evidence which is available suggests that the possibility is unlikely, despite the fact that seasonal and between tree differences in phloem chemistry have a major influence on aphid growth and survival.

First, natural populations on trees are only very loosely aggregated, are widely dispersed throughout the canopy and are maintained at very low population densities (a maximum of four per shoot) by a combination of predation and naturally occurring differences in plant chemistry (Kidd, 1985, 1989). This is in contrast to *C. pini* in which localized aggregations may contain over 100 aphids on an area of equivalent size. In this species the likelihood of an induced response is highly dependent on prevailing localized densities. If the same principle applies to *C. pinea*, it seems unlikely that peak densities of four aphids per shoot would be enough to trigger defensive changes in the tree. Secondly, the hypothesis that the aphids may trigger a rapid defensive reaction which is sustained over a long period, in this case three years, predicts a relatively high peak followed by a rapid drop in numbers the following year (*Figure 3b*). In practice, however, a high peak may be succeeded by an equally high peak the following year, which is not consistent with the hypothesis. Lastly, the fact that the predator–aphid–induced defence interaction appears to be unstable over a wide range of parameter space, suggests that predation and induced defences are unlikely to be operating together in this system. Under the narrow range of conditions where a stable interaction is possible (*Figure 5*), the cycles generated are characterized by even more sharply defined peaks than in the case of induced defences acting alone. This underlines strongly the argument

326   NEIL A.C. KIDD

developed in the previous paragraph. Thus, the available evidence, however incomplete, points to a single factor, predation, generating long-term population cycles in *C. pinea*.

The present modelling exercise has also provided insights of general relevance to recent hypotheses that herbivore-induced plant defences may cause population cycles (e.g. Haukioja, 1980). This may be true when such defences act in isolation from predators, as the present model confirms, but when herbivore-induced defences act in combination with predators a sustainable interaction between predator and prey may be difficult to maintain. Obviously, more detailed modelling of this three-way interaction is needed to provide finer definition of its stability characteristics but, in the meantime, it is clear that future speculations on the impact of induced defences on herbivore population dynamics should not ignore the complicating effects of predation.

Acknowledgements

The work on pine aphids has been funded by the Natural Environment Research Council. I am grateful to Gavin Lewis, Sara Smith, Clive Carter and Mark Jervis for numerous productive discussions which have helped to clarify my ideas.

References

BARLOW, N. D. and DIXON, A. F. G. (1980). *Simulation of Lime Aphid Population Dynamics*. Wageningen: Pudoc.
BEDDINGTON, J. R., FREE, C. A. and LAWTON, J. H. (1976). Concepts of stability and resilience in predator–prey models. *Journal of Animal Ecology* 45, 791–817.
HASSELL, M. P. (1978). *The Dynamics of Arthropod Predator-Prey Systems*. Princeton NJ: Princeton University Press.
HAUKIOJA, E. (1980). On the role of plant defences in the fluctuation of herbivore populations. *Oikos* 35, 202–214.
KIDD, N. A. C. (1977). The influence of population density on the flight behaviour of the lime aphid, *Euallipterus tiliae*. *Entomologia Experimentalis et Applicata* 22, 251–261.
KIDD, N. A. C. (1982). Predator avoidance as a result of aggregation in the grey pine aphid, *Schizolachnus pineti*. *Journal of Animal Ecology* 51, 397–412.
KIDD, N. A. C. (1985). The role of the host plant in the population dynamics of the large pine aphid, *Cinara pinea*. *Oikos* 44, 114–122.
KIDD, N. A. C. (1989). The large pine aphid on Scots pine in Britain. In: *Dynamics of Forest Insect Populations*, pp. 111–128 (ed. A. A. Berryman). New York: Plenum Press.
KIDD, N. A. C. (1990). Population dynamics of the large pine aphid, *Cinara pinea*: simulation of field population. (manuscript submitted to *Res. Pop. Ecol*)
KIDD, N. A. C. and TOZER, D. J. (1985). Distribution, survival and hatching of overwintering eggs in the large pine aphid, *Cinara pinea* (Mordv.) (Hom., Lachnidae). *Zeitschrift für Angewandte Entomologie* 100, 17–23.

WELLINGS, P. W. and DIXON, A. F. G. (1987). Sycamore aphid numbers and population density. III. The role of aphid-induced changes in plant quality. *Journal of Animal Ecology* **56**, 161–171.

29

Understanding the Impact of Natural Enemies on Spruce Aphid Populations through Simulation Modelling

STEPHEN CRUTE AND KEITH DAY

Department of Biological and Biomedical Sciences, University of Ulster, Coleraine, BT52 1SA, Northern Ireland, UK

Introduction

Elatobium abietinum (Walk.) is a pest of Sitka spruce (*Picea sitchensis* [Bong.] Carr.) plantations (Bejer-Petersen, 1962; Bevan 1966; Carter, 1977) and, through direct feeding and resultant defoliation, presents an economic loss to the forest industry (Bevan, 1966) .

Past work on the aphid has centred on its biology (Cunliffe, 1924; Dumbleton, 1932; Fisher, 1982) and description of typical population dynamics observed in the UK (Day 1984, 1986). More recently the focus has changed to the effect of host plant chemistry on aphid performance (Carter and Nichols, 1988) .

The aphid's population dynamics have been outlined by Day (1984) and Dixon (1985) and show a characteristic pattern of growth in aphid numbers in spring to a mid-June maximum followed by a population crash to very low levels in the summer months. Possible reasons for this decline have been proposed and include predation (Hanson, 1952; Hussey, 1952; Bejer-Petersen, 1962), alate production and migration (Bevan, 1966; Carter and Cole, 1977), and summer starvation (Parry, 1974). Day (1984) supported Parry and Powell's (1977) findings that changes in fecundity may be responsible for population collapse. The most frequently accepted view is that changing levels of amino acids in the host sap are responsible for these patterns.

Population Dynamics of Forest Insects
© Intercept Ltd, PO Box 716, Andover, Hampshire, SP10 1YG, UK

The current study is an attempt to integrate past work and, using simulation models, endeavours to provide a fuller understanding of the aphid's biology and ecology. Similar modelling approaches have been used by Barlow and Dixon (1980) on lime aphids, by Carter (1985) on cereal aphids and Kidd (1985) on pine aphids.

It is often assumed that the effects of summer predation on subsequent aphid levels can only be negligible because aphid numbers are generally low in summer. This paper presents results of simulation runs examining the separate and combined effects of Hemerobiids and Syrphids on *E. abietinum* populations and compares predictions from the model with observed field data. Coccinellids were not found in the field samples and are consequently not included in the predation complex.

Methods

MODEL DEVELOPMENT AND DESCRIPTION

Model building has been based upon published data (e.g. Powell and Parry, 1976; Day, 1984, 1986; Fisher, 1987), unpublished data and current laboratory experiments. A VAX Fortran 77 (extended) simulation model implemented on a VAX 8700 has been built and used to test theories of green spruce aphid population dynamics (Crute, in prep.). *Figure 1* shows a simplified relational diagram of the modelled system.

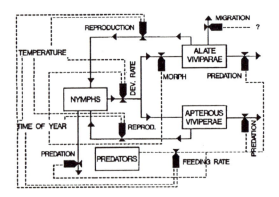

Figure 1. Simplified relational diagram of components in green spruce aphid simulation model. Solid lines represent flow of material around and out of the system, dashed lines are factors having a influence on material flow.

The model is of the discrete deterministic box-car type using temperature as the driving variable. Either daily max/min or hourly temperatures can be used. The latter were used in model runs whose results are presented in this paper. An iteration step of one day is currently used and output is Julian time based whilst aphid development is based on accumulated physiological time (cf. Carter, 1985). The model can take into account a single year or a continuous period for as many years as climatic data is available.

Amino acid levels in needle sap are considered crucial determinants of aphid performance and seasonal population change (Parry, 1974). Results presented here are from a model version which equates levels of phenylalanine with a developmental factor which either increases or decreases the rate of aphid development depending upon time of year. A small amount of additional mortality in the form of summer starvation (Parry, 1979) is also incorporated into this model.

The model is based on the following assumptions:

1. Aphid development is linearly temperature dependent under the field conditions experienced in 1988.
2. Threshold temperatures for aphid development vary with the age of aphids from 6.2 to 4.0 °C (Crute, unpublished data).
3. Lower limits of population size due to predation are set at 0.1 aphids per 100 needles.
4. No net immigration or emigration occurs locally.
5. All aphids are alike in developmental schedules and all are equally liable to predation.
6. Predator numbers which are based upon bi-weekly samples are linearly related between sample dates.
7. Predator feeding rates are dependent upon mean daily temperatures calculated from mean hourly temperatures and use functions derived from current experimental work (Crute, unpublished).

FIELD DATA

Field data are based on a upland forest, Loughermore Forest, Co. Derry at an altitude of 300 m. The sample trees were planted in 1970 and had attained a height of 7 m. Samples were taken weekly during winter and late summer months and bi-weekly during spring and early summer. Sampling was stratified with 25 or 30 shoots of approximately 150 mm length taken from each of three levels in the crown. Aphids and their predators were extracted and counted from shoot material soon after collection. The main predators were Neuroptera: Hemerobiidae (chiefly *Hemerobius micans* Olivier) and Diptera: Syrphidae. Results presented are the means of the combined samples. Nocturnal sampling was conducted during periods when predators were present.

Results

Figures 2–5 present comparisons of model predictions, under varying conditions with actual field data for 1988. *Figure 2* shows output from the model using all subroutines and includes both predator types. The model predictions exhibit a good fit to field data, both in the pattern of the dynamics and the size of the population.

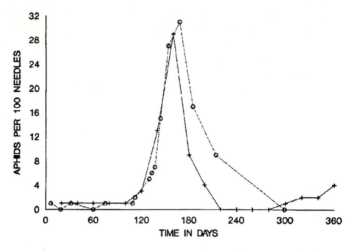

Figure 2. 1988 field data (—–o—–) and predictions of model (— + —) with effects of both Syrphids and Hemerobiids included in the model.

The predicted summer maximum occurs slightly before the actual peak and predictions are advanced compared to field values in the second part of the year. A second small rise in aphid numbers in autumn was predicted but was not observed to such a degree in the field.

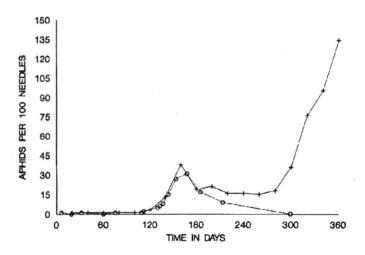

Figure 3. 1988 field data (—–o—–) and predictions of model (— + —), with no predators modelled.

In *Figure 3* the role of natural enemies is removed altogether. A slightly higher value of summer peak population is predicted with a larger population existing during the summer months with a very large increase in October to give an end of year population of 140 aphids per 100 needles.

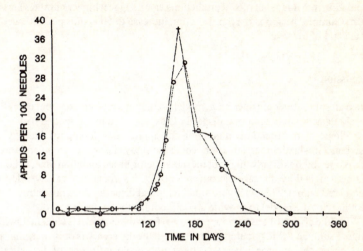

Figure 4.   1988 field data (—–o–—) and predictions of model (— + —), with only Syrphids included in the model.

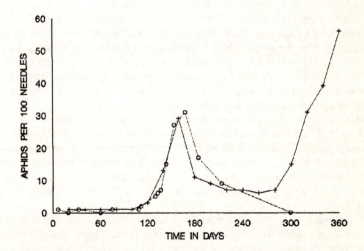

Figure 5.   1988 field data (—–o–—) and predictions of model (— + —), with only Hemerobiids included in the model.

The effects of removal of Hemerobiids from the system are shown in *Figure 4*. This gives a rise in predicted summer peak numbers above both field data and model predictions using both species of predators *(Figure 2)*. Beyond the summer peak the pattern of predictions is similar to *Figure 2*.

The absence of Syrphids from the system *(Figure 5)* has no effect on the summer population maximum but model predictions again suggest higher numbers surviving the summer period and a rapid autumn increase to 60 aphids per 100 needles by the end of the year.

Discussion

When all subroutines are operating, model predictions give an excellent overall fit to field data. A slight temporal disparity is experienced in the second half of the year. Some of the functions modelling the aphid's developmental biology are based on a lowland Sitka plantation whilst the study site was in an upland forest. This is one obvious disadvantage of the model being based on Julian time, although it is outweighed by the ease of conceptualizing daily results over physiological time-based output. This relatively small temporal disparity does not detract from the ability of the model to provide a good tool for theory testing.

When the effects of predators are removed from the system it is clear that, whilst a higher summer aphid population is expected, the difference is only marginal and it is during the autumn where the most noticeable influence of predation is observed. *Figure 3* indicates that a population explosion would occur and we could reasonably expect considerable damage to the host plants. Carter (1977) showed that defoliation of Sitka spruce during winter months causes greater losses in shoot growth than summer defoliation. This would indicate that a healthy predator population whilst being unable to exert any substantial control on summer peaks in normal years plays a vital role in maintaining the autumnal peak in numbers. It is interesting to note that in Iceland where the aphid causes considerable damage it is the autumn population which is the largest (Ottoson, pers. comm.). Further studies could reveal if this is due to a low mid-summer predation rate.

By studying the individual effects of the two main aphidophagous predators found in the system at Loughermore it is possible to demonstrate their relative importance and their impact on aphid population numbers.

Hemerobiids, mainly *Hemerobius micans* in this case, appear earlier in the season than Syrphids. Predators were first observed at the end of May, when the aphid population was already expanding rapidly and these predators were unable to cause any significant reduction in the aphid peak numbers This was in a year with a very mild winter and spring, however, and it seems likely that in years where the overwintering population is low, the effect of Hemerobiids will be more important.

The impact of Syrphid larvae which generally arrive at about the time of the summer peak in aphid numbers is seen to be significant. They reduce the summer population to almost undetectable levels and this has a considerable effect on the number of aphids produced in the autumnal growth phase. Benestad (1970) states that syrphid larvae are able to withstand long periods without feeding and personal observations support this view. This makes syrphid larvae ideal summer predators of the green spruce aphid which continues to reproduce during the summer months but at a lower rate than at other times of the year due to the unsuitability of the host plant sap (Parry, 1974).

Large-scale chemical control of the green spruce aphid is uneconomic but if any form of treatment was considered on nursery trees the timing of pesticide application would be crucial so as to minimize the risk of mortality to natural predators. To maintain maximum needle retention, Carter and Cole (1977) suggest spraying between autumn and late winter. Both Bevan (1966) and Parry (1977) recommend early April applications for control of the spring damage, spraying at this time should avoid killing potential aphid predators. The model could be converted to a predictive form and used to give an indication of years with possible high levels of autumn damage based on late summer aphid population size. Pesticide application would take place once Syrphid larvae were no longer vulnerable.

The usefulness of predators, particularly of Syrphids, needs further investigation for those areas where high autumn numbers regularly occur. Syrphid adults require pollen prior to egg production (Dixon, 1959) and personal observation would suggest that whilst some aphidophagous species are present during early May, eggs do not appear until some time afterwards. Early flowering, pollen-rich plants in forest rides may therefore be beneficial.

In conclusion, it has been shown that the model is providing an ideal tool for furthering our knowledge of part of the green spruce aphid's ecology as well as highlighting areas in which our knowledge can be improved. It is suggested that the role of predators is an important one, especially in determining aphid numbers during late summer and autumn and therefore the numbers of aphids that enter the winter period.

Acknowledgements

SJC acknowledges continuing financial support from the Dept. of Agriculture (NI). We would like to thank Joy Canning for counting needles and Sam Smyth for assistance in the field.

References

BARLOW, N. D. and DIXON, A. F. G. (1980). *Simulation of Lime Aphid Population Dynamics*. Wageningen: Pudoc.

BEJER-PETERSEN, B. (1962). Peak years and the regulation of numbers in the aphid *Neomyzaphis abietina* Walker. *Oikos* **16**, 155–168.

BENESTAD, E. (1970). Food consumption at various temperature conditions in larvae of *Syrphus corollae* (Fabr.) (Dipt., Syrphidae). *Norsk Entomologist Tidsskrift* **17**, 87–91.

BEVAN, D. (1966). The green spruce aphid *Elatobium (Neomyzaphis) abietinum* Walker. *Scottish Forestry* **20**, 193–201.

CARTER, C. (1977). *Impact of Green Spruce Aphid on Growth.* Forestry Commission Research and Development Paper. 116. London: HMSO.

CARTER, C. and COLE, J. (1977). Flight regulation in the green spruce aphid (*Elatobium abietinum*). *Annals of Applied Biology* **86**, 137–151.

CARTER, C. and NICHOLS, J. F. A. (1988). *The Green Spruce Aphid and Sitka Spruce Provenances in Britain.* Forestry Commission Occasional Paper. No 19. Edinburgh.

CARTER, N. (1985). Simulation modelling of the population dynamics of cereal aphids. *Biosystems* **18**, 111–119.

CUNLIFFE, N. (1924). Notes on the biology and structure of *Myzaphis abietina*, Walker. *Quarterly Journal of Forestry* **18**, 133–141.

DAY, K. R. (1984). The growth and decline of a population of the green spruce aphid *Elatobium abietinum* during a three year study, and the changing pattern of fecundity, recruitment and alary polymorphism in a Northern Ireland forest. *Oecologia* **64**, 118–124.

DAY, K. R. (1986). Population growth and spatial patterns of spruce aphids (*Elatobium abietinum*) on individual trees. *Journal of Applied Entomology* **102**, 505–515.

DIXON, A. F. G. (1985). *Aphid Ecology*,157 pp. Glasgow and London: Blackie.

DIXON, T. J. (1959). Studies on oviposition behaviour of Syrphidae (Diptera). *The Transactions of the Royal Entomological Society of London* **111**, 57–80.

DUMBLETON, L. J. (1932). Report on spruce aphis investigation for the year ending December 1930. *New Zealand Journal of Science and Technology* **13**, 207–220.

FISHER, M. (1982). Morph determination in *Elatobium abietinum* (Walk.) the green spruce aphid. Unpublished PhD thesis, University of East Anglia.

FISHER, M. (1987). The effect of previously infested spruce needles on the growth of the green spruce aphid, *Elatobium abietinum*, and the effect of the aphid on the amino acid balance of the host. *Annals of Applied Biology* **111**, 33–41.

HANSON, H. S. (1952). The green spruce aphis, *Neomyzaphis abietina* Walker. *Report on Forest Research 1951*, 98–104.

HUSSEY, N. W. (1952). A contribution to the bionomics of the green spruce aphis, *Neomyzaphis abietina* Walker. *Scottish Forestry* **6**, 121–130.

KIDD, N. A. C. (1985). The role of the host plant in the population dynamics of the large pine aphid *Cinara pinea*. *Oikos* **44**, 114–122.

PARRY, W. H. (1974). The effects of nitrogen levels in Sitka spruce needles on *Elatobium abietinum* populations in north-eastern Scotland. *Oecologia* **15**, 305–320.

PARRY, W. H. (1978). A reappraisal of flight regulation in the green spruce aphid, *Elatobium abietinum*. *Annals of Applied Biology* **89**, 9–14.

PARRY, W. H. (1979). Summer survival of the green spruce aphid, *Elatobium abietinum*, in North East Scotland. *Oecologia* **27**, 235–244.

PARRY, W. H. and POWELL, W. (1977). A comparison of *Elatobium abietinum* populations on Sitka spruce trees differing in needle retention during aphid outbreaks. *Oecologia* **27**, 239–252.

30
Synchronous Fluctuations in Spatially Separated Populations of Cyclic Forest Insects

DAVID A. BARBOUR

Forest Insect Surveys, 62 North Gyle Loan, Edinburgh, EH12 8LD, Scotland, UK

Introduction

Myers (1988) reviewed the general characteristics of cyclic populations of forest Lepidoptera, with particular reference to 18 pest species in North America and Europe. Among the most notable common characteristics was the tendency for populations to fluctuate in synchrony over large geographic areas, although with some variability particularly in the timing of population upsurges. Particularly good examples of this phenomenon include *Oporinia autumnata* Borkhausen (Tenow, 1972), *Orgyia pseudotsugata* McDunnough (Harris, Dawson and Brown, 1985), *Zeiraphera diniana* Guenée (Baltensweiler and Fischlin, 1988), and several species of Lepidoptera and Hymenoptera (Klimetzek 1979, and ch. 1 this volume). For one of the most important pests, the pine looper *Bupalus piniaria* (L.), Klimetzek, in his historical analysis for North Bavaria and the Rheinpfalz, found that the occurrence of outbreak populations was highly significantly correlated across three adjacent areas spanning about 240 km ($r = 0.37 - 0.44$, $P<0.001$). Between North Bavaria and the Rheinpfalz, separated by some 320 km the correlation was almost as strong ($r = 0.21 - 0.38$, $P<0.001$).

In Britain, *B. piniaria* is an occasional pest of Scots pine (*Pinus sylvestris*) plantations. Patterns of population fluctuation have been analysed by Barbour (1980, 1988). A few populations showed cyclic fluctuations similar to those in Germany but, in general, cycles were not well synchronized between different forests. Good examples are Culbin Forest and Roseisle Forest in north Scotland. These two large pine forests on coastal sand-dune sites are separated by less than

5 km at the closest point. *B. piniaria* populations in both showed cyclic fluctuations of similar period and amplitude (*Figure 1*). However, for most of the time from 1954 to the present the fluctuations have been out of synchrony, sometimes almost opposite in phase. The calculated correlation between these two series is not significantly different from zero ($r = 0.007$). The question arises: why is there a conspicuous lack of synchrony in the Scottish populations in contrast to the pattern of geographical synchrony seen in the German ones?

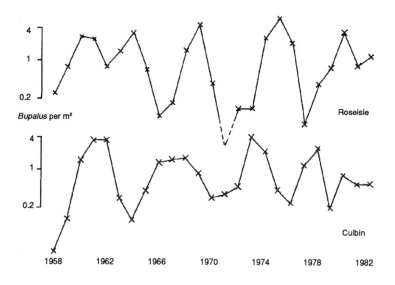

Figure 1.   Population fluctuations in *Bupalus piniaria* the pine looper moth in two closely adjacent forests in north Scotland in 1958–82.

Population cycle synchronization hypotheses

A commonly cited explanation for the synchronization of insect outbreaks over large areas has been the effect of weather. It has been suggested that cycle synchronization might be brought about by catastrophic weather conditions (occurring perhaps only once in many generations) which would bring populations to extreme low levels over a wide area (Moran, 1953, Berryman, 1981). Cyclic populations throughout the area would then 'start again' from a common starting point. However, there seems to be no good published example of such a weather factor affecting a cyclic insect population. Theoretically, too, the explanation seems unlikely, as different populations would be reduced to different degrees and there is no specific 'floor' in population density from which the new increases would begin. Martinat (1987) in a wide-ranging review of climate and weather effects in forest insect outbreaks, concluded that a major influence of weather in the population dynamics of forest insects has seldom been unequivocally demonstrated.

An alternative explanation for synchronization of population cycles is that of intermigration. In periodic events such as locust plagues, mass migrations are easily observed and so are readily accepted as being a factor in the synchronization of spatially separated populations (Rankin and Singer, 1984). In forest insect populations, large-scale emigration is more seldom witnessed, although it has been documented for *Tortrix viridana* L.(Du Merle and Mazet 1985), *Z. diniana* (Vaclena and Baltensweiler, 1978), and for *Choristoneura fumiferana* Clemens (Greenbank, Schaffer and Rainey, 1980). In the majority of cases, however, dispersal between populations seems to occur on a relatively small scale, and intuitively it might seem unlikely that low levels of population interchange could significantly affect synchronization. The existence of two forests only 5 km apart whose populations cycled quite independently would seem to confirm this.

However, the key to the problem of synchronization or the lack of it may well lie in the amplitude of cycles in the different populations. The amplitude of population cycles in some German pine looper populations is close to the highest that has been recorded for any insect population (Williamson, 1972). The amplitude of cycles in Britain is much lower, just as the severity of the insect as a pest is less. This can be quantified using 'Williamson's Index' (Williamson, 1972, 1984) which is the antilog of three times the standard deviation of population size measured logarithmically. This gives a measure of the multiplicative range of the population between peaks and troughs of the cycle (*Table 1*). Suppose two neighbouring populations of the insect whose cycles are out of phase to the extent that the population maximum of one coincides in time with the population minimum of the other. In one case the ratio of the two populations is approximately 2000:1, in the other the ratio is only 40:1 (*Table 1*). The effect of a proportionally small dispersal of moths from the higher to the lower population may be very different in the two cases.

Table 1. Amplitude of Population Cycles of *Bupalus piniaria* in Germany and Britain (data from Varley, 1949 and Forestry Commission, unpublished)

B. piniaria population	Williamson's Index
Letzlingen, 1881–1940	2240
Roseisle Forest 1958–82	71
Culbin Forest 1958–82	39

A Model

A simple model may be used to compare the properties of high and low amplitude cycles with respect to intermigration and synchronization. A cyclic population is modelled by the equation

$$R_t = N_{t+1} - N_t = -0.75N_{t-1}$$

where N is the logarithm of log population density, R is the logarithm of population change and the equilibrium density is assumed to be 1.0. This model generates cycles with a period of 6.6 generations, which will gradually damp to equilibrium unless a stochastic term is added. A stochastic term ε, representing the random effects of weather and other density-independent factors, maintains the amplitude of the cycles, such that amplitude is directly proportional to the variability of the stochastic term. Thus

$$R_t = -0.75N_{t-1} + \varepsilon_t$$

By varying the standard deviation of the stochastic term we can increase or decrease the amplitude of the resultant cycles without changing their other mathematical properties.

If two cyclic populations develop independently, influenced by different random factors, their cycles, though similar in amplitude and period, will rapidly tend to get out of phase. In some time periods, cycles may quite incidentally run together in phase, at other times they may drift apart. We now consider to what extent two such populations could be synchronized by a small intermigration of individuals from one population to the other.

The model considers only the simple case of two nearby but distinct populations, fluctuating cyclically and each generation exchanging one-tenth of the individuals from each population. This could represent, for example, the population of two neighbouring forests on opposite sides of a valley where an essentially random displacement of individuals (by wind, etc.) caused a small amount of interchange in each generation.

An example of a simulation using this model is shown in *Figure 2*, using a stochastic term with mean zero and S.D. = 0.225. Williamson's Index (mean of two populations; *WI*) was 10. The two populations developed largely independently, peaking at different times, and the correlation coefficient between the two series was $r = 0.295$. This r-value with 50 observations is barely significant ($P < 0.05$).

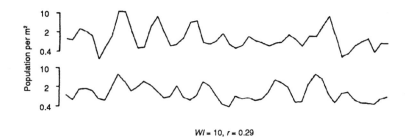

$WI = 10$, $r = 0.29$

Figure 2. A 50-generation simulation of two populations of a forest insect with a *low*-amplitude cyclic fluctuation, exchanging 10% of individuals from each population in each generation.

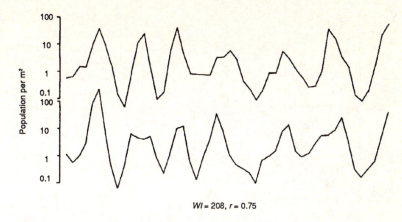

$WI = 208, r = 0.75$

Figure 3. A 50-generation simulation of two populations of a forest insect with a *high*-amplitude cyclic fluctuation, exchanging 10% of individuals from each population in each generation.

A similar simulation giving a higher-amplitude fluctuation is shown in *Figure* 3. Here the stochastic term was given a mean of zero and S.D. = 0.45. In this case WI was 208. The two populations showed much greater synchrony in movement, often peaking in the same years and the correlation coefficient between the two series was $r = 0.750$. This r-value with 50 observations is highly significant ($P<0.001$).

The stochastic input to this model results in very variable performance in terms of amplitude of the cycles. Twenty-five simulations were run with ε given a standard deviation of 0.225 (as in *Figure 2*) and 25 more with ε given a standard deviation of 0.45 (as in *Figure 3*). The results were analysed in relation to cycle amplitude, measured by the *WI*, and degree of synchronization, measured by the correlation coefficient r. The results are illustrated in *Figure 4*. Considering each group separately, within each there was a strong relationship between degree of synchronization and amplitude of the cycles. Between the two groups, mean cycle amplitude changed from WI = 12 to WI = 182 and mean correlation between the population pairs changed from $r = 0.381$ to $r = 0.671$.

In summary, for two sets of simulations differing in a single parameter, the degree of correlation between population pairs both within and between sets was strongly related to amplitude of the fluctuations. Put more simply, the wider the amplitude of the cycles the more closely the populations became synchronized.

The two sets of simulation results shown in *Figure 4* may well fit a single curve, characteristic for the 10% rate of population exchange, and which is likely to be robust to different mathematical forms of the cyclic population model. Different rates of population exchange would result in different curves, but the general finding is established that surprisingly low amounts of intermigration can synchronize cycles which are intrinsically of high amplitude.

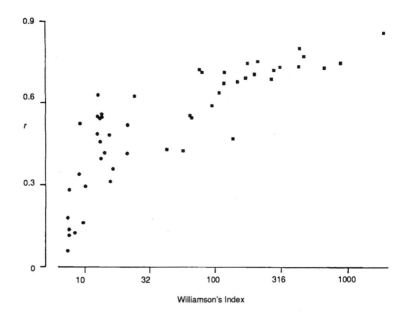

Figure 4. Relationship of the degree of synchronization (measured by correlation coefficient r) and the amplitude of fluctuations (measured by Williamson's Index) in 50 simulations of paired cyclic populations in which 10% of individuals from each population are exchanged in each generation. (●) low-amplitude cycles (as shown in *Figure 2*); (■) high-amplitude cycles (as shown in *Figure 3*).

Discussion

Royama (1979) in a theoretical investigation of the effects of adult dispersal on local population dynamics, including effects on population cycles, concluded that density-independent dispersal (as discussed above) had no significant effect on population cycles except for some increase in the amplitude of fluctuations. However, his model took no account of interaction between neighbouring populations, treating the dispersal purely as a loss of individuals from a single population. It was shown that density-dependent dispersal, by contrast, would reduce both the amplitude and period of population cycles.

The addition of density-dependent dispersal to the present model should increase synchronization for high-amplitude cycles, in which emigration is enhanced at population peaks, but should not influence the synchronization of low-amplitude cycles, in which populations seldom rise to the level where emigration becomes significant. This is probably a realistic scenario for many forest pests, where emigration occurs chiefly as a response to the depleted food resources in heavily defoliated forests (Berryman, 1988).

Fischlin (1983), modelling dispersal effects in Z. *diniana,* demonstrated that synchronization could be achieved by low levels of dispersal in the very high-amplitude cycles shown by the larch budmoth in its optimum zone. The present findings generalize that result, to show how synchronization may be critically dependent on the amplitude of the cycles, and so may be almost absent in low-amplitude fluctuations such as those of the British *B. piniaria* populations.

The hypothesis may be of very general application to those animal populations that show cyclic fluctuation patterns. For example, some small mammals (Finerty, 1980), game birds (Tapper, 1988), and locusts (Waloff, 1976) have comparable cyclic dynamics and we would expect to see a similar trend towards closer synchronization where cycles are of highest amplitude.

References

BALTENSWEILER, W. and FISCHLIN, A. (1988). The larch budmoth in the Alps.In: *Dynamics of Forest Insect Populations,* pp. 331–351 (ed. A. A. Berryman). New York: Plenum Press.

BARBOUR, D. A. (1980). Population dynamics of the pine looper moth *Bupalus piniaria* (L.) (Lepidoptera, Geometridae) in British pine forests. Unpublished PhD thesis, University of Edinburgh.

BARBOUR, D. A. (1988). The pine looper in Britain and Europe. In: *Dynamics of Forest Insect Populations,* pp. 291–308 (ed. A. A. Berryman). New York: Plenum Press.

BERRYMAN, A. A. (1981). *Population Systems. A General Introduction,* 222 pp. New York: Plenum Press.

BERRYMAN, A. A. (1988). *Dynamics of Forest Insect Populations: Patterns, Causes, Implications,* 603 pp. New York: Plenum Press.

DU MERLE, P. and MAZET, R. (1985). Piégeage sexuel de *Tortrix viridana* (Lep.,Tortricidae) en montagne méditerranéenne. I. Époque de vol et dispersion des adultes. *Zeitschrift für Angewandte Entomologie* **100**, 146–163.

FINERTY, J.P. (1980). *The Population Ecology of Cycles in Small Mammals,* 234 pp. New Haven and London: Yale University Press.

FISCHLIN, A. (1983). Modelling of alpine valleys, defoliated forests, and larch bud moth cycles: the role of moth migration. In: *Mathematical Models of Renewable Resources,* vol. II, pp. 102–104 (ed. R. Lamberson). Proceedings of the second Pacific Coast Conference on Mathematical Modelling of Renewable Resources. University of Victoria, Victoria BC, Canada.

GREENBANK, D. O., SCHAFFER, G. W. and RAINEY, R. C. (1980). Spruce budworm (Lepidoptera: Tortricidae) moth flight and dispersal: New understanding from canopy observations, radar, and aircraft. *Memoirs of the Entomological Society of Canada* **110**, 1–49.

HARRIS, J. W. E., DAWSON, A. F. and BROWN, R. G. (1985). The Douglas-fir tussock moth in British Columbia 1916–1984. *Canadian Forest Service, Pacific Forest Research Centre Information Report BC-X-268.*

KLIMETZEK, D. (1979). Insekten-Grosschadlinge an Kiefer in Nordbayern und der

Pfalz: Analyse und Vergleich 1810–1970. *Freiburger Waldschutz-Abhandlungen* **2**, 1–173.

MARTINAT, P. (1987). The role of climatic variation and weather in forest insect outbreaks. In: *Insect Outbreaks*, pp. 241–268 (eds. P. Barbosa and J. C. Schultz). San Diego: Academic Press.

MORAN, P. A. P. (1953). Statistical analysis of the Canadian Lynx cycle, 2. *Australian Journal of Zoology* **1**, 291–298.

MYERS, J. (1988). Can a general hypothesis explain population cycles of forest Lepidoptera? *Advances in Ecological Research* **18**, 179–242.

RANKIN, M. A. and SINGER, M.C. (1984). Insect movement: mechanism and effects. In: *Ecological Entomology*, pp. 185–216 (eds C. B. Huffaker and R. L. Rabb). New York: Wiley Interscience.

ROYAMA, T. (1979). Effect of adult dispersal on the dynamics of local populations of an insect species: a theoretical investigation. In: *Proceedings of the Second IUFRO Conference on Dispersal of Forest Insects: Evaluation, Theory and Management Implications* (eds A. A. Berryman and L. Safranyk). Conference Office, Cooperative Extension Service, Washington State University, Pullman, Washington DC.

TAPPER, S. C. (1988). Population changes in gamebirds. In: *Ecology and Management of Gamebirds*, pp. 18–47 (eds P. J. Hudson and M. R. W. Rands). Oxford: Blackwell Scientific.

TENOW, O. (1972). The oubreaks of *Oporinia autumnata* Bkh. and *Operophthera* spp, (Lep., Geometridae) in the Scandinavian mountain chain and northern Finland 1862–1968. *Zoologiska Bidrag Fran Uppsala*, Supplement 2.

VACLENA, K. and BALTENSWEILER, W. (1978). Untersuchungen zur Dispersionsydynamik des Grauen larchenwicklers, *Zeiraphera diniana* Gn. (Lep., Tortricidae): 2. Das Flugverhalten der Falter im Freiland . *Mitteilungen der Schweizerischen Entomologischen Gesellschaft* **51**, 59–88.

VARLEY, G. C. (1949). Population changes in German forest pests. *Journal of Animal Ecology* **18**, 117–122.

WALOFF, Z. (1976). Some temporal characteristics of Desert Locust plagues. *Antilocust Memoir* **13**. London: Centre for Overseas Pest Research.

WILLIAMSON, M. (1972). *The Analysis of Biological Populations*. London: Edward Arnold.

WILLIAMSON, M. (1984). The measurement of population variability. *Ecological Entomology* **9**, 239–241.

31
Modelling the Dynamics of Larch Casebearer

GARRELL E. LONG

Department of Entomology, Washington State University, Pullman WA 99164-6432, USA

Introduction

The adult larch casebearer (LCB), *Coleophora laricella* Hubner, is a silver-grey moth 5–7 mm in length. The larvae mine needles of larch (*Larix* spp.) worldwide. Apparently of European origin, this defoliator was first recorded in the US near Northampton, Mass. in 1886 and had spread west to Minneapolis by 1950 (Webb, 1953). In the spring of 1957, specimens collected just south of St. Maries, Idaho, were identified, and a survey indicated that about 44 000 ha of forest was infested. The population spread rapidly and uniformly until virtually all larch in the intermountain regions of eastern Washington, Oregon, and northern Idaho had some LCB. Extrapolation of data on its rate of spread suggests that LCB arrived in Idaho in 1953 (Long, 1977).

BIOLOGY OF LARCH CASEBEARER

Female LCB lay about 50 eggs each, half of which are female. In late June eggs are laid on the larch needles, usually no more than one per needle. Under high population pressure, females may deposit eggs on needles already occupied. I have found as many as five eggs on a single needle but this is extremely rare. It is

Population Dynamics of Forest Insects
© Intercept Ltd, PO Box 716, Andover, Hampshire, SP10 1YG, UK

generally agreed that no more than one neonate will survive the competition for space and food within the needle. On hatching, the first instar mines directly into the needle and begins to feed. As it grows and moults to the second instar, pieces of needle epidermis, connected by silk, are used to fashion a protective case when the larva becomes too large to remain inside the needle. In this externally feeding habit, the LCB larva consumes the needle parenchyma through a hole cut in the epidermis.

As November approaches, the larva moults to third instar, migrates toward the tip of a branch, and fastens itself to the base of a spur shoot with silk. The LCB overwinters as a third instar. Diapause is not obligatory under laboratory conditions. In the field, diapause is terminated about the same time the larch buds burst in April.

The major defoliation damage generally results from the spring feeding fourth instars. Larvae usually consume distal portions of the needles first. Depending on larval density, only a portion of each needle is consumed. In light infestations, this habit leaves a respectable proportion of each needle in the fascicle; in heavy infestations, so much of the needle complement may be consumed that the larch generates a new crop, or second flush, of foliage.

During the early 1960s, when the interaction between western larch and LCB was new, LCB densities were often high, as much as 3.6 larvae per spur shoot. Since then, defoliation has been severe at times, but neither direct tree mortality nor LCB densities in excess of 0.8 larvae per spur shoot has been noted. This pattern is similar to that reported in the eastern US and suggests that one or more compensatory mechanisms are beginning to moderate the effects of the defoliator on the larch trees.

BIOLOGY OF THE NATURAL ENEMIES

Over 30 species of native parasites have been recovered from LCB in the northwest. Two of these, *Spilochalcis albifrons* (Walsh), a chalcid wasp, and *Mesopolobus verditer* (Norton), a pteromalid wasp, are recovered consistently and in sufficient numbers to attract interest as potential biocontrol agents. To enhance the possibilities of natural control individuals of the braconid species *Agathis pumila* (Ratzburg) were collected from Rhode Island, shipped to Spokane and released at five locations near St. Maries Idaho in 1960. This host-specific parasite has been credited with natural control of LCB in Europe and eastern US and Canada. Larch branches infested with LCB were later removed from these sites and taken to other sites in the Pacific northwest where *A. pumila* had not yet been released.

The female *A. pumila* parasitizes only needle-mining (first and second instars) larvae. There is little development of the parasite larva inside the LCB until the following spring. About the time that unparasitized larvae are getting ready to pupate, the parasite larvae begin to develop rapidly; at the same time the LCB larval development is arrested. Adult parasites begin emerging in early June but peak emergence activity occurs in early to mid-July.

In 1972 LCB larvae and pupae parasitized by *Chrysocharis laricinellae* (Ratzburg) were imported to the northwestern US from Austria and England. This is a eulophid parasite which attacks only case-bearing stages of LCB. It was not thought to be host specific, but in the northwestern US it has three generations per year and the timing of these is such that no alternative host seems to be required. At this time, *C. laricinellae* seems to be more abundant than *A. pumila* in the northwest. Several years ago the reverse was true, but I have found no evidence that competitive displacement was occurring between these two species.

POPULATION DYNAMICS

Parasitoid and larch casebearer densities have been monitored annually at several sites in northern Idaho since parasitoids were released on the sites in 1974. I showed (Long, 1988) that during the late 1970s northern Idaho populations of LCB were limited by food availability. Densities fluctuated around a theoretical carrying capacity of about 0.35 pupating larvae per spur shoot. During the 1980s, population densities dropped precipitously to less than 0.02 pupating larvae per spur shoot (*Figure 1*). Although such low density estimates were not precise, there was some indication that population densities were fluctuating around a value of about 0.01 pupating larvae per spur shoot. This occurred at about the same times that parasitism rates increased sharply suggesting that host-specific parasitoids were effectively regulating larch casebearer population densities. I further suggested that this lower value represented a second theoretical equilibrium density maintained by natural enemies of the LCB, so that an unstable third equilibrium, or escape threshold, must exist with an intermediate value.

In this paper I compare the relationships among densities of the host LCB, exotic parasitoids, and native parasitoids when host LCB densities are above, or below, this presumed escape threshold to seek further evidence for or against the hypotheses that introduced parasitoids are controlling densities of northern Idaho

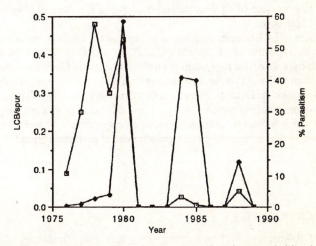

Figure 1. Numbers of larch casebearer larvae per spur shoot (open symbols) and proportion parasitized by exotic parasitoids (closed symbols) between 1976 and 1989 at a site near Priest Lake, Idaho.

populations of LCB, and to clarify the structure of the host-parasitoid communities. Finally, I show a simple model based on these results which could be used to predict changes in LCB densities over short periods.

Methods

Since 1975 LCB pupating larvae have been collected annually from 10 larch trees on each of five sites in northern Idaho. Two additional sites were sampled less often during the period. The numbers of larvae per spur shoot were noted, and all larvae returned to the laboratory for emergence of adult LCB or adult parasitoids. Any LCB cases from which no insect emerged within one month were cleared in sodium hydroxide solution and examined under a stereomicroscope to identify its occupant (Ismail and Long, 1982).

Before statistical analysis all observations were sorted into two groups: those associated with LCB densities above 0.08 larvae per spur shoot and those associated with LCB densities below 0.08 larvae per spur shoot. This distinction seems to separate the population dynamic behaviours into those associated with the upper and lower presumed LCB equilibria, respectively.

Values for the intrinsic rate of increase, the natural logarithm of annual replacement rate R, were calculated for LCB and parasitoid population densities, N_t at time t, as

$$\ln R = \ln(N_{t+1}/N_t)$$

and compared with densities of each other population by simple linear regression. A regression was considered significant when the ratio of regression sum of squares to error sum of squares exceeded tabulated F values with appropriate degrees of freedom.

Densities of LCB are pupating larvae per spur shoot and may include parasitized larvae. Densities of parasitoids are adult, or pharate adult, parasitoids per LCB. I acknowledge that regression of ratios on one another may show increased levels of correlation when the denominators are similar. However, this does not seem an appropriate forum to debate whether dynamics of biological populations include autoregressive functions, nor do I find this assumption damaging to the analysis which follows. Finally, stepwise linear regression was used to combine significant effects in order to predict growth rates of each population.

Results and discussion

HIGH LCB DENSITIES

Above LCB densities of 0.08 larvae per spur shoot, four significant interactions are apparent in the host–parasitoid community (*Table 1*). As noted previously LCB

Table 1. Potential interactions among host and parasitoids when host density is greater than 0.08 larvae per spur shoot. The intrinsic rate of increase, $\ln R$, of the source population were compared by simple linear regressions with densities of the populations indicated resulting in the sign of the relationship, coefficient of determination and the F statistic (* P <0.05). Populations analysed were larch casebearer (LCB), the exotic parasitoids *A. pumila* (Ap) and *C. laricinellae* (Cl), and the native parasitoids (Natives)

$\ln R$	Density	Sign	r^2	$F(1,14)$
LCB	LCB/Spur	−	0.49	11.5*
LCB	Ap/LCB	−	0.64	17.0*
LCB	Cl/LCB	−	0.09	1.5
LCB	Natives/LCB	−	0.08	1.2
Ap	LCB/Spur	+	0.13	2.2
Ap	Ap/LCB	−	0.13	2.1
Ap	Cl/LCB	−	0.40	9.4*
Ap	Natives/LCB	−	0.42	1.0
Cl	LCB/Spur	+	0.12	1.9
Cl	Ap/LCB	+	0.25	4.7*
Cl	Cl/LCB	−	0.12	1.9
Cl	Natives/LCB	+	0.06	0.9
Natives	LCB/Spur	−	0.14	2.3
Natives	Ap/LCB	+	0.02	0.3
Natives	Cl/LCB	+	0.16	2.7
Natives	Natives/LCB	+	0.01	0.1

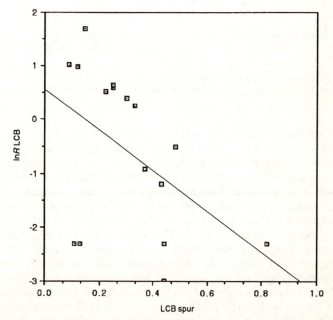

Figure 2. The relationships between intrinsic rate of increase of larch casebearer, $\ln R$ LCB, and its density, LCB/Spur, when larch casebearer densities exceed 0.08 larvae per spur shoot.

rate of increase is positive when LCB densities are below the carrying capacity of about 0.35 larvae per spur shoot and negative above this value (*Figure 2*).

The only parasitoid population that significantly influenced the LCB rate of increase was *A. pumila*. The relationship is illustrated in *Figure 3*, and shows a fairly consistent depression of LCB growth rates above about 15% parasitism (0.15 *A. pumila* per LCB).

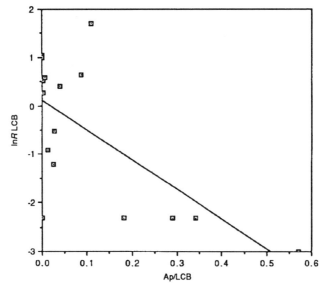

Figure 3. Depression of larch casebearer population growth rates, lnR LCB, by density of the parasitoid *A. pumila*, Ap/LCB, when larch casebearer densities exceed 0.08 larvae per spur shoot.

Apparently the introduced parasitoids *A. pumila* and *C. laricinellae* interact (*Table 1*). Quednau (1970) suggested that *C. laricinellae* might be hyperparasitic on *A. pumila*, but Ismail and Long (1982) were unable to detect evidence of hyperparasitism empirically. The results here suggest significant detriment to *A. pumila* rates of increase result from increasing densities of *C. laricinellae*, and rates of increase for *C. laricinellae* are enhanced by higher densities of *A pumila*.

Figure 4 shows that the effect of *C. laricinellae* on *A. pumila* results from only one, possibly spurious, observation, while *Figure 5* shows that rates of increase of *C. laricinellae* are improved when parasitism by *A. pumila* exceeds about 10%. As noted earlier, *C. laricinellae* attacks only casebearing stages of LCB and parasitism by *A. pumila* delays LCB development during pupation. Parasitized LCB are still available after non-parasitized LCB adults have eclosed. Thus parasitism of LCB by *A. pumila* increases the host resources available to *C. laricinellae*.

But what of the reverse effect? Percentage parasitism by *C. laricinellae* is generally quite low compared to that of *A. pumila*, so that the probability that a *C.*

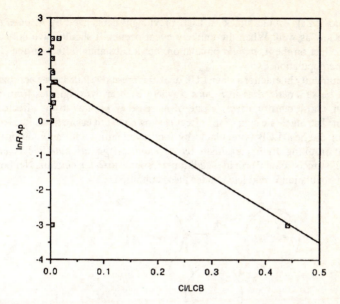

Figure 4. Possible competitive effect between the introduced parasitoids, *A. pumila* (Ap) and *C. laricinellae* (Cl) when larch casebearer densities exceed 0.08 larvae per spur shoot.

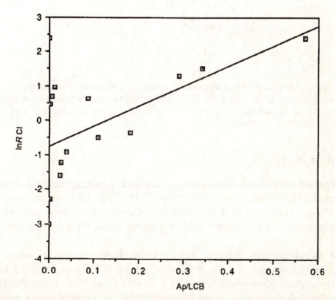

Figure 5. Mutualistic effect of the introduced parasitoid, *A. pumila* (Ap) on rate of increase of *C. laricinellae* (Cl) when larch casebearer densities exceed 0.08 larvae per spur shoot.

laricinellae female would attack an LCB previously parasitized by *A. pumila* should be low as well. When the unlikely event occurs, it should have only a marginal effect on the *A. pumila* population, but a substantial effect on the *C. laricinellae* population.

The community interactions when LCB densities are above 0.08 larvae per spur shoot can be summarized as in *Figure 6* where each arrow originates from a population whose density affects a rate of increase and points to the affected population. The algebraic sign of the effect is shown next to each arrow. Rate of increase of the host LCB is regulated by food availability with an additional decrement resulting from parasitism by *A. pumila*. *Agathus pumila* has an abundant host resource and provides additional resource to *C. laricinellae*. Neither parasitoid population is regulated under these conditions.

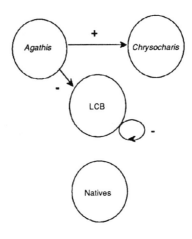

Figure 6. Significant host–parasitoid community interactions when larch casebearer densities exceed 0.08 larvae per spur shoot. Each arrow originates from a population whose density affects the rate of increase of the population to which it points. The algebraic sign of the effect is shown next to each arrow.

LOW LCB DENSITIES

At LCB densities below 0.08 larvae per spur shoot, only two significant interactions are apparent in the host-parasitoid community (*Table 2*). LCB rate of increase is still significantly impacted by *A. pumila*, but, as shown by the relative values for coefficient of determination, less reliably so (*Figure 7*) than when LCB densities are higher.

When *A. pumila* are rare relative to the host LCB, but nonetheless present on a site, it appears that density of the parasitoid is very likely to increase during the following generation. However, as parasitism exceeds about 10%, *A. pumila* rate of increase drops dramatically (*Figure 8*). Presumably this is due to overexploitation of a rare host by the parasitoid. This is in sharp contrast to the previous cases when LCB hosts were abundant. *Figure 9* shows the community organization

Table 2. Potential interactions among host and parasitoids when host density is less than 0.08 larvae per spur shoot. The intrinsic rate of increase, ln$R$, of the source population were compared by simple linear regressions with densities of the populations indicated resulting in the sign of the relationship, coefficient of determination and the F statistic (*$P <0.05$). Populations analyzed were larch casebearer (LCB), the exotic parasitoids *A. pumila* (Ap) and *C. laricinellae* (Cl), and the native parasitoids (Natives)

ln$R$	Density	Sign	R^2	F(1,39)
LCB	LCB/Spur	–	0.07	1.2
LCB	Ap/LCB	–	0.26	13.8*
LCB	Cl/LCB	–	0.01	0.4
LCB	Natives/LCB	–	0.01	0.3
Ap	LCB/Spur	+	0.09	3.7
Ap	Ap/LCB	–	0.24	11.1*
Ap	Cl/LCB	–	0.06	2.5
Ap	Natives/LCB	+	0.03	1.3
Cl	LCB/Spur	+	0.08	3.5
Cl	Ap/LCB	+	0.00	0.0
Cl	Cl/LCB	+	0.09	4.0
Cl	Natives/LCB	+	0.10	3.5
Natives	LCB/Spur	+	0.09	3.9
Natives	Natives/LCB	+	0.13	0.0
Natives	Ap/LCB	+	0.02	0.8
Natives	Cl/LCB	–	0.02	0.9

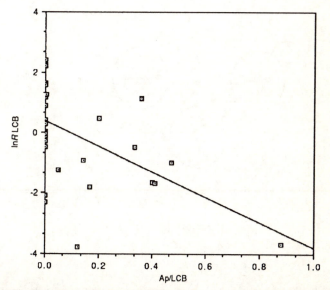

Figure 7. Depression of larch casebearer population growth rates, ln$R$ LCB, by density of the parasitoid *A. pumila*, Ap/LCB, when larch casebearer densities are less than 0.08 larvae per spur shoot. Note that while the range on the abscissa is similar to that in *Figure 3*, the variation about the regression line is much higher.

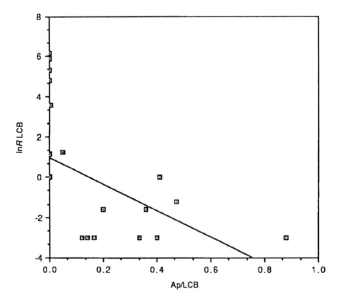

Figure 8.   When densities of the introduced parasitoid *A. pumila* were near zero, its rate of increase was positive and parasitism increased the following year. However, parasitism above 10% apparently leads to overexploitation of hosts and, therefore, negative rates of increase.

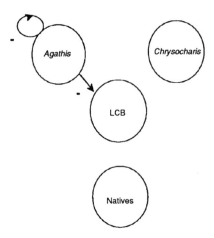

Figure 9.   Significant host–parasitoid community interactions when larch casebearer densities are less than 0.08 larvae per spur shoot. Each arrow originates from a population whose density affects the rate of increase of the population to which it points. The algebraic sign of the effect is shown next to each arrow.

when LCB densities are less that 0.08 larvae per spur shoot. Rate of increase for the host LCB is regulated only by the parasitoid *A. pumila* which is regulated in turn by host availability.

Predictive Model

Stepwise linear regression showed that for high LCB densities

$$\ln R \text{ LCB} = 1.2 - 3.9 \text{ LCB/Spur} - 4.1 \text{ Ap/LCB} - 6.8 \text{ Cl/LCB}$$

The inclusion of *C. laricinellae* in this function is not statistically significant by itself but none the less increased the coefficient of determination from 45% to 71% in the third step of the regression. No other population improved the predictive capacity of the equation, nor was any other population predictable.

Stepwise linear regression for low LCB densities yielded the two functions

$$\ln R \text{ LCB} = 0.5 - 3.9 \text{ Ap/LCB}$$

and

$$\ln R \text{ Ap} = 0.9 - 6.5 \text{ Ap/LCB}$$

with coefficients of determination 22% and 24%, respectively. Again, no other populations improved the predictive capacity of the equations, nor were any other populations predictable.

Conclusion

From these analyses it is inferred that, in the geographic region covered by these data, LCB and its associated parasitoids exhibit different community structures, and therefore different dynamic behaviours, depending on whether LCB densities are relatively high or low. Native parasitoids are unimportant to community dynamics as these have alternative hosts and are facultative parasitoids on LCB. The introduced parasitoid *C. laricinellae* has a measurable, but relatively unimportant impact on population dynamics of LCB. It may have an alternative host. The introduced parasitoid *A. pumila* does not effectively utilize dense populations of LCB. It is adapted to finding and colonizing rare aggregations of LCB, and this suggests that the parasitoid is host specific.

Based on anecdotal evidence from Europe, the United Kingdom, Canada, and the eastern United States, and the information presented here, I would conclude that LCB is not an eruptive species, and that population densities will remain low unless natural controls are drastically reduced .

References

ISMAIL, A. B. and LONG, G. E. (1982). Interactions among parasites of the larch casebearer (Lepidoptera: Coleophoridae) in northern Idaho. *Environmental Entomology* **11**, 1242–1247.

LONG, G. E. (1977). Spatial dispersion in a biological control model for larch casebearer *(Coleophora laricella)*. *Environmental Entomology* **6**, 843–852.

LONG, G. E. (1988). The larch casebearer in the Intermountain Northwest. In: *Dynamics of Forest Insect Populations*, pp. 233–242 (ed. A. A. Berryman). New York: Plenum Press,.

QUEDNAU, F. W. (1970). Competition and co-operation between *Chrysocharis laricinellae* and *Agathis pumila* on larch casebearer. *Canadian Entomologist* **102**, 602–612.

WEBB, F.E. (1953). An ecological study of the larch casebearer, *Coleophora laricella* (Lepidoptera: Coleophoridae). PhD thesis, University of Michigan, Ann Arbor.

32
Models of the Spatial Dynamics of Epidemic Gypsy Moth Populations

ANDREW M. LIEBHOLD* AND JOSEPH S. ELKINTON**

*USDA Forest Service, Northeastern Forest Fxperiment Station, Morgantown, West Virginia, USA
**Department of Entomology, University of Massachusetts, Amherst, Massachusetts, USA

Introduction

The gypsy moth, *Lymantria dispar* (L.), was introduced to North America in 1869 near Boston, Massachusetts. Since that time it has expanded its range by 300 miles as far as Ohio, West Virginia and North Carolina. Within the area where gypsy moth is permanently established, populations typically are unstable, often fluctuating through changes of several orders of magnitude in density over a period of only three or four years.

Many state and federal agencies are developing new approaches for managing gypsy moth populations. At the core of modern gypsy moth integrated pest management systems are grids of monitoring stations deployed over large geographic areas (Ravlin, Bellinger and Roberts, 1987; Reardon *et al.*, 1987). The approach is to detect new, rising populations and then supress these populations in order to prevent regional outbreaks. In order to evaluate treatment decisions, there is a need for models that predict the large-scale spatial dynamics of gypsy moth populations.

Unfortunately, relatively little is known about spatial patterns of gypsy moth outbreak development. Valentine and Houston (1979) and Campbell (1976) hypothesized that outbreaks may be initiated by immigration of larvae from nearby outbreak areas. Campbell (1973, 1976) concluded that outbreaks were likely to

Population Dynamics of Forest Insects
© Intercept Ltd, PO Box 716, Andover, Hampshire, SP10 1YG, UK

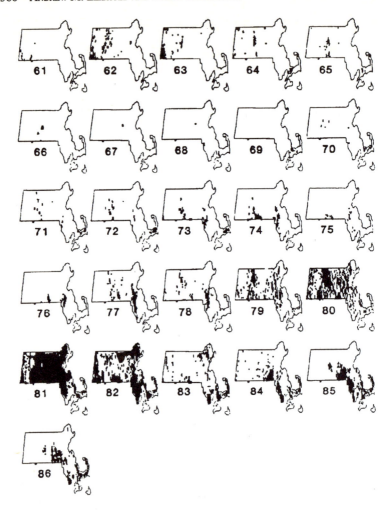

Figure 1.   Defoliation incidence for each 1.9 x 1.9 km cell in Massachusetts, 1961–86.

persist from one year to the next when insect densities ranged widely among subpopulations in a region and conversely that when numerical variability was minimal among subpopulations, outbreaks were likely to decline. Several new statistical techniques have been developed for characterizing static and dynamic spatial patterns. In this study we applied several such statistical methods to characterize historical defoliation records and to simulate hypothesized mechanisms of outbreak spread.

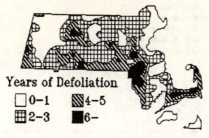

Figure 2. Spatial contour of defoliation frequency in Massachusetts, 1961–86.

Description of the data

The Massachusetts Department of Environmental Management monitors gypsy moth defoliation annually using aerial sketch maps. Maps are sketched during a series of flights over the state in late July, when defoliation is at its peak. Composite 1:760 320 maps covering the entire state from the years 1961–86 were digitized into a 65 x 101 matrix of cells (*Figure 1*). Each cell represented a 1.9 x 1.9 km area. Each cell was coded as either: (1) containing defoliation; (2) undefoliated; (3) not part of the state.

Defoliation frequencies for each cell were summed from 1961 to 1986. From the original 65 x 101 matrix of defoliation frequencies, a 25 x 25 matrix was interpolated using a modified inverse-distance weighting function (Sampson, 1975). This new grid matrix was then used in a piecewise Bessel interpolation algorithm to generate defoliation-frequency contour intervals (*Figure 2*).

Transition models

Two-state transition models (Parzen, 1962) were used to describe the transition of cells from 0 (undefoliated) to 1 (defoliated) and from 1 to 0. Separate transition probabilities were calculated to compare cells in different regions (previously identified as groups of areas operating in synchrony (Liebhold and Elkinton, 1989)) in the 65 x 101 matrix. We computed separate probabilities for cells with different numbers of defoliated adjacent cells [(adjacency was defined using a Queen's move definition of adjacency (Cliff and Ord, 1973)]. All regions exhibited similar trends in defoliation transition probabilities with respect to the defoliation status of adjoining cells. Probabilities of defoliation initiation increased as the proportion of defoliated neighbours increased (*Figure 3a*). This represents what is referred to as the 'focal area' phenomenon where visible defoliation begins at some localized area and then this area grows larger through time (Chugunin, 1949; Valentine and Houston, 1979; Wallner, 1987). These results confirm this pattern; however, they do not provide evidence of a causal connection between defoliation in the localized area and subsequent defoliation in surrounding areas (see below). The probability of defoliation termination decreased as the proportion of defoliated neighbours increased (*Figure 3b*).

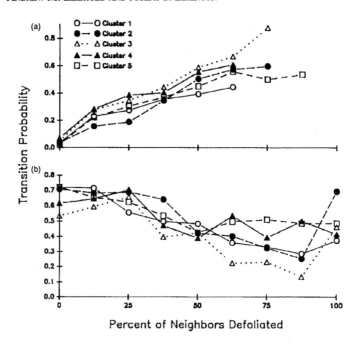

Figure 3. Probabilities of transition to a different defoliation state with varying numbers of defoliated neighbours. (a) transition from non-defoliated to defoliated; (b) transition from defoliated to non-defoliated.

Autocorrelation analysis

Spatial autocorrelation is a method of describing spatial patterns by quantifying the correlation between spatially adjacent points (Sokal and Oden, 1978; Cliff and Ord, 1973). We applied spatio-temporal autocorrelation analysis (STAA) to the 65 x 101 x 26 matrix of defoliation data. STAA is used to evaluate the relationship between adjacent points, where adjacency is described in both space and time. Results from STAA can be useful for understanding the spread of epidemics through space (Cliff and Ord, 1981; Reynolds *et al.*,1988). We restricted adjacency definitions to separate cardinal directions (Oden and Sokal, 1986). This directional STAA was thus used to statistically evaluate the extent to which defoliation spread in one cardinal direction. Qualitative analysis of other gypsy moth defoliation maps suggested that changes in the spatial distribution of defoliation between years often is unidirectional due to the wind-borne disperal of first instars (Anderson and Gould, 1974).

Directional STAA yielded nearly identical correlograms in all four cardinal directions (*Figure 4*). Cells which were within 6 time-space cells were significantly autocorrelated. These results suggest that defoliation in a given cell may have had an effect on the condition of spatially adjacent cells in the future. However, it does not indicate that there was a unifying directional component to the spread of gypsy moth defoliation. Instead, the lack of a significant effect of

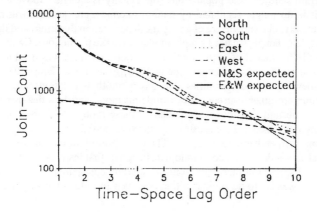

Figure 4. Directional spatio-temporal autocorrelation correlograms (Reynolds 1988; Oden and Sokal, 1986) based on defoliation join-counts from 1961 to 1986. Join-counts expected under a random distribution through time and space were the same for both north–south and east–west adjacency definitions.

cardinal direction on directional STAA suggests that, at least in Massachusetts, outbreaks radiate outward in all directions.

Simulation of the focal-area phenomenon

The so called 'focal-area' hypothesis (Chugunin, 1949; Valentine and Houston, 1979; Wallner, 1987) discussed earlier postulates that gypsy moth outbreaks begin in small specific areas (foci) and expand outward into larger, concentric zones in successive years. These foci are characterized by forest stands growing on poor sites such as ridgetops, and frequently are subject to drought (Houston and Valentine, 1977, 1985). This hypothesis has major implications for managing gypsy moth populations because it suggests that suppressing defoliating populations in one area will prevent their expansion into surrounding areas (Leonard, 1971). It has been proposed that focal areas, 'epicentres' or 'refugia' play similar roles in the initiation of region-wide epidemics of a variety of forest insects (Wallner, 1987).

The casual connection between outbreaks in foci and subsequent outbreaks in surrounding areas has not been demonstrated. Most evidence for the spread of gypsy moth defoliation from foci comes from the examination of defoliation maps in historical records. We present here a simple simulation model based on the assumption of the focal-area theory that we have used to determine if the theory can explain these historical patterns of defoliation.

Proponents of the focal-area theory hypothesize that the immigration of large numbers of first instar gypsy moths into areas near foci upsets an assumed endemic equilibrium state of local populations (the density is elevated above a point at which it is no longer regulated by natural enemies, and an outbreak ensues). Unfortunately, our knowledge of gypsy moth population dynamics and dispersal is incomplete; hence, we used simple models to make predictions. Campbell and Sloan's (1978) model of gypsy moth population dynamics was used to predict generation-to-generation changes in densities (*Figure 5*). The Campbell and Sloan replacement rate model assumes that gypsy moth populations are bimodal: populations are either in an endemic or epidemic mode, and densities are regulated about equilibrium densities within each of these modes. Campbell and Sloan (1978) hypothesized that some catastrophic event caused populations to make the transition from endemic to epidemic. The other portion of our simululation, the dispersal submodel, assumed a random diffusion of first instars such that only 5% of larvae dispersed farther than 50 m. As a crude approximation, this dispersal magnitude was set to be similar to that observed in field studies by Mason and

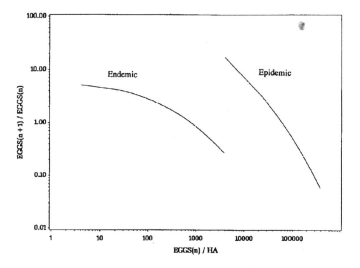

Figure 5.   Bimodal density-dependent replacement-rate model developed by Campbell and Sloan (1978). Epidemic model: $\log(\text{DENSITY}(n+1)/\text{DENSITY}(n)) = 2.28 - 0.016(\log(\text{DENSITY}(n)))^3$. Endemic model: $\log(\text{DENSITY}(n+1)/\text{DENSITY}(n)) = 0.733 - 0.021(\log(\text{DENSITY}(n)))^3$.

McManus (1981), and Minott (1922). The two submodels (Campbell and Sloan's replacement rate model and our dispersal model) were applied over a hypothetical

25 x 25 array of 0.0625 ha cells for three generations. The dispersal submodel moved first instars among cells, and the Campbell and Sloan model predicted the next year's egg density based on the post-dispersal density for each cell.

In the simulation, each cell started with an initial density of 2000 eggs/ha (typical endemic density; roughly equivalent to 5–10 egg masses/ha) except the centre cell started with 200 000 eggs/ha (typical epidemic density; roughly equivalent to 1000-2000 egg masses/ha). This cell represented a hypothetical focal-area. Over the course of the next three years, the defoliation spread: larvae that dispersed out of cells harbouring epidemic densities elevated populations in nearby endemic cells above the threshold density (10 000 eggs/ha) and, therefore, caused them to enter the epidemic mode (*Figure 6*). The simulation was repeated using different

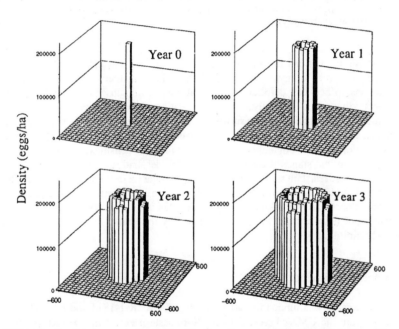

Figure 6. Predicted gypsy moth egg densities from focus simulation. Dispersal submodel parameterized such that 5% of larvae dispersed beyond 50 m. Replacement rate submodel parameterized with a threshold density of 10 000 eggs/ha. Values on X and Y axes are expressed in metres. Densities shown on Z axis are expressed as eggs/ha.

parameters for the dispersal function and different threshold values which were used in the replacement rate model. Growth of the outbreak area was relatively slow (*Table 1*). Even under conditions in which the dispersal rate was set unrealistically high (50% of insects dispersing beyond 50 m) and the threshold density was set unrealistically low (5000 eggs/ha), the outbreak grew to an area of only approximately 50 ha. In contrast, historical records show that in three years outbreaks typically grow to several hundred or thousands of hectares in size.

Table 1.   Defoliated areas predicted under focus simulation

Proportion Dispersing Beyond 50 m	Threshold Density*	Predicted area (ha) defoliated		
		Year 1	Year 2	Year 3
5	10 000	1.2	4.0	8.3
50	10 000	5.9	14.2	30.7
5	5 000	1.2	5.9	11.0
50	5 000	5.9	21.6	47.2

* Threshold determines which density-dependent replacement model to use (*Figure 5*).
Expressed as numbers of eggs per ha.

These simulations do not support the hypothesis that dispersal causes the observed spread of gypsy moth defoliation. Instead, there may be no causal relationship between defoliation in 'focal areas' and subsequent defoliation in surrounding areas. Generally, there is a gradation of stand susceptibility from the center of 'foci' out through the surrounding region (Houston and Valentine, 1985). When some exogenous factor(s), such as weather, triggers the release of endemic populations, 'focus' populations may reach defoliating densities first. In adjoining, less susceptible stands it may take longer for defoliating densities to build. In distant, more resistant stands it may take even longer for defoliation to appear. The result would be an apparent spread of defoliation in concentric areas about foci without necessarily invoking an explanation that involves dispersal. Instead, this pattern of defoliation is more likely caused by the spatial distribution of stand susceptibility.

Acknowledgements

This research was funded in part by grants 85-CRCR-1-1814 and 87-CRCR-1-2498 from the USDA Forest Service, Northeastern Forest Experiment Station, Gypsy Moth Research & Development Program. Competitive Research Grants Programme and Funds from the USDA.

References

ANDERSON, J. F. and GOULD, S. W.(1974), Defoliation in Connecticut 1969–1974 tabulated by use of the geo-code. *Connecticut Agricultural Experiment Station* Bulletin 749.
CAMPBELL, R. W. (1973). Numerical behavior of a gypsy moth population system. *Forest Science* **19**, 162–167.
CAMPBELL, R. W. (1976). Comparative anaylsis of numerically stable and violently

fluctuating gypsy moth populations. *Environmental Entomology* 5, 1218–1224.

CAMPBELL, R.W. and SLOAN, R. J. (1978). Numerical bimodality among North American gypsy moth populations. *Environmental Entomology* 7, 641–646.

CHUGUNIN, Y. (1949). Focal periodicity of gypsy moth outbreaks. *Zoologicheskiy Zhurnal* 28, 431–438.

CLIFF, A. D. and ORD, J. K. (1973). *Spatial Autocorrelation*. London: Pion Press.

HOUSTON, D. R. and VALENTINE, H. T. (1977). Comparing and predicting forest stand susceptibility to gypsy moth. *Canadian Journal of Forest Research* 7, 447–461.

HOUSTON D. R. and VALENTINE, H. T. (1985). Classifying forest susceptibility to gypsy moth defoliation. Washington DC: *US Department of Agriculture Agricultural Handbook* 542.

LEONARD, D. E. (1971). Air-borne dispersal of larvae of the gypsy moth and its influence on concepts of control. *Journal of Economic Entomology* 64, 638–640.

LIEBHOLD, A. M. and ELKINTON, J. S. (1989). Characterizing spatial patterns of gypsy moth, *Lymantria dispar* (Lepidoptera: Lymantriidae), regional defoliation. *Forest Science* 35, 557–568.

MASON, C. J. and McMANUS, M. L. (1981). Larval dispersal of the gypsy moth. In: *The Gypsy Moth: Research Toward Integrated Pest Management.*, pp. 161–202 (eds C.C. Doane and M.L. McManus) Washington US Department of Agriculture Forest Service Technical Bulletin 1584.

MINOTT, C. W. (1922). The gipsy moth on cranberry bogs. Washington DC: *US Department of Agriculture Bulletin* 1093.

ODEN, N. L. and SOKAL, R. R. (1986). Directional autocorrelation: an extension of spatial correlograms to two dimensions. *Systematic Zoology* 35, 608–617.

PARZEN, E. (1962). *Stochastic Processes*. Oakland, California: Holden Day.

RAVLIN, F. W., BELLINGER, R. G. and ROBERTS, E. A. (1987). Gypsy moth management programs in the United States: status, evaluation, and recommendation. *Bulletin of the Entomological Society of America* 33 90–98.

REARDON, R., MCMANUS, M., KOLODNY-HIRSCH, D., TICHENOR, R., RAUPP, M., SCHWALBE, C., WEBB, R. and MECKLEY, P. (1987). Development and implementation of a gypsy moth integrated pest management program. *Journal of Arboriculture* 13, 209–216.

REYNOLDS, K. M., BULGER, M. A., MADDEN, L. V. and ELLIS, M. A. (1988). Analysis of epidemics using spatio-temporal autocorrelation. *Phytopathology* 78, 240–246.

SAMPSON, R. J. (1975). Surface II graphics system. *Kansas Geological Survey Series on Spatial Analysis* No. 1.

SOKAL, R. R. and ODEN, N. L. (1978). Spatial autocorrelation in biology 1. Methodology. *Biological Journal of the Linnean Society* 10, 199–228.

VALENTINE, H. T. and HOUSTON, D. R. (1979). A discriminant function for identifying mixed-oak stand susceptibility to gypsy moth defoliation. *Forest Science* 25, 468–474.

WALLNER, W. E. (1987). Factors affecting insect population dynamics. Differences between outbreak and non-outbreak species. *Annual Review of Entomology* 32, 317–340.

33
Modelling Douglas-fir Tussock Moth Population Dynamics: The Case for Simple Theoretical Models

ALAN A. BERRYMAN,* JEFFREY A. MILLSTEIN** AND
RICHARD R. MASON†

*Department of Entomology, Washington State University, Pullman, WA 99164
USA
**Applied Biomathematics, Inc., 100 North Country Rd., Setauket, NY 11733
USA
†Forestry and Range Sciences Laboratory, La Grande, OR 97850 USA

Introduction

Insect population modelling was born in the 1960s, grew up in the 1970s, and has
matured in the 1980s. Forest entomologists have played a leading role in this
evolution.

The first attempts to model insect population dynamics occurred at the end of
the famous 'Green River Project' on the eastern spruce budworm, *Choristoneura
fumiferana* (Morris, 1963). At this time, entomologists were preoccupied with
statistical methods, and the model reflected this preoccupation. The spruce
budworm model had the following structure

$$N(t) = N(t-1) R \prod_{i=1}^{I} S(i) \qquad (1a)$$

$$S(i) = a(i) + \sum_{j=1}^{J} b(i,j) X(i,j) \qquad (1b)$$

Population Dynamics of Forest Insects
© Intercept Ltd, PO Box 716, Andover, Hampshire, SP10 1YG, UK

where $N(t)$ is the density of a particular stage, say eggs, at time t, R is the reproductive output per individual, and $S(i)$ is the probability of survival over the ith stage of development ($0 \le S \le 1$). The survival and reproductive components were defined by multiple regression, equations (lb), on a set of independent variables $X(i, j), j = 1, 2, \ldots J$, representing variables such as weather, new foliage production, etc. Thus, the spruce budworm model is structured on the life cycle of the insect and rests on the assumptions of statistical theory. As far as we are aware, this model has never been used to predict budworm outbreaks, and has yielded few insights into the forces determining budworm abundance.

The next attempt at insect population modelling was with the winter moth, *Operophthera brumata* (Varley, Gradwell and Hassell, 1973). The winter moth model can be written

$$n(t) = n(t-1) + r - \sum_{i=1}^{I} k(i) \qquad (2)$$

where $n(t)$ is the logarithm of adult winter moth density at time t, r is a reproductive constant, and $k(i)$ is the k-value, the mortality of the ith stage of development. Mortalities were either assumed to be constants, input variables, or were computed as functions (submodels) of other variables. Three submodels involved density-dependent interactions with parasitoids and predators.

The winter moth and spruce budworm models are similar, in some respects, both being structured on the insect life cycle. However, the winter moth model also contains explicit feedback structures in its density-dependent predation and parasitism submodels, and rests more on ecological than statistical theory; that is, the model was built around the concepts of density-dependent and density-independent processes. Although we are unaware of the winter moth model being used to predict population fluctuations in Britain, it has provided valuable inferences on the role of parasitoids in the biological control of the moth in North America (Varley, Gradwell and Hassell 1973, Hassell, 1978).

In the early seventies, insect population modellers shifted their attention to the methods of operations research, systems analysis and simulation. Once again, forest entomologists were at the front of this new move, with detailed simulation models being constructed for the fir engraver beetle (Berryman and Pienaar, 1974), the eastern spruce budworm (Holling, Jones and Clark 1976), and the Douglas-fir tussock moth (Brookes, Stark and Campbell, 1978). Then, as the term 'systems analysis' caught hold and spread into the agricultural sector, a proliferation of highly detailed simulation models followed, many retaining the life cycle structure of their predecessors, but few the ecological concepts of the winter moth model.

In the seventies, new ideas of engineering control theory, cybernetics and non-linear dynamics, forming what had come to be called general systems theory, also began to penetrate entomology. This perspective drew attention to the feedback structure of population systems and the importance of feedback in determining the dynamics and stability of natural systems (Berryman, 1981). In effect this was a return to ecological principles, for the concept of density-induced negative feedback is synonymous with the ecological idea of density-dependence (Berryman, Stenseth and Isaev, 1987; Berryman, 1989). This return to ecological theory in forest insect modelling has resulted in much more simple models which seem to provide better predictions and more powerful insights than their highly detailed counterparts. In this paper we will attempt to demonstrate this somewhat counter-intuitive notion as exemplified by population dynamics models for the Douglas-fir tussock moth, *Orgyia pseudotsugata*.

Detailed simulation models

In the early seventies, outbreaks of the Douglas-fir tussock moth flared up in many areas of western North America. Commercial and government organizations reacted to this 'threat' to the timber resources in the usual manner - kill the insects! However, since the last outbreak in the sixties (the 10-year cycle again), Rachael Carson's 'Silent Spring' had been published and the newly formed Environmental Protection Agency had banned DDT, the only pesticide known to control effectively the tussock moth. The timber concerns applied for special use permission for DDT and, as part of the special use package, the Forest Service was instructed to mount a large-scale 'system' study to develop alternative control methods and an 'integrated' approach to tussock moth management. This research and development project started in 1974, perhaps the last time that DDT was sprayed on a forest insect in the USA, and ended four years later (Brookes, Stark and Campbell, 1978).

A major challenge to the scientists of the project was to develop a model of Douglas-fir tussock moth population dynamics that could be linked to the Forest Service's 'Stand Prognosis Model' and used to make forest management and pest control decisions. The programme took the view that a highly detailed system simulation model should be built, and this was eventually achieved at considerable cost (Colbert, Overton and White, 1979). This large model (50 plus parameters) faithfully mimics the 1972–74 outbreak but cannot be used to predict the occurrence or intensity of future tussock moth outbreaks. Neither has it yielded any valuable insights into the forces that regulate tussock moth populations or which are responsible for the periodic outbreaks (Vezina and Peterman, 1985). The model is useful, however, for projecting long-term stand trajectories into the future under the assumptions of time-dependent outbreak probabilities and that all outbreaks will follow the trend of the seventies.

Simple theoretical models

During the early phases of the tussock moth project, some scientists urged that parallel studies be initiated on other modelling approaches, particularly k-value models and simple theoretically orientated synoptic models. Although programme managers decided not to fund such studies, some independent work was done along these lines. One attempt was the simple time-delayed 'logistic' model proposed by Berryman (1978)

$$N(t) = N(t-1) \, exp\{A[1 - N(t-2)/K\,]\} \tag{3}$$

where A is the maximum per capita rate of increase and K is the equilibrium density or 'carrying capacity'. With parameter values roughly estimated from data on a tussock moth outbreak in Arizona, this model predicts the following:

1. Outbreaks will occur every nine years or so, a prediction that is in line with the historical facts (Brookes, Stark and Campbell, 1978).
2   The intensity of outbreaks, as represented by the amplitude of the cycles, should be related to site, stand and/or climatic variables. Experimental studies seem to support this prediction (Stoszek *et al.*, 1981).
3. Outbreak frequencies can be synchronized by widespread climatic events and desynchronized by local catastrophies. This prediction was tested by the eruption of Mount St. Helens which deposited an insecticidal ash over the forests to the north of Moscow, Idaho in May 1980. This resulted in heavy mortality to the newly emerged tussock moth larvae (Mason *et al.*, 1984) and apparently delayed the next cycle which peaked in 1985 in this region, two years later than in other areas. Whether a widespread climatic fluctuation, such as an extremely cold winter, or tussock moth larval dispersal (ch. 32, this volume), will bring these populations back into phase will have to await the test of time.
4. The cyclic dynamics of tussock moth populations are probably driven by numerical interactions with the host plant or with its predators and/or parasitoids. Studies on sparse tussock populations seem to support the latter hypothesis (Dahlsten *et al.*,1977; Mason and Torgersen, 1987).

These are rather rich insights from such a simple model. However, the model also predicts that tussock moth populations will invariably reach levels of visible defoliation, around 300 small larvae per 10 m² of foliage. Thus the model failed to predict that tussock moth populations would not attain damaging levels in the mid-eighties; i.e. tussock moth densities in the eighties did not exceed visible defoliation levels in most forested areas. In fact, a spray project against the tussock moth in Northern Idaho in 1986 was cancelled when populations that were expected to reach damaging levels suddenly declined. The main problems with the simple 'logistic' model proposed by Berryman (1978) are the crude parameter estimates made from limited data and the simplifying assumptions of classical linear 'logistic' theory. In the meantime, however, more extensive data series have been published (Mason and Wickman, 1988), and 'logistic' theory has been generalized to include non-linear negative feedback and more than one domain of attraction (Berryman, 1981, 1987; Berryman and Millstein, 1988; Berryman, Stenseth and Isaev, 1987).

The generalized 'logistic' model can be written

$$N(t) = N(t - 1) \, exp[\, R + V(O,S)\,] \qquad (4a)$$

$$R = A\{1 - [N(t - T)/K]^Q\} \qquad (4b)$$

where R is the per capita rate of increase, expressed as a function of A, the maximum per capita rate of increase, T, the time-delay in the negative feedback,

K, the equilibrium density, and Q, a coefficient of non-linearity. The R-function is convex when $Q>1$ and concave when $Q<1$ (*Figure 1a*). Stochastic variability due to density-independent factors is introduced by a random variable V which has a

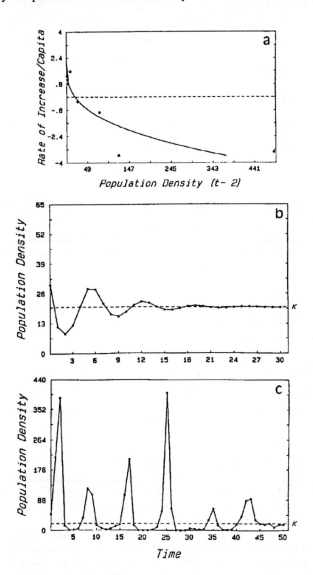

Figure 1. (a) Non-linear time-delayed 'logistic' model fit to the Sierra Nevada Douglas-fir tussock moth population (*Table 1*). (b) Steady-state behaviour of the Sierra Nevada model under constant environmental conditions. (c) Behaviour of the Sierra Nevada model in a variable environment ($S = 0.856$).

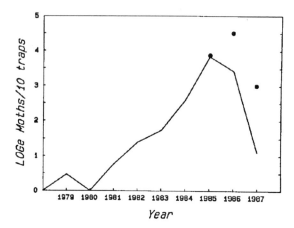

Figure 2.   Natural logarithm of the average number of male Douglas-fir tussock moths caught in pheromone baited traps from 1978 to 1987. Notice the suppression of captures in 1980, the year Mount St. Helens erupted, and that densities did not exceed the level of visible defoliation during the cycle (around in $300 = 5.7$). Solid circles represent predictions of tussock moth densities made with the Sierra Nevada model (*Figure 1a* and *Table 1*).

Table 1.   Parameter values of the non-linear 'logistic' model of equation (4b) estimated from four Douglas-fir tussock moth data series*

Population	N	A	K	Q	T	r
Arizona	10	1.50	160	0.31	2	0.81
Sierra Nevada	10	2.00	175	0.33	2	0.83
Blue Mountains	17	1.00	75	0.33	2	0.61
Oregon Cascades	12	2.00	30	0.20	2	0.34

* Population densities adjusted to $10 \ m^2$ of foliage

mean of zero and standard deviation S, equation (4a).

When this model is fit to Douglas-fir tussock moth data (Berryman and Millstein 1988), we obtain concave R-functions (*Table 1, Figure 1a*). These models all generate damped-stable dynamics in constant environments (*Figure 1b*), but cycle continuously with variable amplitude in changing environments (*Figure 1c*). The models predict that non-outbreak tussock moth populations will be found in relatively stable (consistent) environments while outbreak populations will be associated with less stable (variable) environments. The model of the Californian Sierra Nevadas population (*Table 1, Figure 1a*) predicted that tussock moth densities in Northern Idaho would not attain damaging levels in the mideighties (*Figure 2*). Although these predictions were quantitatively imprecise, they were qualitatively accurate in the sense that they predicted no damage to the forest.

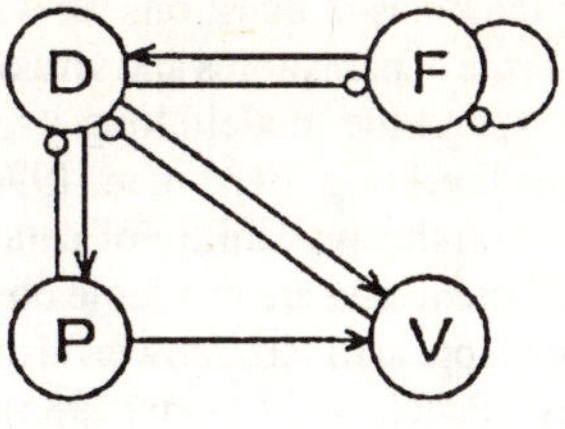

$$\begin{array}{c c c c c} & F & D & P & V \\ F & \begin{pmatrix} -FF & -FD & 0 & 0 \\ D & +DF & 0 & -DP & -DV \\ P & 0 & +PD & 0 & 0 \\ V & 0 & +VD & +VP & 0 \end{pmatrix} \end{array}$$

$F_1 = -\alpha_{FF} = -1$

$F_2 = -[\alpha_{FD}\alpha_{DF} + \alpha_{DV}\alpha_{VD} + \alpha_{DP}\alpha_{PD}] = -3$

$F_3 = -[\alpha_{VP}\alpha_{DV}\alpha_{PD} + \alpha_{FF}\alpha_{DV}\alpha_{VD} + \alpha_{FF}\alpha_{DP}\alpha_{PD}] = -3$

$F_4 = -\alpha_{FF}\alpha_{DV}\alpha_{PD}\alpha_{VP} = -1$

$F_1F_2 + F_3 = \alpha_{FF}\alpha_{FD}\alpha_{DF} - \alpha_{VP}\alpha_{DV}\alpha_{PD} = 0$

Figure 3. Loop analysis of the interactions between populations of fir tree foliage (F), Douglas-fir tussock moths (D), insect parasitoids (P) and viral pathogens (V). This system is stable but can exhibit limit cycles.

In contrast, other forecasting techniques such as pheromone traps and larval sampling predicted economic damage in 1986, a prediction which precipitated the spray project that was eventually terminated prematurely. Our simple model also predicts that tussock moth populations North of Moscow will reach peak densities earlier than expected in the 1990s. This is because cycles with lower amplitudes should have higher frequencies (see Figure 1 in Berryman, 1978). Thus, the next peak should occur in the early 1990's in Montana, Eastern Oregon and Idaho south of Moscow (e.g. 1990–92) but two years later in Idaho north of Moscow (e.g. 1992–94). These predictions will also be subjected to the test of time.

Expanding the model

Although the simple 'logistic' model can provide useful predictions and valuable insights when fit to real data, it has a fundamental limitation in that it is a single-species model and, therefore, can only answer questions about single-species populations. In other words, equation (4) can only address questions concerning the tussock moth and cannot be used to explore, for instance, the role of parasitoids or viruses in the dynamics of tussock moth populations. To ask such questions, the model must be expanded to include these variables (*Figure 3*).

Before adding dimensions to the model, however, we must first decide on the direction in which we wish to expand. For instance, do we want to add detail about the insect's life cycle, as in the spruce budworm and winter moth models (expanded life cycle), or would it be better to add functions describing the interactions with the host plant, parasitoids and/or virus populations (expanded trophic levels)?

Expansion of the model should depend on the kinds of questions we wish it to answer. For example, if we are interested in the role of parasitoids and virus disease in tussock moth dynamics, then we should expand the model along its trophic dimension, building equations for each trophic level (e.g. Berryman, 1990). The success of this exercise, however, will depend on the availability of data. In the case of the tussock moth, unfortunately, insufficient data are available on parasitoid and virus populations. However, we can still appeal to some powerful methods of qualitative analysis, such as 'loop analysis' (Puccia and Levins, 1985).

Loop analysis is based on the principles of feedback and mathematical stability. This approach allows one to determine the qualitative stability (or instability) of a system from its feedback structure, and can also identify the conditions under which cyclical dynamics are to be expected. Using loop analysis, Millstein (1988) analysed all possible kinds of interactions between the tussock moth and its host plant, parasitoid and virus populations. He found that the system was always stable in the absence of virus but could exhibit cyclical dynamics when parasitoids were involved in spreading virus infections; i.e. when there is a positive effect of parasitoids on the virus (*Figure 3*). This result is very interesting, and perhaps counterintuitive, suggesting that the virus is responsible for the wide amplitude outbreak cycles of the Douglas-fir tussock moth and that the virus, rather than the insect, is the pest. Empirical data seem to support this contention because virus has never been isolated from sparse (non-outbreak) tussock moth populations (Mason and Torgersen, 1987). This inference, gleaned from a simple qualitative model, has sweeping implications for pest control and biotechnology because spraying virus formulations, particularly the more virulent (bioengineered) strains, can destabilize otherwise stable insect–plant–parasitoid systems.

Although we cannot test the virus/instability hypothesis quantitatively with the single-species tussock moth model, we can test it in principle on models of similar cyclical forest insects. We chose to do this with a two-species model of the interaction between populations of the blackheaded budworm, *Acleris variana*, and its parasitoids (Berryman, 1990):

$$Rb = 2.152 - 0.003\,B(t) - 4.131\,P(t)/B(t) \tag{5a}$$

$$Rp = 2.441 - 5.623\,P(t)/B(t) \tag{5b}$$

where Rb and Rp are the per-capita rates of increase of budworm and parasitoids, respectively, $B(t)$ and $P(t)$ are their respective population densities, and the parameters are estimated by multiple linear regression from data in Morris (1959). The correlation coefficients for both regressions were highly significant [r(budworm) = 0.92; r(parasitoids) = 0.95].

Simulation experiments can be performed by calculating the per capita rates of increase for each population in each generation [equation (5)], then computing their population densities in the next generation with equation (4a). Simulation in constant environments ($S = 0$) demonstrates that this model is damped-stable (*Figure 4a*).

The effect of virus disease on budworm populations was modelled by a simple threshold function

$$Rv = 0.6\,Rb, \text{ when } B(t) > 300 \tag{6a}$$

$$Rv = 0.2\,Rb, \text{ when } B(t\text{-}1) > 300 \tag{6b}$$

In other words, virus disease causes 40% mortality to budworm larvae surviving parasitism when the initial budworm densities exceed the transmission threshold

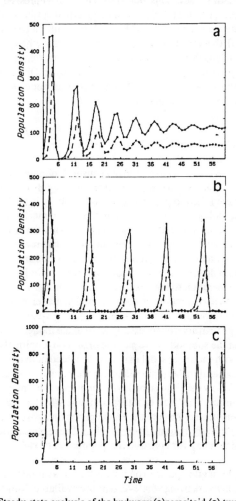

Figure 4. (a) Steady-state analysis of the budworm(●)parasitoid (o) two-species model, described by equation (5) showing damped-stability in a constant environment ($S = 0$). (b) Analysis of the budworm(●)parasitoid(o)model in the presence of a viral pathogen described by equation (6) showing continuous outbreaks with period around 10 years, also in a constant environment ($S = 0$). (c) Analysis of the budworm–virus interaction described by equations (5) and (6) in the absence of parasitoids showing continuous outbreaks of budworm with a period of 4 years ($S = 0$, $P(1) = 0$).

of 300 larvae while, in the next generation, 80% of the surviving larvae are killed by virus infections. Inclusion of this two-step virus epizootic destabilizes the budworm–parasitoid model and precipitates continuous outbreak cycles with period of about 10 years (*Figure 4b*). This analysis seems to support the hypothesis of virus-induced instability.

Finally, we could ask the model if virus epizootics alone are capable of driving 10-year tussock moth cycles. This question has been addressed previously by Anderson and May (1980) and Vezina and Peterman (1985) with conflicting results. This problem can be examined by setting the initial parasitoid density to zero in equation (5), thereby eliminating the parasitoid from the system. Under this condition, cycles of much higher amplitude and shorter frequency are obtained (*Figure 4c*). Thus, parasitoids act to lower the frequency and intensity of outbreaks while virus has the opposite effect. Again, the virus appears to be the main destabilizing component in this system.

Strategy of model building

Simple theoretical models seem to provide useful qualitative predictions of Douglas-fir tussock moth population dynamics while highly detailed mechanistic models seem to have failed to perform as expected. Why is this?

In the past, entomologists may have felt that detailed mechanistic models that faithfully describe the events and processes characterizing the biology of a species would provide more accurate population predictions. The reason that they do not may be associated with the fact that many biological events are not relevant to the ecological processes determining population dynamics. For instance, population change from one generation to the next is dependent on the average reproduction and survival over that generation. It does not really matter in what stage of the insect's life cycle the mortality occurs. Thus, structuring a population model on the life cycle of the organism adds unnecessary detail to the model, detail that can give rise to errors when the submodels are parameterized on empirical data. The error terms associated with each submodel may then be added or even multiplied when the model is run. In addition, large simulation models are often so complicated that it is impossible to follow their causal pathways or discover the major processes that govern their dynamics. This lack of transparency is why large complex models rarely provide valuable insights into the mechanisms that govern the dynamics of a system (Lee, 1975; Berryman and Brown, 1981). These observations suggest the following optimal strategy for constructing population models:

1. Build a simple single-species (one state variable) model around a sound theoretical framework using whatever data are available. Non-linear 'logistic' (Berryman and Millstein, 1988) and autoregressive time-series (Royama, 1977, 1981) analyses are alternative approaches that have their separate advantages and disadvantages.

2. Analyse the behaviour and structure of the model and try to identify the mechanisms responsible for the observed dynamics (Berryman, 1987, Berryman and Millstein, 1988).

3. Identify the kinds of questions that you wish to ask of the model and then decide on a strategy for model expansion.
4. Expand the dimensions of the model by fitting theoretically sound equations to available data (Berryman, 1990) or to new data obtained by carefully designed experiments.

An approach such as this should help to minimize costs and, at the same time, maximize the predictive and interpretive capabilities of insect population models.

References

ANDERSON, R. M. and MAY, R. M. (1980). Infectious diseases and population cycles of forest insects. *Science* **210**, 658–661.

BERRYMAN, A. A. (1978). Population cycles of the Douglas-fir tussock moth (Lepidoptera: Lymantriidae): the time-delay hypothesis. *Canadian Entomologist* **110**, 513–518.

BERRYMAN, A. A. (1981). *Population Systems: A General Introduction.* New York: Plenum Press.

BERRYMAN, A. A. (1987). The theory and classification of outbreaks. In: *Insect Outbreaks*, pp. 3–30 (eds P. Barbosa and J. C. Schultz). New York: Academic Press.

BERRYMAN, A. A. (1989). The conceptual foundations of ecological dynamics. *Bulletin of the Ecological Society of America* **70**, 234–240.

BERRYMAN, A. A. (1990). *Population Analysis System: POPSYS Series 2 User Manual.* Pullman WA: Ecological Systems Analysis.

BERRYMAN, A. A. and BROWN, G. C. (1981). The habitat equation: a useful concept in population modelling. In: *Quantitative Population Dynamics*, pp. 11–24 (eds G. C. Chapman and V. F. Gallucci). Burtonsville MD: International Cooperative.

BERRYMAN, A. A. and MILLSTEIN, J. A. (1988). *Population Analysis System: POPSYS Series 1 User Manual.* Pullman WA: Ecological Systems Analysis.

BERRYMAN, A. A. and PIENAAR, L. V. (1974). Simulation: A powerful method for investigating the dynamics and management of insect populations. *Environmental Entomology* **3**, 199–207.

BERRYMAN, A. A., STENSETH, N. C. and ISAEV, A. S. (1987). Natural regulation of herbivorous forest insect populations. *Oecologia* **71**, 174–184.

BROOKES, M. H., STARK, R. W. and CAMPBELL, R. W. (eds) (1978). *The Douglas-fir Tussock Moth: A Synthesis.* Washington DC: *United States Department of Agriculture Technical Bulletin* 1585.

COLBERT, J. J., OVERTON, W. S. and WHITE, C. (1979). Documentation of the Douglas-fir tussock moth outbreak population model. Washington DC: *United States Department of Agriculture General Technical Report* PNW-89.

DAHLSTEN, D. L., LUCK, R. F., SCHLINGER, E. I., WENZ, J. M. and COPPER, W. A. (1977). Parasitoids and predators of the Douglas-fir tussock moth, *Orgyia pseudotsugata* (Lepidoptera: Lymantriidae), in low to moderate populations in central California. *Canadian Entomologist* **109**, 727–746.

HASSELL, M. P. (1978). *The Dynamics of Arthropod Predator-Prey Systems.* Princeton, NJ: Princeton University Press.

HOLLING, C. S., JONES, D. D. and CLARK, W. C. (1976). Ecological policy design: a case study in forest and pest management. In: *Pest Management*, pp. 13–90 (eds G. A. Norton and C. S. Holling). New York: Pergamon Press.

LEE, D. B. (1975). Requiem for large-scale models. *Simuletter* **6** (2/3), 16–29.

MASON, R. R. and TORGERSEN, T. R. (1987). Dynamics of a nonoutbreak population of the Douglas-fir tussock moth (Lepidoptera: Lymantriidae) in Southern Oregon. *Environmental Entomology* **16**, 1217–1227.

MASON, R. R. and WICKMAN, B. E. (1988). The Douglas-fir tussock moth in the Interior Pacific Northwest. In: *Dynamics of Forest Insect Populations: Patterns, Causes, Implications*, pp. 179–209 (ed. A. A. Berryman). New York: Plenum Press.

MASON, R. R., WICKMAN, B. E. and PAUL, H. G. (1984). Effect of ash from Mount St. Helens on survival of neonate larvae of the Douglas-fir tussock moth. *Canadian Entomologist* **116**, 1145–1147.

MILLSTEIN, J. A. (1988). The population dynamics of the Douglas-fir tussock moth: a study of multiple interactions. PhD Thesis, Washington State University, Pullman, WA.

MORRIS, R. F. (1959). Single-factor analysis in population dynamics. *Ecology* **40**, 580–588.

MORRIS, R. F. (ed.) (1963). Dynamics of epidemic spruce budworm populations. *Memoirs of the Entomological Society of Canada* **31**, 1–332.

PUCCIA, C. J. and LEVINS, R. (1985). *Qualitative Modeling of Complex Systems.* Cambridge MA: Harvard University Press.

ROYAMA, T. (1977). Population persistence and density dependence. *Ecological Monographs* **47**, 1–35.

ROYAMA, T. (1981). Fundamental concepts and methodology for the analysis of animal population dynamics, with particular reference to univoltine species. *Ecological Monographs* **51**, 473–493.

STOSZEK, K. J., MIKA, P. G., MOORE, J. A. and OSBORNE, H. L. (1981). Relationships of Douglas-fir tussock moth defoliation to site and stand characteristics in Northern Idaho. *Forest Science* **27**, 431–442.

VARLEY, G. C., GRADWELL, G. R. and HASSELL, M. P. (1973). *Insect Population Ecology: An Analytical Approach.* Oxford: Blackwell Scientific.

VEZINA, A. and PETERMAN, R. M. (1985). Tests of the role of a nuclear polyhedrosis virus in the population dynamics of its host, Douglas-fir tussock moth, *Orgyia pseudotsugata* (Lepidoptera: Lymantriidae). *Oecologia* **67**, 260–266.

34

Population Dynamics of the Eastern Spruce Budworm: Inferences and Model Performance Examination using Survey Data

RICHARD A. FLEMING

Forest Pest Management Institute, Forestry Canada, Box 490, Sault Ste. Marie, ON P6A 5M7, Canada

Introduction

The eastern spruce budworm, *Choristoneura fumiferana* (Clem.) (Lepidoptera: Tortricidae), is the most destructive defoliator in North America's boreal forests. It attacks spruce, *Picea* spp., as well as balsam fir, *Abies balsamea* (L.) Mill.; during uncontrolled budworm outbreaks, dense, mature fir stands sustain severe damage and tree mortality can approach 100%.

The economic importance of budworm outbreaks has motivated a number of intensive, innovative modelling efforts to improve our understanding of the budworm–forest system (e.g. Antonovsky *et al.*, 1990; Gage and Sawyer, 1979; Jones, 1977; Kemp, Nyrop and Simmons, 1980; Ludwig, Jones and Holling, 1978; Royama, 1984; Watt, 1964). In two cases (Jones, 1979; Stedinger, 1984), such modelling efforts have produced detailed syntheses of theories of budworm population dynamics to guide the design of better management policies (Baskerville, 1976).

Due to the costs and difficulties of data collection, most studies of insect population dynamics are based on rather small plots. However, in a small plot, insect dispersal from surrounding areas can seriously distort estimates of population change. In addition, a few scattered plots usually cannot represent the range of environmental and geographical conditions experienced by the population across its natural habitat. Serious bias can arise when the results of small-scale population dynamics studies are transferred to larger scales (as happens almost routinely in developing management strategies for forest pests).

This paper presents some results from a larger study (Fleming, Shoemaker and Stedinger, 1983, 1984; Fleming and Shoemaker, 1989) which used Maine Forest Survey (MFS) data to describe the population dynamics of the eastern spruce budworm across the state of Maine. These data were collected during a prolonged outbreak (Trial, 1980) when a major suppression programme was under way. The paper begins with a description of the survey methods and their limitations. This first section discusses the spatial variability of the budworm population density. Such information is essential for using the results of small-scale population dynamics studies at the larger scales of management planning. The paper continues with a short presentation of some general inferences about budworm population dynamics in Maine. After these preliminaries, we summarize a test of the two large-scale simulation models which were developed to design budworm control strategies (Jones, 1979; Stedinger, 1984).

Survey methods

The spruce budworm overwinters as a second instar larva. It emerges from hibernation in early May and passes through four feeding instars until pupating in late June. Mating, moth dispersal, and oviposition (of masses of about 20 eggs each) occur in July. The eggs hatch by mid-August and the newly hatched, non-feeding larvae immediately search the branches for overwintering sites. Spraying occurs in late May (during the fourth instar) in some areas (depending on political, environmental, and logistical concerns) where high budworm populations and heavy defoliation existed in the previous year.

The surveys are used to identify these areas of threatened forest. In early August, shortly after oviposition is completed, the MFS estimates the defoliation of needles produced that spring and the budworm population density (in egg masses/m² of fir foliage) at approximately 1000 points in Maine's 28 000 km² Spruce-Fir Protection District.

LIMITATIONS

For our purposes, the survey data was limited by its poor spatial resolution and by its discrete measure of budworm population density. The analysis focused on the relationship between past and present conditions (in terms of defoliation or egg-mass density) of small forest areas. Unfortunately, the surveys seldom collected data at the same sampling points in consecutive years; therefore, when considering relationships involving survey data from consecutive years, we enlarged the spatial scale of the analysis to coincide with the 8.0 x 6.4 km 'blocks' of forest used

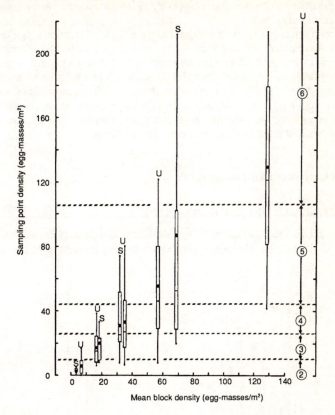

Figure 1. Box plots showing the distribution of egg-mass densities among sampling points within a forest block and the block means. The top and bottom edges, the enclosed circle, and the horizontal interior line (when visible), of the rectangle in each box plot respectively indicate the distribution's 75th and 25th percentiles, its mean, and its median. Vertical lines extend upward and downward to the 95th and 5th percentiles, respectively. The ranges of density classes 2-6 used in sequential sampling are shown to the right of the horizontal axis. Class 1 densities range from 0-0.05 masses/m^2. Only unsprayed (U) and sprayed (S) blocks with over 3 sampled points are included. Sample sizes exceed 30 points. (After Fleming, Shoemaker and Stedinger, 1984)

to map the surveys. Blocks with both sprayed and unsprayed sample points were excluded from the analysis whenever they occurred. However, since only one point was sampled in 60% of the block-years, getting block estimates by averaging over all sampled points was often no more precise than sampling a single randomly selected point within a block. *Figure 1* shows the variability encountered in blocks with over three sampled points. It represents a likely upper limit to the sampling error from estimating block conditions with a single sample point because sampling density was greatest (one point every 12 km^2) in areas where the greatest variability was expected.

Another limitation concerns the sequential sampling used to estimate egg-mass densities. A sample consists of one mid-crown branch from each of four dominant or co-dominant balsam firs. Branches are searched one by one for egg-masses until the density falls into a predetermined class (*Figure 1*) with a preselected level of confidence (Morris, 1954). This procedure, however, allows mean densities at different sampling points to be estimated with varying confidence. Therefore, rather than use average densities as a continuous variable with inconsistent precision, we deal directly with the discrete density classes.

General inferences based on survey data

PREDICTABILITY OF FOREST CONDITIONS

Figure 2 shows the distribution of egg densities (in year t+1) among blocks associated with each initial density class (the density of eggs recorded the previous

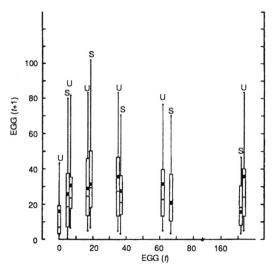

Figure 2. The relationship in the survey data between budworm densities (egg masses/m^2) in consecutive years, EGG(t) and EGG(t+1), for forest blocks which were not sprayed in year t, and sprayed (S) or not (U) in the following year, t+1. (The caption to *Figure 1* explains the box plots.) Sample sizes exceed 25 blocks. (After Fleming, Shoemaker and Stedinger, 1984)

year, year *t*). Each box plot occurs above the class average of its corresponding initial density. Although general trends in the means and medians are evident, the large scatter along the y-axis shows that knowing only a block's density class is a poor basis for predicting that block's future density. The degree to which this scatter is caused by averaging sampling point data over MFS survey blocks (as explained above), and the degree to which it is inherent in the budworm–forest system due to variability in geography, local weather and forest conditions is unknown. Taking survey samples in consecutive years from the same sampling points would help clarify this question and better define the relationships shown in *Figure 2*.

LONG-TERM DAMAGE

Since it takes about four or five consecutive years of almost 100% defoliation of new needles to kill a balsam fir (MacLean, 1980), the presence of budworm-killed trees is evidence of long-term damage to a stand. Fleming, Shoemaker and Stedinger (1983) used such data from the MFS surveys to study the effects of long-term damage. They showed that unsprayed stands with little long-term damage had smaller mean densities (by about 15 masses/m^2) than heavily damaged stands which had lost the same percentage of their current foliage (and therefore presumably had comparable larval densities). In fact, the differences between corresponding means were so great that their 95% confidence intervals did not overlap. This corresponds to $P \ll 0.05$ (Jones, 1984).

YEAR-TO-YEAR VARIABILITY

The relationships between the past and present conditions of forest blocks showed large variation from year to year. For instance, for all density classes in the previous year, unsprayed blocks had significantly higher ($P \ll 0.05$) mean densities in 1979 than in 1980. Attempts to relate this year to year variability to variations in the total area sprayed (which might affect moth immigration) and to a 'warm-dry' day index (Miller, 1971) were unsuccessful (Fleming, Shoemaker and Stedinger, 1983).

GEOGRAPHY

To test for a geographic influence, Maine's Spruce-Fir Protection District was split into north-west and south-east regions based on climate and past spray history. Among unsprayed sampling points with less than 70% defoliation, southeastern stands had lower ($P \ll 0.05$) densities than northwestern stands sustaining comparable defoliation. This could be due to the (presumably) large populations of dispersing moths in the northwest where Maine's outbreak was centred. There was no evidence of a geographic effect on the efficacy of operational spraying.

Evaluation of model performance

Testing models, in general, is an essential part of the scientific method and provides a means of refining the theories and understanding on which the models are based. Testing management-orientated models has an additional function. It indicates the reliability of policies recommended by these models and thus helps ensure that the considerable expense of putting management policy into operational practice is justified.

Both of the large-scale simulation models for decision support in budworm management (Jones, 1979; Stedinger, 1984) comprise four major submodels: one for the spray policy, one for the harvest policy, a third for the weather, and one describing budworm–forest dynamics. Proper testing of this last submodel (essentially a dynamic life table driven by forest age, an indicator of host quality) requires that the other three submodels and the initial conditions be representative of the situation in Maine. Otherwise, one cannot tell which submodel is involved when differences between model output and the survey data arise. When information gaps (e.g. the absence of observations on foliage condition) or model structure (e.g. the limited spatial resolution) precluded the precise representation of Maine conditions, we used assumptions made by the only modeller (Stedinger, 1984) to specifically apply his model to Maine. Therefore, in the simulations, stands were even-aged and harvested at age 60, and weather patterns were representative of northern Maine. In addition, we used a stochastic spray policy submodel which targetted modelled blocks of forest in a particular condition (with respect to budworm density and defoliation) for spraying as often as the MFS sprayed surveyed blocks in the same condition.

In each run, initial transient effects damped out within the first ten simulated years. In the following 100 simulated years of each run, the budworm density and defoliation for each of 80 modelled blocks of forest were recorded, giving 8000 block-years of results for both models. Subsequent analysis determined the relationships between block conditions in consecutive years in the models. These relationships, inherent in the model output, were compared with corresponding relationships determined from the survey data.

We concentrate on short-term evaluation of the models for two reasons. First, the developers of the models depended largely on long-term qualitative evaluations for establishing the reliability of their models (Baskerville, 1976; Stedinger, 1984). Hence, the evaluation described in here provides a different perspective. Secondly, the short history of MFS surveys precludes a meaningful long-term evaluation of the models.

Because differences between the output of the two models were generally relatively small compared to the differences between the survey data and the output of either model, we concentrate on the latter differences. Fleming and Shoemaker (1989) relate differences between the output of the two models to their detailed structure.

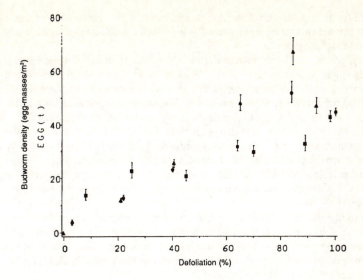

Figure 3. Plots of budworm density (egg masses/m²) against the average of the defoliation class (%) at sampling points which were not sprayed. Vertical lines encompass the 95% confidence intervals about each mean. The boxes, circles and triangles correspond respectively to the surveys, Stedinger's (1984) model, and Jones's (1979) model. Sample sizes exceed 35 points. (After Fleming and Shoemaker, 1989.)

PERFORMANCE IN UNSPRAYED AREAS

Here we focus on budworm–forest dynamics in blocks of forest unsprayed in consecutive years. In this, model components describing the budworm mortality caused by operational insecticide application are of little importance and can be (temporarily) ignored.

Figure 3 compares model output and the survey data in terms of the relationship between the defoliation of new needles (i.e. short-term damage) at an average unsprayed sampling point (*x*-axis) and the budworm density there later that summer. Since defoliation and density were observed together at each sampling point, the sampling error inherent in averaging over sampling points to estimate block conditions is not a factor here. Discrepancies between the model output and survey results are clear: in just four of 12 comparisons do the confidence intervals from the model output and corresponding survey results overlap. There also appear to be qualitative differences in the trends. Budworm densities from the model output increase steadily with increasing defoliation but drop between 85 and 100%. In contrast, the survey data show a slower increase but one which is maintained over the entire range of observed defoliation.

The models' apparent underestimation of budworm densities at low levels of defoliation and overestimation at high levels, could be due to underrating the smoothing effect of moth dispersal. In the models, greater tendencies for moths to disperse (or to oviposit after dispersing) would increase the transfer of egg masses (due to the net flux of moth dispersal) from heavily defoliated to less heavily defoliated sampling points.

Using graphs similar in form to *Figure 3*, we also examined the relationship between egg masses/m² in an unsprayed block of forest (*x*-axis) and the defoliation in that block in the following summer. The models appeared to underestimate ($P \ll 0.05$) defoliation at densities below 25 masses/m². Dispersal is not a factor because it occurs in a different part of the lifecycle (i.e. before oviposition).

Sampling error, however, is involved. With observations at only one or two points in most surveyed blocks, some estimates of extremely high or extremely low budworm densities are likely for blocks with average densities. Since average defoliation is generally expected next year in blocks with average densities, this produces bias towards the overall mean defoliation at extreme egg-mass densities in the survey results.

Sampling error alone, however, cannot explain the models' apparent underestimation of defoliation at low densities: there is no evidence of the concomitantly expected overestimation of defoliation at high egg-mass densities. Therefore, the most comprehensive explanation for these results is unrealistically slow population increase at low densities in the models.

This apparent deficiency of the models could be due to difficulties in representing the non-linear population dynamic relationships observed in small (0.1 km²) experimental plots (Morris, 1963) at the models' block size of 93–173 km². As *Figure 1* shows, there can be considerable variability, even within 50 km². To account for this variability in the models, one takes the expected value of the small scale process distributed over space (O'Neill, 1988). This amounts to integrating the budworm–forest dynamics submodel across the frequency distributions representing the spatial variability of the relevant independent variables (e.g. budworm densities) in a model block. Due to the particular non-linearities of the budworm dynamics in the models, a pocket of moderate budworm density within a block can more than compensate for a predominance of low-density areas (Fleming and Shoemaker, 1989). Thus, incorporating the effects of small-scale spatial variability in the models' budworm population dynamics could increase population growth rates at low densities in the models, and hence produce better agreement between model output and the survey results.

PERFORMANCE IN SPRAYED AREAS

Figure 4 illustrates the relationship between budworm densities in successive years for blocks unsprayed one year and sprayed the next. Apparently the models overestimate the effectiveness of operational spray programmes at all but the highest densities.

Although the models' 90% spray mortality rate is typical of small-scale insecticide trials, there are two reasons to suspect it overrates operational rates. First, the spruces are a major component of Maine's forest protection area and insecticide applications are usually much less effective on spruce, even in small scale trials (Trial and Devine, 1982). Moreover, there is an increasing body of evidence (Webb and Irving, 1983) that spruce supports a substantial proportion of

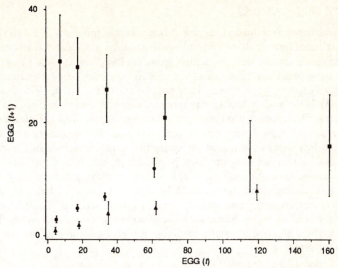

Figure 4. The relationships between budworm densities (egg masses/m²) in consecutive years, EGG(*t*) and EGG(*t*+l), for forest blocks which were not sprayed initially, year *t*, but were sprayed in the following year, *t*+l. Vertical lines encompass the 95% confidence intervals about each mean. Sample sizes exceed 25 blocks for the surveys (boxes) and 15 for the models, except near EGG(*t*)=115 where they are 5 and 4, respectively, for Stedinger's (circles) and Jones's (triangles). (After Fleming and Shoemaker, 1989.)

the budworm population. Secondly, it is inherently harder to maintain good application procedures and good timing over the expansive areas sprayed operationally than over small trial plots. Poor weather, logistical problems and asynchrony in larval and shoot development all pose difficulties. The criteria for timing sprays are, first, that balsam fir shoots have expanded enough to give an adequate spray target and, secondly, that the larvae are in their vulnerable fourth instar. Often these events don't occur simultaneously, and even when they do, shoot development on spruce is usually slower than on balsam fir. Thus, it is almost impossible to achieve 90% larval mortality with operational spraying of large multispecies forests.

Comparison of results from sprayed blocks (e.g. *Figure 4*) with corresponding results from unsprayed blocks, uncovered another difference between the models' output and the survey data. The models suggest that spraying is most effective (as measured by the relative change in corresponding means) in stands that were moderately infested (i.e. 20–60 masses/m², 20–60% defoliation) in the previous year. On the other hand, the survey data (e.g. *Figure 2*) corroborate Miller's (1971) report that spraying makes the largest improvement in heavily infested blocks. Graphs of defoliation against the egg-mass density in a block the previous summer, showed similar differences between the models' output and the survey data, and thus indicate that dispersal alone cannot explain these differences. Rather, unrealistically slow population increase from low budworm densities in the models may best explain these differences.

POLICY IMPLICATIONS

The performance examination suggested two potential problems in the models: inflated rates of insecticide-caused budworm mortality, particularly at low densities, and unrealistically slow population growth at low densities. A key question is how these problems might affect the models' common recommendation of aggressively spraying low density local populations (Baskerville, 1976; Stedinger, 1984). In the models, low density spray policies keep budworm densities low enough that: (1) maximum population growth rates are not realized; (2) infestations do not reach damaging levels even in later years; and (3) contamination of other stands through dispersal is limited. Adjusting the models to include smaller rates of pesticide-caused mortality and faster population growth, both at low densities, would increase the cost/benefit ratio for low-density spraying. Immediately after spraying, modelled populations would be higher (because of reduced spray mortality) and growing more quickly than in earlier versions of the models. Spraying would, therefore, be needed more often to maintain low densities. This would produce higher costs and thus favour a higher threshold as the optimal density for spraying.

Conclusions

Survey data can provide a useful and unique perspective for studying the population dynamics of the eastern spruce budworm. In an analysis of the Maine Forest Service surveys conducted from 1975 to 1980, relationships between the mean conditions (i.e. budworm density or defoliation) of small forest areas in consecutive years were defined with enough confidence to determine general trends. However, the conditions observed from one year to the next in a particular forest block were extremely variable and depended on weather, geographical location, the block's recent spray history, and the amount of long-term damage to the stand.

The survey data also provided a basis for examining the performance of the two large-scale management-orientated simulation models (Jones, 1979; Stedinger, 1984). Throughout this examination, however, the inherent complexities of comparing data and models that function on fundamentally different spatial scales were a major problem. Consequently, sampling error imparted some bias in most of our model output-survey data comparisons. Nevertheless, certain patterns were clear. In the models, spray mortality rates are inflated, particularly at low budworm densities, and population growth rates underestimated, also at low densities. As a result, the models tend to underestimate both the budworm density of the economically optimal spray threshold, and the cost/benefit ratio of the low-density thresholds they recommend.

These results highlight a more general concern in studying forest-insect population dynamics on a large scale. In the budworm models studied here, results based on small experimental plots were applied at much larger spatial scales, apparently without adjusting for the effects of spatial variability. This contributed to inflated rates of spray mortality and to underestimated rates of budworm population growth at low densities in the models. No doubt, such difficulties with spatial scale are not unique to the models examined here. Difficulties can arise whenever the results of small-scale population dynamics studies are transferred to larger scales without accounting for the effects of spatial heterogeneity at that larger scale.

Acknowledgements

Many people and organizations helped with this work. Particular thanks are due Henry Trial, Jr., of the Maine Forest Service, Chris Shoemaker and Jery Stedinger of Cornell University, and Buzz Holling of the University of Florida. This research was supported in part by a grant from the Eastern CANUSA Spruce Budworm Program of the USDA Forest Service.

References

ANTONOVSKY, M. Y., FLEMING, R. A., KUZNETSOV, Y. A. and CLARK, W. C. (1990). Forest-pest interaction dynamics: the simplest mathematical models. *Theoretical Population Biology* **37**(2), (in press).

BASKERVILLE, G. L. (1976). *Report of the Task-Force for Evaluation of Budworm Control Alternatives.* Fredericton, New Brunswick, Canada: New Brunswick Department of Natural Resources. 210 pp.

FLEMING, R. A. and SHOEMAKER, C. A. (1989). *Evaluation of Models of Spruce Budworm–Forest Dynamics: Comparison of Predictions to Regional Field Data.* File Report No. 99. Sault Ste. Marie, Ontario: Forest Pest Management Institute, Forestry Canada. 40 pp.

FLEMING, R. A., SHOEMAKER, C. A. and STEDINGER, J. R. (1983). An analysis of the regional dynamics of unsprayed spruce budworm (*Lepidoptera: Tortricidae*) populations. *Environmental Entomology* **12**, 707–713.

FLEMING, R. A., SHOEMAKER, C. A. and STEDINGER, J. R. (1984). An assessment of the impact of large scale spraying operations on the regional dynamics of spruce budworm (*Lepidoptera: Tortricidae*) populations. *Canadian Entomologist* **116**, 633–644.

GAGE, S. H. and SAWYER, A. J. (1979). A simulation model for eastern spruce budworm populations in a balsam fir stand. *Modeling and Simulation* **10**, 1103–1113.

JONES, D. (1984). Use, misuse, and role of multiple-comparison procedures in ecological and agricultural entomology. *Environmental Entomology* **13**, 635–649.

JONES, D. D. (1977). Catastrophe theory applied to ecological systems. *Simulation* **29**, 1–15.

JONES, D. D. (1979). The budworm site model. In: *Pest Management*, pp. 9–155 (eds G. A. Norton and C. S. Holling). Oxford: Pergamon Press.

KEMP, W. P., NYROP, J. P. and SIMMONS, G. A. (1980). Simulation of the effects of stand factors on spruce budworm (*Lepidoptera: Tortricidae*) larval redistribution. *Great Lakes Entomologist* **13**, 81–91.

LUDWIG, D., JONES, D. D. and HOLLING, C. S. (1978). Qualitative analysis of insect outbreak systems: the spruce budworm and the forest. *Journal of Animal Ecology* **47**, 315–332.

MACLEAN, D. A. (1980). Vulnerability of fir-spruce stands during uncontrolled spruce budworm outbreaks: a review and discussion. *Forestry Chronicle* **56**, 213–221.

MILLER, C. A. (1971). The spruce budworm in eastern North America. In: *Proceedings, Tall Timbers Conference on Ecological Animal Control by Habitat Management*, pp. 169–177. 25–27 Feb. 1971.

MORRIS, R. F. (1954). A sequential sampling technique for spruce budworm egg surveys. *Canadian Journal of Zoology* **32**, 302–313.

MORRIS, R. F. (1963). The dynamics of epidemic spruce budworm populations. *Memoirs of the Entomological Society of Canada* **31**, 1–332.

O'NEILL, R. V. (1988). Hierarchy theory and global change. In: *Scales and Global Change*, pp. 29–56 (eds T. Rosswall, R. G. Woodmansee and P. G. Riser). New York: John Wiley.

ROYAMA, T. (1984). Population dynamics of the spruce budworm, *Choristoneura fumiferana*. *Ecological Monographs* **54**, 429–462.

STEDINGER, J. R. (1984). A spruce budworm-forest model and its implications for suppression programs. *Forest Science* **30**, 597–615.

TRIAL, H., JR. (1980). A cartographic history of the spruce budworm in Quebec, Maine and New Brunswick 1970–1980. *Maine Forest Review* **13**, 3–7.

TRIAL, H., JR. and DEVINE, M. E. (1982). *Spruce Budworm in Maine: 1982*. Technical Report No. 18. Orono, Maine: Entomology Division, Maine Forest Service. 122 pp.

WATT, K. E. F. (1964). The use of mathematics and computers to determine optimal strategies and tactics for a given insect pest control problem. *Canadian Entomologist* **96**, 202–220.

WEBB, F. E. and IRVING, H. J. (1983). My fir lady. *Forestry Chronicle* **59**, 118–122.

Index